Advances in Nutrient Dynamics in Soil - Plant System for Improving Nutrient Use Efficiency

Advances in Nutrient Dynamics in Soil - Plant System for Improving Nutrient Use Efficiency

Editors

R. Elanchezhian

A.K. Biswas

K. Ramesh

A.K. Patra

ICAR-Indian Institute of Soil Science

Nabi Bagh, Berasia Road, Bhopal – 462038

Madhya Pradesh (India)

NEW INDIA PUBLISHING AGENCY

New Delhi – 110 034

NEW INDIA PUBLISHING AGENCY
101, Vikas Surya Plaza, CU Block, LSC Market
Pitam Pura, New Delhi 110 034, India
Phone: + 91 (11) 27 34 17 17 Fax: + 91 (11) 27 34 16 16
Email: info@nipabooks.com
Web: www.nipabooks.com

Feedback at feedbacks@nipabooks.com

ISBN: 978-93-85516-96-2

Citation: *Elanchezhian, R., Biswas, A.K., Ramesh, K. and Patra, A.K. Eds. (2017). Advances in Nutrient Dynamics in Soil - Plant System for Improving Nutrient Use Efficiency, Published by New India Publishing Agency, New Delhi, 404 pages.*

Composed, Designed and Printed in India

त्रिलोचन महापात्र, पीएच.डी.
एफ एन ए, एफ एन ए एस सी, एफ एन ए ए एस
सचिव एवं महानिदेशक
TRILOCHAN MOHAPATRA, Ph.D.
FNA, FNASc, FNAAS
SECRETARY & DIRECTOR GENERAL

भारत सरकार
कृषि अनुसंधान और शिक्षा विभाग एवं
भारतीय कृषि अनुसंधान परिषद
कृषि एवं किसान कल्याण मंत्रालय, कृषि भवन, नई दिल्ली 110 001

GOVERNMENT OF INDIA
DEPARTMENT OF AGRICULTURAL RESEARCH & EDUCATION
AND
INDIAN COUNCIL OF AGRICULTURAL RESEARCH
MINISTRY OF AGRICULTURE AND FARMERS WELFARE
KRISHI BHAVAN, NEW DELHI 110 001
Tel.: 23382629; 23386711 Fax: 91-11-23384773
E-mail: dg.icar@nic.in

Foreword

The factor productivity in Indian agriculture is on a steady decline since the advent of green revolution. Though intensive cultivation of land with fertilizer inputs have enhanced the production of agricultural crops, it has put immense pressure on scarce natural resources like soil, posing a threat to future agricultural production systems. Over the last four to five decades, soil health has been declining at faster rate with higher rates of erosion, declining nutrient use efficiency, loss of soil biota and degradation of land due to environmental pollution. Under such scenario, improvement in nutrient use efficiency of crops is essentially required to sustain crop productivity in the country. This warrants a thorough understanding of the nutrient dynamics in soil-plant system, besides characterization of plant physiological, genetic and molecular biological basis of nutrient uptake.

The book titled "Advances in Nutrient Dynamics in Soil-Plant System for Improving Nutrient use Efficiency" with 31 chapters gives a comprehensive understanding of the subject covering different aspects of nutrient use efficiency in agroecosystems and different agro-climatic conditions. I am sure, this book will be highly useful to researchers, teachers and students to understand, further research and define crop combinations for enhanced nutrient use efficiency and sustainable agriculture.

I congratulate all the editors for this commendable job.

Dated the 21st November, 2016 **(T. MOHAPATRA)**
New Delhi

Preface

Mineral nutrients such as nitrogen, phosphorus, potassium, calcium, magnesium, sulphur, and other micronutrients are essential for plant growth and food production. They ultimately contribute towards adequate nutrition for human beings. Presently, we face a glaring contrast of insufficient use of nutrients on one hand and excessive use on another. Nutrient Use Efficiency (NUE) represents a key indicator to assess progress towards better nutrient management. A goal for a 20% relative improvement in NUE by 2020 would lead to an annual saving of around 20 million tons of nitrogen, and equate to an initial estimate of improvement in human health, climate and biodiversity worth around $170 billion per year.

Humans have been altering the world's biogeochemical cycles for many millennia to ensure food and energy security. Many of these anthropogenic activities modified the nutrient cycles of major and micro nutrients of the world. The scale of these changes has massively accelerated since the industrial revolution throwing the equilibrium into disarray. The rates of anthropogenic carbon dioxide and other green-house gas emissions have increased substantially since 1750 (IPCC, 2007). The greenhouse gases include both methane, especially from fossil fuel sources and livestock, and nitrous oxide, which is particularly emitted from agricultural soils.

While recent trends in nutrient consumption are relatively stable in developed countries, growing human population and rising per capita meat based food consumption as a result of increasing incomes are together causing a rapid increase in nutrient consumption in transitional and developing countries including India and China. It is anticipated that these developing countries may account for three-fourths of global nutrient consumption by 2050. Indiscriminate and imbalanced use of nutrients has created a web of pollution at the global level. There are major problems associated with high levels of nutrient use, especially in Americas, Europe and Asia. The efficiency of nutrient use is very low, on an average 70-80% of N and 25-75% of P consumed end up lost to the environment, wasting the energy used to prepare them, and causing environmental pollution.

Oversupply of nutrients or imbalance between nutrients also reduces the efficiency of nutrient use. In addition, insufficient uses of nutrients lead to land degradation. Biological nitrogen fixation and manure recycling are key local nutrient sources which are not always optimally exploited. The inability to match crop harvests with a sufficient nutrient return leads to depletion of nutrients and organic matter, reducing soil quality and increasing the risk of land degradation through erosion and of agricultural incursion into virgin ecosystems. Shortages of water and other nutrients such as sulphur, zinc, molybdenum, etc. can limit N and P use efficiency, preventing the best use being made of these major nutrients.

This book comprises of 31 chapters under four major themes viz. 1) Introduction and fundamentals of soil plant atmosphere continuum and nutrient use efficiency; 2) Soil physical, chemical, biological and agronomic management for improving NUE; 3) Plant physiological, genetic & molecular biological basis for improving nutrient uptake & use efficiency and 4) Climate change aspects related to soil and plant systems for improving NUE.

We sincerely thank all the authors for their contributions to this book. We want to express our gratitude to Secretary DARE & DG ICAR, DDG (NRM), ICAR and ADG (SWM), ICAR for their constant encouragement in the field of nutrient use efficiency research. We thankfully acknowledge ICAR, New Delhi for sponsoring a short course on Advances in nutrient use efficiency, through which we could able to generate and collate information related to nutrient use efficiency of crops for this book.

We hope, this book on "Advances in nutrient dynamics in soil-plant system for improving nutrient use efficiency" will be highly useful to students, researchers, scientists, land managers, policy planners and administrators for an in-depth understanding for improving nutrient use efficiency.

Editors

List of Contributors

Adhikari, T.
Principal Scientist (Soil Science)
ICAR-Indian Institute of Soil Science
Nabi Bagh, Berasia Road, Bhopal – 462038

Asit Mandal
ICAR-Indian Institute of Soil Science
Nabi Bagh, Berasia Road, Bhopal – 462 038

Ajay
Principal Scientist (Plant Physiology)
ICAR-Indian Institute of Soil Science
Nabi Bagh, Berasia Road, Bhopal – 462038

Bhadana, V.P.
Senior Scientist (Genetics)
Indian Institute of Rice Research
Rajendranagar, Hyderabad
Telangana – 500030

Biswas, A.K.
Principal Scientist & HOD
(Soil Chemistry & Fertility)
ICAR-Indian Institute of Soil Science
Nabi Bagh, Berasia Road, Bhopal – 462038

Brajendra
Indian Institute of Rice Research
Rajendranagar, Hyderabad
Telangana – 500030

Chaudhary, R.S.
Principal Scientist & HOD (Soil Physics)
ICAR-Indian Institute of Soil Science
Nabi Bagh, Berasia Road, Bhopal – 462038

Dotaniya, M.L.
Scientist (Soil Science)
ICAR-Indian Institute of Soil Science
Nabi Bagh, Berasia Road, Bhopal – 462038

Elanchezhian, R.
Principal Scientist (Plant Physiology)
ICAR-Indian Institute of Soil Science
Nabi Bagh, Berasia Road, Bhopal – 462038

Hati, K.M.
Principal Scientist (Soil Science)
ICAR-Indian Institute of Soil Science
Nabi Bagh, Berasia Road, Bhopal – 462038

H.K. Mahadevaswamy
Indian Institute of Rice Research
Rajendranagar, Hyderabad
Telangana – 500030

Jha, Pramod
Senior Scientist (Soil Science)
ICAR-Indian Institute of Soil Science
Nabi Bagh, Berasia Road, Bhopal – 462038

Kundu, S.
Principal Scientist (Soil Science)
ICAR-Indian Institute of Soil Science
Nabi Bagh, Berasia Road, Bhopal – 462038

Lakaria, B.L.
Principal Scientist (Soil Science)
ICAR-Indian Institute of Soil Science
Nabi Bagh, Berasia Road, Bhopal – 462038

Manna, M.C.
Principal Scientist & HOD (Soil Biology)
ICAR-Indian Institute of Soil Science
Nabi Bagh, Berasia Road, Bhopal – 462038

Manoj Kumar
ICAR- National Institute of High Security
Animal Diseases, Anand Nagar
Bhopal – 462022

Meena, B.P.
Scientist (Agronomy)
ICAR-Indian Institute of Soil Science
Nabi Bagh, Berasia Road
Bhopal – 462038

Mohanty, M.
Senior Scientist (Soil Science)
ICAR-Indian Institute of Soil Science
Nabi Bagh, Berasia Road
Bhopal – 462038

Neenu, S.
Scientist (Soil Science)
ICAR-Indian Institute of Soil Science
Nabi Bagh, Berasia Road
Bhopal – 462038

Neeraja, C.N.
Principal Scientist (Genetics)
Indian Institute of Rice Research
Hyderabad

P. Dey
Principal Scientist & PC (STCR)
ICAR-Indian Institute of Soil Science
Nabi Bagh, Berasia Road
Bhopal – 462038

P.K. Tiwari
ICAR-Indian Institute of Soil Science
Nabi Bagh, Berasia Road
Bhopal – 462 038

Pandey, Renu
Senior Scientist (Plant Physiology)
Division of Plant Physiology
IARI, New Delhi – 110012

Patra, A.K.
Director
ICAR-Indian Institute of Soil Science
Nabi Bagh, Berasia Road
Bhopal – 462038

P. Senguttuvel
Indian Institute of Rice Research
Rajendranagar, Hyderabad
Telangana – 500030

Rajukumar, K
Senior Scientist (Veterinary Pathology)
ICAR-National Institue for High Security Animal Diseases, Anand Nagar, Bhopal

Ramesh, K.
Principal Scientist (Agronomy)
ICAR-Indian Institute of Soil Science
Nabi Bagh, Berasia Road, Bhopal – 462038

Rashmi, I.
Scientist (Soil Science)
ICAR-Indian Institute of Soil Science
Nabi Bagh, Berasia Road, Bhopal – 462038

R.H. Wanjari
ICAR-Indian Institute of Soil Science
Nabi Bagh, Berasia Road
Bhopal – 462038

R.M. Sundaram
Indian Institute of Rice Research
Rajendranagar, Hyderabad
Telangana – 500030

S. Rajendiran
Scientist (Soil Science)
ICAR-Indian Institute of Soil Science
Nabi Bagh, Berasia Road, Bhopal
Madhya Pradesh – 462038

S. Srivastava
ICAR-Indian Institute of Soil Science
Nabi Bagh, Berasia Road
Bhopal – 462 038

S. Kota
Indian Institute of Rice Research
Rajendranagar, Hyderabad
Telangana – 500030

Saha, J.K.
Principal Scientist & HOD (Env. Soil Science)
ICAR-Indian Institute of Soil Science
Nabi Bagh, Berasia Road, Bhopal – 462038

Shirale, A.O.
Scientist (Soil Science)
ICAR-Indian Institute of Soil Science
Nabi Bagh, Berasia Road , Bhopal – 462038

Singh, Muneshwar
Principal Scientist & PC (LTFE)
ICAR-Indian Institute of Soil Science
Nabi Bagh, Berasia Road
Bhopal – 462038

Sinha, N.K.
Scientist (Soil Science)
ICAR-Indian Institute of Soil Science
Nabi Bagh, Berasia Road
Bhopal – 462038

Somasundaram, J.
Senior Scientist (Soil Science)
ICAR-Indian Institute of Soil Science
Nabi Bagh, Berasia Road
Bhopal – 462038

Steven Reeves
Soil Processes, Department of Science
Information Technology and Innovation
(DSITI)
Ecosciences Precinct, Boggo Road
Dutton Park, Brisbane, Queensland
Australia

Subba Rao, A.
Ex. Director
ICAR-Indian Institute of Soil Science
Bhopal – 462038

Thakur, J.K.
Scientist (Agricultural Microbiology)
ICAR-Indian Institute of Soil Science,
Nabi Bagh, Berasia Road
Bhopal – 462038

Vassanda Coumar, M.
Scientist (Soil Science)
ICAR-Indian Institute of Soil Science
Nabi Bagh, Berasia Road
Bhopal – 462038

Voleti, S.R.
Principal Scientist (Plant Physiology)
Indian Institute of Rice Research
Rajendranagar, Hyderabad
Telangana – 500030

Contents

Advances in Nutrient Dynamics in Soil - Plant System, pp 1-8

Editors: R. Elanchezhian, A.K. Biswas, K. Ramesh and A.K. Patra

1

Nutrient Management and Nutrient Use Efficiency: Indian Perspective

A.K. Patra and R. Elanchezhian

ICAR-Indian Institute of Soil Science, Nabi Bagh, Bhopal – 462038, India

Plants require altogether 17 essential elements to nourish themselves and to complete their reproduction and life cycle. Among them nine elements are required in large amounts for their growth, the so called macro nutrients like C, H, O, N, P, K, Ca Mg and S. The former three elements (C, H and O) are available to the plants through air and water over which we have relatively little control either in their application rates or uptake into plant system. Of the remaining macronutrients N, P and K are being utilized in large amounts by the farmers as fertilizers globally for the nutrient supplementation of the crop over the past half century. Presently the modern agricultural practices feed more than 6 billion people of the world. This was possible due to the green revolution technologies which have doubled the global cereal production in the past five decades (Fig. 1) mainly from the increased crop yields resulting from greater inputs of fertilizer, water and pesticides, new crop strains, and other

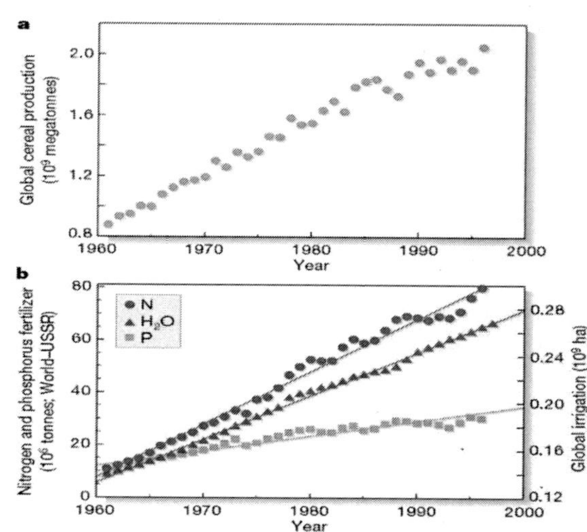

Fig. 1: Global cereal production (a) and nitrogen and phosphorus fertilizer use (b) (Adapted from Tilman *et al.*, 2002)

technologies (Tilman *et al.,* 2002). On a global scale, agricultural production systems are divided into two major regions viz. 1) regions where nutrients are used in excess and 2) regions where nutrient inputs are insufficient to provide adequate crop production. However, among the developing nations, India and China were found to be the major consumers of fertilizers of N, P and K.

Intensive agricultural production system is dependent on addition of nitrogenous fertilizers, especially industrially produced ammonia and nitrate. However, in some regions of the world, crop production is still constrained by too little application of fertilizers. Without the use of synthetic fertilizers, world food production could not have increased at the rate it did and more natural ecosystems would have been converted to agriculture. Between 1960 and 1995, global use of nitrogenous fertilizers have increased sevenfold, while phosphorus 3.5-fold. Moreover both are expected to increase another threefold by 2050 unless there is a substantial increase in fertilizer efficiency (Tilman *et al.,* 2002). Hence nutrient use efficiency (NUE) of fertilizers is an important criterion for improving the input use efficiency and reducing the burden on land resources. Nutrient use efficiency can be improved mainly by fertilizer management as well as by soil- and plant-water management. The sole aim of nutrient use is to increase the overall performance of crop and cropping systems by providing economically optimum nourishment to the crop with minimal nutrient losses from the field.

At global level, the temporal trends in NUE vary from region to region. For N, P and K, partial nutrient balance (ratio of nutrients removed by crop harvested to fertilizer nutrients applied) and partial factor productivity (crop production per unit of nutrient applied) for Africa, North America, and Europe are showing an increasing trends, while the reverse for Latin America, India, and China (Fixen *et al.,* 2014). This has become a serious challenge to developing countries like India and China, which are the major consumers of those macro elemental fertilizers, wherein the response ratio is declining steadily and rapidly. Current scenario of the nutrient use efficiency of major nutrients and the causes for low efficiency are enumerated in Table 1.

Table 1: Nutrient use efficiency of major nutrients and the causes for their low efficiency

Nutrient	Efficiency (%)	Cause of low efficiency
Nitrogen	30-50	Immobilization, volatilization, denitrification, Leaching
Phosphorus	15-20	Fixation in soils Al – P, Fe – P, Ca – P
Potassium	70-80	Fixation in clay - lattices
Sulphur	8-10	Immobilization, Leaching with water
Micro nutrients (Zn, Fe, Cu, Mn, B)	1-2	Fixation in soils

In India, similar to global trends, the food grain production has steadily increased while the use of nitrogenous fertilizers also increased tremendously (Fig. 2 and Fig. 3).

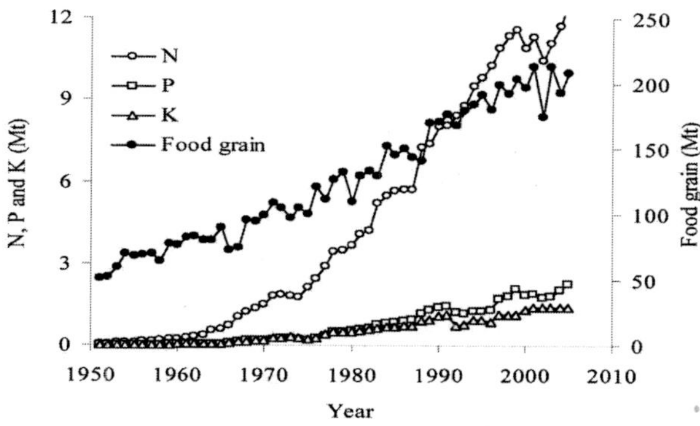

Fig. 2: Food grain production and fertilizer use in India

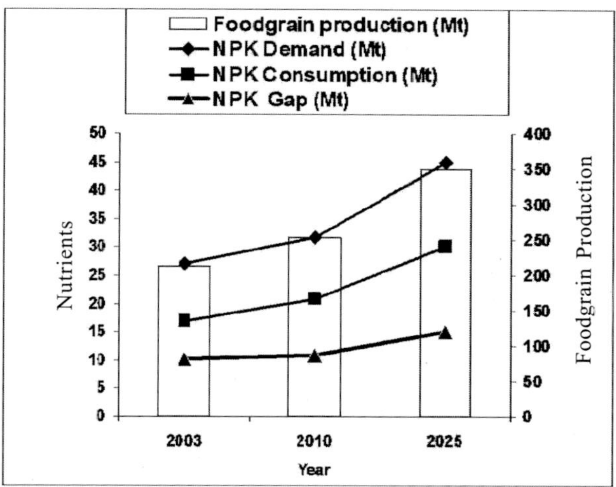

Fig. 3: Projected food grain production and fertilizer demand and consumption in India

India is the second largest producer and consumer of nitrogenous fertilizers and the second largest consumer of phosphatic fertilizers in the world. China and USA are the other two countries competing with India in terms of production and consumption of nitrogenous, and phosphatic fertilizers. India is totally dependent on imports for potassic fertilizers, imports large amounts of phosphatic fertilizers and the raw materials required for their production. In terms of overall consumption, India has attained a level of 52.5 million tons of fertilizers of various types which is represented by a consumption of 24.4 million tons of nutrients comprising 16.7 million tons of nitrogen, 5.6 million tons of phosphorus and 2.1 million tons of potash during 2013-14 (Table 2). The average fertilizer consumption per hectare is 144.14 kg of NPK which is, however, far less than China (289 kg),

Bangladesh (197 kg) and many European countries as per the compendium on soil health released by DAC, New Delhi (2012). The fertilizer consumption of different states of India is given in Fig. 4.

Table 2: Consumption of major chemical fertilizers (N, P and K) in India since 2001-02 (in lakh tons)

Year	Nitrogen (N)	Phosphate (P)	Potash (K)	Total (N+P+K)
2001–02	113.10	43.82	16.67	**173.60**
2002–03	104.74	40.19	16.01	**160.94**
2003–04	110.77	41.24	15.98	**167.99**
2004–05	117.13	46.24	20.61	**183.98**
2005–06	127.23	52.04	24.13	**203.40**
2006–07	137.73	55.43	23.35	**216.51**
2007-08	144.19	55.15	26.36	**225.70**
2008-09	150.90	65.06	33.13	**249.09**
2009-10	155.80	72.74	36.32	**264.86**
2010-11	165.58	80.50	35.14	**281.22**
2011-12	173.00	79.14	25.75	**277.90**
2012-13	168.21	66.53	20.62	**255.36**
2013-14	167.50	56.33	20.99	**244.82**

Source: DAC Annual report 2014-15

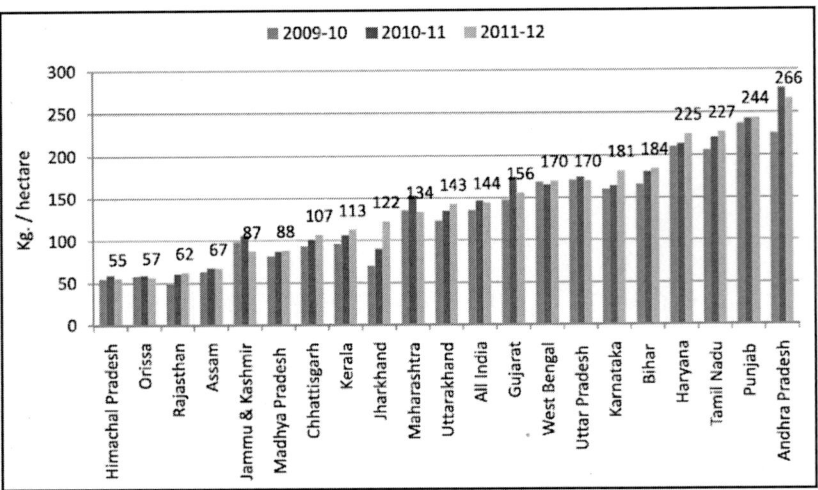

Fig. 4: Fertilizer NPK consumption across different states of India (DAC Annual report 2014-15)

Besides, the fertilizer response ratio is declining since the last five decades (Fig. 5) in India which indicates that the nutrient use efficiency is reducing drastically and point toward the poor soil health.

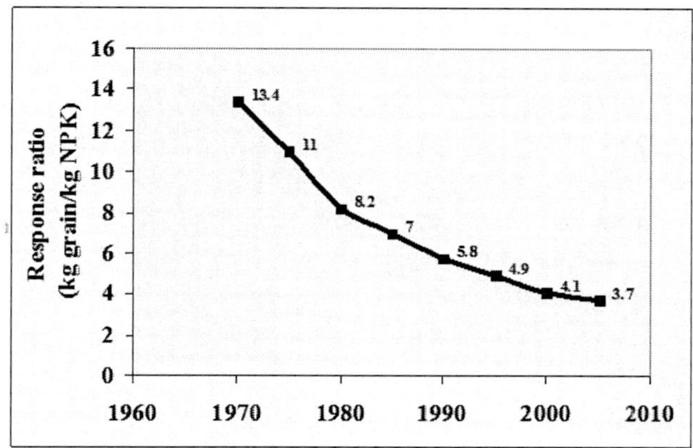

Fig. 5: Declining fertilizer NPK response ratio in India

The NPK ratio of 4:2:1 is considered desirable for the major crops except pulses. In India, it is estimated that by the year 2025, total food grain requirement will be ~300 million tonnes (mt) and to achieve this, about 13.1 mt of P_2O_5 would be required (based on current usage pattern) besides 22.4 mt of nitrogen (N) and potassium (K) with an average N: P_2O_5: K_2O ratio of 4: 2: 1 (Elanchezhian *et al.,* 2015). However it is glaringly far from being balanced nourishment across India (Table 3). Except the states in south and east zone which are more or less closer to the ideal 4:2:1 ratio of balanced fertilization, the ratio was greatly skewed towards nitrogenous and phosphatic fertilizers in north and west zone. The all India average is thus 6.7:3.1:1, which gives tremendous scope for balanced fertilization at national level. However, the fertilizer dose has to be worked out based on soil analysis to find out (i) available nutrient status of the soils and (ii) the crop requirement of the nutrients; the difference of the two (ii – i) is the required fertilizer dose for a given crop.

Table 3: NPK consumption ratios across India

Year	2007-08			2008-09			2009-10			2010-11			2011-12		
Zone	N	P	K	N	P	K	N	P	K	N	P	K	N	P	K
SOUTH	2.3	1.0	1.0	2.2	1.1	1.0	2.4	1.1	1.0	2.7	1.3	1.0	3.7	1.7	1.0
EAST	5.7	2.2	1.0	3.9	1.7	1.0	3.5	1.6	1.0	4.4	2.0	1.0	5.9	2.2	1.0
WEST	11.9	5.0	1.0	10.4	5.2	1.0	8.5	4.3	1.0	9.3	4.9	1.0	13.4	6.8	1.0
NORTH	20.8	5.6	1.0	15.9	4.9	1.0	11.5	3.7	1.0	12.3	4.0	1.0	17.6	5.7	1.0
INDIA	5.5	2.1	1.0	4.6	2.0	1.0	4.3	2.0	1.0	4.7	2.3	1.0	6.7	3.1	1.0

South Zone comprises- AP, TN, Karnataka, Kerala & Pondicherry; East zone comprises Bihar, Jharkhand, WB, Orissa & N.E states; West zone comprises - Gujarat, Maharastra, Rajasthan Madhya pradesh, Chattisgharh, Goa; North zone comprises Haryana, Punjab, UP, Uttaranchal, HP, J&K

Source: Indian Fertilizer Scenario 2013

Inherently, Indian soils show deficiency of primary nutrients (Nitrogen and Phosphorous) as well as secondary nutrients (Sulphur, Calcium and Magnesium) and micro nutrients (Boron, Zinc, Copper and Iron) in most parts of the country (Table 4 and Table 5). In a comprehensive study carried out by ICAR-IISS through their All India Coordinated Research Project on STCR and Micronutrients, revealed the deficiency status of macro and micro nutrients at all India level, which has become a limiting factor in increasing crop productivity. Such soils require balanced and integrated management of nutrients for sustaining the soil fertility in the long run. Earlier, Tandon (2007) has reported nutrient mining of soils by crops leading to negative balance of major nutrients (Table 6).

Table 4: Extent of macro nutrient deficiency in India

Nutrient	No. of samples analyzed	Percent of samples by category		
		Low	Medium	High
Nitrogen	3,650,004	63	26	11
Phosphorus	3,650,004	42	38	20
Potassium	3,650,004	13	37	50
Sulphur	27,000	40	35	25

Source: AICRP on STCR and MSNPE, IISS

Table 5: Extent of micronutrient deficiency in India

Nutrient	No. of samples analyzed	Percent of samples deficient
Zn (DTPA extract)	97,464	43.0
Fe (DTPA extract)		12.1
Cu (DTPA extract)		5.4
Mn (DTPA extract)		5.5
B (Hot water soluble)	73360	18.3

Source: AICRP on MSNPE, IISS; Shukla *et al.*, (2014)

Table 6. Nutrient mining in soils of India (mt)

Nutrient	Gross balance			Net balance		
	Addition	Removal	Balance	Addition	Removal	Balance
N	10.9	9.6	1.3	5.5	7.7	-2.2
P_2O_5	4.2	3.7	0.5	1.5	3.0	-1.5
K_2O	1.4	11.6	-10.2	1.0	7.0	-6.0
Total	16.5	24.9	-8.4	8.0	17.7	-9.7

Source: Tandon (2007)

Management practices for improving nutrient use efficiency

Depletion in soil organic carbon is leading to poor fertilizer use efficiency (FUE) of the soil which on an average is estimated to be 33% for N; 15% for P; 20% for K and 3-5% for micronutrients as against 50% for N; 30% for P and 50% for K with the best management practices. Intensive agriculture, while increasing food production, has at the same time caused second generation problems in respect of nutrient imbalance including greater mining of soil nutrients, depletion of soil fertility, emerging deficiencies of secondary and micronutrients, decline in the water table and quality of water, decreasing organic carbon content, and overall deterioration in soil health. Hence the following nutrient management practices and remedial measures are advocated for improving nutrient use efficiency.

Integrated nutrient management: Government is promoting Integrated Nutrient Management (INM), advocating soil test based balanced and judicious use of chemical fertilizers in conjunction with organic sources of nutrients for improving soil fertility. Introduction of customized fertilizers on the basis of soil testing and the agronomic multi-locational trials which are crop specific and area specific are recommended. Promotion of INM which includes soil test based balanced and judicious use of chemical fertilizers in conjunction with bio-fertilizers, and organic manures like FYM, compost, vermi-compost, green manure, fruit and vegetable waste compost, MSW compost etc.; use of complex fertilizers (NPK) and customized fertilizers which are considered to be agronomically superior and more balanced fertilizers in place of straight fertilizers; use of fertilizers fortified with micro-nutrients; use of Bio-fertilizers - phosphate solubilizing bacteria; *Azospirillum*, *Azotobacter*, and *Rhizobium*; potash mobilizing bio-fertilizers which can supplement upto 20-25% of chemical fertilizers (NPK).

STCR based fertilizer recommendations: It is an inductive approach, a refined method of fertilizer recommendation for varying soil test values developed by All India Coordinated Research Project Soil Test Crop Response (AICRP-STCR) located at ICAR-IISS Bhopal, for different crops under different agro-ecological sub regions. Soil Test Crop Response studies have used the targeted yield approach to develop relationship between crop yield on the one hand, and soil test estimates and fertilizer inputs, on the other. Considerable agronomic and economic benefits were accrued when farmers applied fertilizer nutrient doses based on soil test. Lately, the calibrations are being developed under integrated supply of organics and fertilizers keeping into account the nutrient contribution of organics, soil and fertilizers. It is evident that STCR based approach of nutrient application has definite advantage in terms of increasing nutrient response ratio over general recommended dose of nutrient application.

Conclusion

In conclusion, it is reiterated that widespread nutrient deficiencies and deteriorating soil health are major causes of low nutrient use efficiency, productivity and profitability of the agricultural production system in India. Hence adoption of site-specific balanced and integrated nutrient management involving major, secondary and micro nutrients, organic manures, bio-fertilizers and suitable amendments for problem soils is very much required. Besides, suitable policy environment for more investments in the fertilizer sector for sustained supplies of fertilizers is needed. Moreover, utilizing all indigenously available nutrient sources to reduce dependence on imports can be explored. Developing new efficient fertilizer products/ approaches through state of art R & D applications is also the need of the hour. This also warrants effective soil testing service to back up precise fertilizer use and creating awareness amongst farmers on benefits of site-specific balanced fertilization and integrated nutrient management.

References

Annual Report (2014-15). Department of Agriculture & Cooperation, Ministry of Agriculture and Farmers' Welfare, GOI, New Delhi.

Annual Report of AICRP on Micro, Secondary Nutrients and Polluted Elements (2014). ICAR-Indian Institute of Soil Science, Bhopal.

Annual Report of AICRP on Soil Test Crop Response (2013). ICAR-Indian Institute of Soil Science, Bhopal.

Compendium on soil health (2012). INM Division, Department of Agriculture & Cooperation, New Delhi.

Elanchezhian, R., Krishnapriya, V., Pandey, R., Subba Rao, A and Abrol, Y.P. (2015). Physiological and molecular approaches for improving phosphorus uptake efficiency of crops, *Current Science*, 108 (7): 1271-1279.

Fixen, P., Brentrup, F., Bruulsema, T., Garcia, F., Norton, R and Zingore, S. (2014). Nutrient/ Fertilizer Use Efficiency: Measurement, Current Situation and Trends. In *Managing Water and Fertilizer for Sustainable Agricultural Intensification*, IFA, IWMI, IPNI and IPI.

Indian Fertilizer Scenario 2013. Department of Fertilizers, Ministry of Chemicals and Fertilizers, Government of India, New Delhi.

Shukla, A.K., Tiwari, P.K. and Prakash, C. (2014). Micronutrients deficiencies vis-a-vis food and nutritional security of India. *Indian Journal of Fertilizers*, Vol. 10(12): 94-112.

Tandon, H.L.S. (2007). Soil Nutrient balance sheets in India: Importance, status, issues and concerns. *Better crops – India*, Nov 7: 15-19.

Tilman, D., Cassman, K.G., Matson, P.A., Naylor, R. and Polasky, S. (2002). Agricultural sustainability and intensive production practices. *Nature,* 418: 671-677.

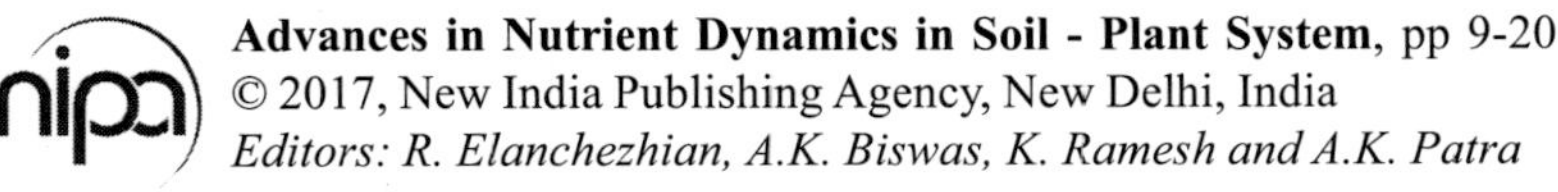

Advances in Nutrient Dynamics in Soil - Plant System, pp 9-20

Editors: R. Elanchezhian, A.K. Biswas, K. Ramesh and A.K. Patra

2

Nutrient Dynamics in Soil-Plant System–NPK

A.K. Biswas and R. Elanchezhian

ICAR-Indian Institute of Soil Science, Nabi Bagh, Bhopal – 462038, India

Soil is a major source of nutrients needed by plants for growth. The three main nutrients are nitrogen (N), phosphorus (P) and potassium (K). Together they make up the trio known as NPK. Other important nutrients are calcium, magnesium and sulfur. Plants also need small quantities of iron, manganese, zinc, copper, nickel, chlorine, boron and molybdenum, known as trace elements because only traces are needed by the plant. The primary nutrients are nitrogen, phosphorus and potassium, because they are required in larger quantities than other nutrients. These three elements form the basis of the N-P-K label on commercial fertilizer bags. As a result, the management of these nutrients is very important. However, the primary nutrients are no more important than the other essential elements since all essential elements are required for plant growth. Remember that the 'Law of the Minimum' tells us that if deficient, any essential nutrient can become the controlling force in crop yield.

Nitrogen

Environmental and economic issues combined have increased the need to better understand the role and fate of nitrogen (N) in crop production systems. Nitrogen is the nutrient most often deficient for crop production in most soils and its use can result in substantial economic return for farmers. However, when N inputs to the soil system exceed crop needs, there is a possibility that excessive amounts of nitrate (NO_3^-) may enter either ground or surface water.

Managing N inputs to achieve a balance between profitable crop production and environmentally tolerable levels of NO_3^- in water supplies should be every grower's goal. The behavior of N in the soil system is complex, yet an understanding of these basic processes is essential for a more efficient N management program.

Nitrogen exists in the soil system in many forms and changes (transforms) very easily from one form to another. The route that N follows in and out of the soil system is collectively called the "nitrogen cycle" and is biologically influenced. Biological processes, in turn, are influenced by prevailing climatic conditions along with the physical and chemical properties of a particular soil. Both climate and soils vary greatly across the agro-ecoregions and affect the N transformations for the different areas.

Sources of N for plant growth

Nitrogen can be supplied for plant growth from several sources

- The atmosphere
- Biological fixation
- Atmospheric fixation
- Precipitation
- Commercial fertilizers
- Soil organic matter
- Crop residues
- Animal manures

Atmospheric N: It is the major reservoir for N in the N cycle (air is 79% N_2 gas). Although unavailable to most plants, large amounts of N_2 can be used by leguminous plants via N fixation. In this biological process, nodule-forming Rhizobium bacteria inhabit the roots of leguminous plants and through a symbiotic relationship convert atmospheric N_2 to a form the plant can use. The amount of N_2 fixed by legumes into usable N can be substantial. Any portion of a legume crop, that is left after harvest, including roots and nodules, supplies N to the soil system. When the plant material is decomposed, N is released. Several non-symbiotic organisms exist that fix N, but N additions from these organisms are quite low. In addition small amounts of N are added to soil from precipitation.

Commercial N fertilizers: They are also derived from the atmospheric N pool. The major step is to combine $N_2$with hydrogen (H_2) to form ammonia (NH_3). Anhydrous ammonia is then used as a starting point in the manufacture of other nitrogen fertilizers. Anhydrous ammonia or other N products derived from NH_3 can then supplement other N sources for crop nutrition.

Organic N sources: Nitrogen can also become available for plant use from organic N sources which must be converted to inorganic forms before they are available to plants. Nitrogen is available to plants as either ammonium (NH_4^+) or nitrate (NO_3^-). Animal manures and other organic wastes can be important sources of N for plant growth. The amount of N supplied by manure will vary with the type of livestock, handling, rate applied, and method of application. Since the N form and content of manures varies widely, an analysis of manure is recommended to improve N management.

Crop residues: Residues from non-leguminous plants also contain N, but in relatively small amounts compared with legumes. Nitrogen exists in crop residues in complex organic forms and the residue must decay (a process that can take several years) before N is made available for plant use.

Soil organic matter: It is also a major source of N used by crops. Organic matter is composed primarily of rather stable material called humus that has collected over a long period of time. Easily decomposed portions of organic material disappear relatively quickly, leaving behind residues more resistant to decay. Decomposition of this portion of organic matter proceeds at a rather slow rate.

Nitrogen transformations

Nitrogen, present or added to the soil, is subject to several changes (transformations) that dictate the availability of N to plants and influence the potential movement of NO_3^- to water supplies.

Organic N that is present in soil organic matter, crop residues, and manure is converted to inorganic N through the process of **mineralization**. In this process, bacteria digest organic material and release ammonium (NH_4^+) nitrogen. Formation of NH_4^+ increases as microbial activity increases. Bacterial growth is directly related to soil temperature and water content. The ammonium supplied from fertilizers is the same as the ammonium supplied from organic matter.

Ammonium-N has properties that are of practical importance for N management. Plants can absorb NH_4^+ -N. Ammonium also has a positive charge and, therefore, is attracted or held by negatively charged soil and soil organic matter. This means that NH_4^+ does not move downward in soils. Nitrogen in the ammonium form that is not taken up by plants is subject to other changes in the soil system.

Nitrification is the conversion of NH_4^+ -N to NO_3^- -N. Nitrification is a biological process and proceeds rapidly in warm, moist, well-aerated soils. Nitrification slows at soil temperatures below 50 degrees F. Nitrate is a negatively charged ion and is not attracted to soil particles or soil organic matter like NH_4^+. Nitrate-N is water soluble and can move below the crop rooting zone under certain conditions.

Denitrification is a process by which bacteria convert NO_3^- to N gases that are lost to the atmosphere. Denitrifying bacteria use NO_3^- instead of oxygen in the metabolic processes. Denitrification takes place where there is waterlogged soil and where there is ample organic matter to provide energy for bacteria. For these reasons, denitrification is generally limited to topsoil. Denitrification can proceed rapidly when soils are warm and become saturated for 2 or 3 days.

A temporary reduction in the amount of plant-available N can occur from **immobilization** (tie up) of soil N. Bacteria that decompose high carbon-low N residues, such as corn stalks or small grain straw, need more N to digest the material than is present in the residue. Immobilization occurs when nitrate and/or ammonium present in the soil is used by the growing microbes to build proteins. The actively growing bacteria that immobilize some soil N also break down soil organic matter to release available N during the growing season. There is often a net gain of N during the growing season because the additional N in the residue will be the net gain after immobilization-mineralization processes.

Nitrogen loss from the soil system

Nitrogen is lost from the soil system in several ways:

- Leaching
- Denitrification
- Volatilization
- Soil erosion and runoff

In contrast to the biological transformations previously described, loss of nitrate by **leaching** is a physical event. Leaching is the loss of soluble NO_3^- as it moves with soil water, generally excess water, below the root zone. Nitrate that moves below the root zone has potential to enter either groundwater or surface water through tile drainage systems.

Coarse-textured soils have a lower water-holding capacity and, therefore, a higher potential to lose nitrate from leaching when compared with fine-textured soils. Nitrate can be leached from any soil if rainfall or irrigation moves water through the root zone.

Denitrification can be a major loss mechanism of NO_3^- when soils are saturated with water for 2 or 3 days. Nitrogen in the NH_4^+ form is not subject to this loss. Management alternatives are available if denitrification losses are a potential problem.

Significant losses from some surface-applied N sources can occur through the process of **volatilization**. In this process, N is lost as the ammonia (NH_3) gas. Nitrogen can be lost in this way from manure and fertilizer products containing urea. Ammonia is an intermediate form of N during the process in which urea is transformed to NH_4^+. Incorporation of these N sources will virtually eliminate volatilization losses. Loss of N from volatilization is greater when soil pH is higher than 7.3, the air temperature is high, the soil surface is moist, and there is a lot of residue on the soil.

Nitrogen can be lost from agricultural lands through **soil erosion and runoff**. Losses through these events do not normally account for a large portion of the soil N budget, but should be considered for surface water quality issues.

Incorporation or injection of manure and fertilizer can help to protect against N loss through erosion or runoff. Where soils are highly erodible, conservation tillage can reduce soil erosion and runoff, resulting in less surface loss of N.

In considering the many transformations and reactions of N in soils, there are some major points to keep in mind. Although N can be added to soil in either organic or inorganic forms, plants take up only inorganic N (that is, NO_3^- and NH_4^+). One form is not more important than the other and all sources of N can be converted to nitrate. Commercial N fertilizers, legumes, manures, and crop residues are all initial sources of NO_3^- and NH_4^+ and once in the plant or in the water supply it is impossible to identify the initial source.

Nitrate is always present in the soil solution and will move with the soil water. Inhibiting the conversion of NH_4^+ to NO_3^- can result in less N loss and more plant uptake; however, it is not possible to totally prevent nitrification. There is no way to totally prevent the movement of some NO_3^- to water supplies, but sound management practices can keep losses within acceptable limits.

Phosphorus

With increasing demand of agricultural production and as the peak in global production will occur in the next decades, phosphorus (P) is receiving more attention as a nonrenewable resource (Cordell *et al.,* 2009; Gilbert, 2009). One unique characteristic of P is its low availability due to slow diffusion and high fixation in soils. All of this means that P can be a major limiting factor for plant growth. Applications of chemical P fertilizers and animal manure to agricultural land have improved soil P fertility and crop production, but caused environmental damage in the past decades. Maintaining a proper P-supplying level at the root zone can maximize the efficiency of plant roots to mobilize and acquire P from the rhizosphere by an integration of root morphological and physiological adaptive strategies. Furthermore, P uptake and utilization by plants plays a vital role in the determination of final crop yield. A holistic understanding of P dynamics from soil to plant is necessary for optimizing P management and improving P-use efficiency, aiming at reducing consumption of chemical P fertilizer, maximizing exploitation of the biological potential of root/rhizosphere processes for efficient mobilization, and acquisition of soil P by plants as well as recycling P from manure and waste. Taken together, overall P dynamics in the soil-plant system is a function of the integrative effects of P transformation, availability, and utilization caused by soil, rhizosphere, and plant processes.

P DYNAMICS IN SOIL

Soil P Transformation

Soil P exists in various chemical forms including inorganic P (Pi) and organic P (Po). These P forms differ in their behavior and fate in soils (Hansen *et al.*, 2004; Turner *et al.*, 2007). Pi usually accounts for 35% to 70% of total P in soil (calculation from Harrison, 1987). Primary P minerals including apatites, strengite, and variscite are very stable, and the release of available P from these minerals by weathering is generally too slow to meet the crop demand though direct application of phosphate rocks (i.e. apatites) has proved relatively efficient for crop growth in acidic soils. In contrast, secondary P minerals including calcium (Ca), iron (Fe), and aluminum (Al) phosphates vary in their dissolution rates, depending on size of mineral particles and soil pH (Pierzynski *et al.*, 2005; Oelkers and Valsami-Jones, 2008). With increasing soil pH, solubility of Fe and Al phosphates increases but solubility of Ca phosphate decreases, except for pH values above 8 (Hinsinger, 2001). The P adsorbed on various clays and Al/Fe oxides can be released by desorption reactions. All these P forms exist in complex equilibria with each other, representing from very stable, sparingly available, to plant-available P pools such as labile P and solution P.

In acidic soils, P can be dominantly adsorbed by Al/Fe oxides and hydroxides, such as gibbsite, hematite, and goethite (Parfitt, 1989). P can be first adsorbed on the surface of clay minerals and Fe/Al oxides by forming various complexes. The nonprotonated and protonated bidentate surface complexes may coexist at pH 4 to 9, while protonated bidentate innersphere complex is predominant under acidic soil conditions (Luengo *et al.*, 2006; Arai and Sparks, 2007). Clay minerals and Fe/Al oxides have large specific surface areas, which provide large number of adsorption sites. The adsorption of soil P can be enhanced with increasing ionic strength. With further reactions, P may be occluded in nanopores that frequently occur in Fe/Al oxides, and thereby become unavailable to plants (Arai and Sparks, 2007).

In neutral-to-calcareous soils, P retention is dominated by precipitation reactions (Lindsay *et al.*, 1989), although P can also be adsorbed on the surface of Ca carbonate (Larsen, 1967) and clay minerals (Devau *et al.*, 2010). Phosphate can precipitate with Ca, generating dicalcium phosphate (DCP) that is available to plants. Ultimately, DCP can be transformed into more stable forms such as octocalcium phosphate and hydroxyapatite (HAP), which are less available to plants at alkaline pH (Arai and Sparks, 2007). HAP dissolution increases with decrease of soil pH (Wang and Nancollas, 2008), suggesting that rhizosphere acidification may be an efficient strategy to mobilize soil P from calcareous soil.

Po generally accounts for 30% to 65% of the total P in soils (Harrison, 1987). Soil Po mainly exists in stabilized forms as inositol phosphates and phosphonates, and active forms as orthophosphate diesters, labile orthophosphate monoesters, and organic polyphosphates (Turner *et al.,* 2002; Condron *et al.,* 2005). The Po can be released through mineralization processes mediated by soil organisms and plant roots in association with phosphatase secretion. These processes are highly influenced by soil moisture, temperature, surface physical-chemical properties, and soil pH and Eh (for redox potential). Po transformation has a great influence on the overall bioavailability of P in soil (Turner *et al.,* 2007). Therefore, the availability of soil P is extremely complex and needs to be systemically evaluated because it is highly associated with P dynamics and transformation among various P pools.

Chemical Fertilizer P in Soil

The modern terrestrial P cycle is dominated by agriculture and human activities (Oelkers and Valsami-Jones, 2008). The concentration of available soil Pi seldom exceeds 10 μM (Bieleski, 1973), which is much lower than that in plant tissues where the concentration is approximately 5 to 20 mM Pi (Raghothama, 1999). Because of the low concentration and poor mobility of plant-available P in soils, applications of chemical P fertilizers are needed to improve crop growth and yield. The major forms of phosphate fertilizers include monocalcium phosphate (MCP) and monopotassium phosphate. Application of MCP can significantly affect soil physicochemical properties. After application to soil, MCP undergoes a wetting process, generates large amounts of protons, phosphate, and DCP, and eventually forms a P-saturated patch (Benbi and Gilkes, 1987). This Pi-saturated patch forms three different reaction zones including direct reaction, precipitation reaction, and adsorption reaction zones. The direct reaction zone is very acidic (pH = 1.0–1.6), resulting in enhanced mobilization of soil metal ions. These metal ions can also react with high concentrations of Pi in the zone thus causing further precipitation of Pi. The amorphous Fe-P and Al-P that thereby form can be partly available to plants. In calcareous soil, new complexes of MCP and DCP can be formed and with time DCP is gradually transformed into more stable forms of Ca phosphates (octocalcium phosphate or apatite). Because the Pi concentration is relatively low, P adsorption by soil minerals is dominant in the outer zone (Moody *et al.,* 1995). In contrast, the application of monopotassium phosphate has little influence on soil physical and chemical properties (Lindsay *et al.,* 1962). Therefore, matching P fertilizer types with soil physical and chemical properties may be an efficient strategy for rational use of chemical fertilizer P.

Manure P in Soil

Manure can be applied to soil to increase P fertility. The total P content in manure is very variable and nearly 70% of total P in manure is labile. In manure, Pi accounts for 50% to 90% (Dou *et al.,* 2000). Manure also contains large amounts of Po, such as phospholipids and nucleic acids (Turner and Leytem, 2004), which can be released to increase soil Pi concentrations by mineralization. Furthermore, small molecular organic acids from mineralization of humic substances in manure can dissolve Ca phosphate, and especially for citrate, it can efficiently weaken the nanoparticle stability of HAP, by controlling the free Ca availability and thereby the nucleation rate (Martins *et al.,* 2008). P adsorption to soil particles can be greatly reduced through applying organic substances. The humic acids contain large numbers of negative charges, carboxyl and hydroxyl groups, which strongly compete for the adsorption sites with Pi. Manure can also change soil pH and thus alter soil P availability. However, mechanisms of manure-induced P transformation processes between Pi and Po in soil still need further investigation.

POTASSIUM

Soil analysis is an important tool when evaluating soil nutrient status: The results of soil tests are frequently taken as a basis for fertilizer recommendations. This is justified in such cases where a correlation exists between soil test results and crop response to fertilizer application. As a rule, the effect of a fertilizer nutrient should be the lower the higher its content in the soil.

As to potassium, however, many cases are known in which no correlation has been found between soil test data and yield response to potash application. In hundreds of trials with rice in India good responses to applied potash were observed on soils testing high in "available" or exchangeable K, but sometimes low or no effects of potash on soils with poor K status.

In extreme cases, even negative correlations may exist between exchangeable K and yield, while the correlation between the K concentration of the soil solution and the yield is positive and highly significant.

K dynamics and K availability

Only a small fraction of the potassium requirements of the plant is attained by direct contact through root interception. The largest proportion of the K needed by the plants has to be transported in the soil to the roots. This transport of potassium ions is an important factor of K availability. It occurs mainly in the soil solution, the liquid phase of the soil, by mass flow (with the water moving to the plant root) and diffusion along a concentration gradient that is built up by the absorbing root. In the immediate vicinity of the roots, the soil solution is rapidly exhausted of nutrients due to removal by the plants. Continuous potassium supply to the growing plant is only ensured when the rate of potassium release to the soil solution and transport to the roots keeps pace with the rate of nutrient uptake.

Clay minerals are the most important source of soil K. They hold the bulk of mobile K and release it when the concentration of the soil solution falls due to plant uptake or to an increase in soil moisture. A good K saturation of the clay minerals results in an equilibrium with a high K concentration of the soil solution, whereas poor K saturation is in equilibrium with low K concentration in the soil solution.

The composition of the soil solution can change rapidly due to variations in soil moisture, nutrient uptake by plants and other factors. Nevertheless, it has been found experimentally that the K concentration of the equilibrated soil solution is a reproducible value if determined at a suitable standard moisture, e.g. field capacity or water saturation. Under these conditions the concentration in the soil solution depends on the K saturation of the inorganic cation exchange capacity (CEC) of the soil. At a given content of exchangeable K a soil with many K adsorbing particles (clay soil) usually has a lower K concentration in the soil solution than a sandy soil with less clay. At equal clay content, the K concentration of the soil solution depends on the nature of the clay minerals.

1. Kaolinitic clay minerals have no interlattice binding sites for potassium and a low cation exchange capacity. They do not hold non-exchangable K. Therefore, they behave similarly to sand and soil organic matter, as far as K dynamics are concerned.
2. Illitic clay minerals, vermiculite and chlorite, on the other hand, adsorb K selectively.
3. The selectivity of montmorillonitic clay minerals (smectite) for potassium is lower than that of illitic but greater than that of kaolinitic clay minerals.
4. Allophanes contain very small amounts of K, but experiments indicate, that K is preferentially adsorbed.

Depending on the degree of K saturation or depletion of these minerals, they will either release K into the soil solution or adsorb it from the solution. Apart from the type of mineral the selectivity of the clay minerals for potassium depends on the site of K adsorption. Planar positions do not represent specific K binding sites. Edge positions bind K more selectively. Inter-lattice positions have the highest K selectivity. Potassium held at inter-lattice positions is generally "non-exchangeable".

Adequate K concentration in the· soil solution (above 0.5 me/1) is attained only when the K selective sites have been saturated with K so that no fixation of K takes place, and a sufficient number of planar positions are occupied by K.

The exchangeable (or "available") potassium, as determined by soil analysis, does not give satisfactory information on the level of the actually available soil potassium unless related to the clay content and to the nature of clay minerals. 15 mg K_2O per 100 g soil (150 ppm) means high availability in a sandy soil but poor saturation in a clay soil containing illitic or montmorillonitic clay minerals. In laboratory tests it is possible to assess the amount of exchangeable potassium which is needed in a heavy soil to raise the soil solution concentration to an adequate level. In a clay loam with predominantly illitic clay minerals this may amount to 60 mg K_2O/100 g soil (600 ppm) or even more. In other words: heavy soils require considerably more fertilizer potassium to attain a high level of K availability than light soils. On the other hand, clay soils possess a better K buffering capacity than concentration of the soil solution at a similar level for a long time. In sandy soils, the soil solution concentration decreases rather fast so that split applications of potash may be expedient.

As the soil solution is depleted of nutrients in the immediate vicinity of plant roots, the disturbance of the equilibrium will result in a release not only of exchangeable K but also of potassium which initially had been non-exchangeable. This release of K from non-exchangeable sources can be considerable. In most cases, however, the rate at which non-exchangeable K is set free is too low to ensure an adequate K supply for high yields. Experiments have shown that yields were lowest on soils where the plants were left to the release of K from non-exchangeable reserves.

Such conditions prevail in many heavy soils which have not received adequate potash applications for decades. These soils do not only strongly bind the fertilizer potassium, but even fix it. Potassium then migrates into the expanded clay minerals which contract, thereby trapping the K ions. Thus K is transferred into a sparingly available form instead of being readily available to the plant. lt goes without saying that under such circumstances the K concentration in the soil solution is too low for optimum plant growth. The lack of yield response to usual potash dressings sometimes observed on soils testing low or medium in “available” K can often be ascribed to the strong fixation of fertilizer potash in the soil. This applies particularly to clay soils, where heavy rates of potash, may be necessary to overcome fixation. When applied at lower rates, band placement is advisable.

As mentioned before, diffusion is one of the major factors of potassium movement in the soil. lt increases with improved K saturation of the clay minerals. When comparing two soils with equal levels of exchangeable K, but different contents of clay, higher diffusion rates are found in the soil with lower clay content because of its higher degree of K saturation and, consequently, higher K concentration in the soil solution.

The diffusion rate also depends on the moisture status of the soil. In laboratory tests it has been shown that K movement is faster in a moist soil than in a dry soil. The influence of soil moisture on K availability has been confirmed in greenhouse trials and in field experiments. The results show that at optimum soil moisture conditions, less K is needed to produce a certain yield level.

In a relatively dry soil more K has to be given in order to overcome the decrease in K mobility. This is of special importance in rainless periods of short duration. Due to impaired movement of nutrients, potash deficiency and corresponding losses in yield and quality will occur in case of insufficient K saturation. In a wet soil additional potash supplies help to counter balance the diminished nutrient uptake capacity of the roots caused by poor aeration and to avoid unfavourable reduction processes in the soil.

References

Arai, Y. and Sparks, D.L. (2007). Phosphate reaction dynamics in soils and soil minerals: a multiscale approach. *Advances in Agronomy,* 94: 135–179.

Benbi, D.K. and Gilkes, R.J. (1987). The movement into soil of P from superphosphate grains and its available to plants. *Fertilizer Research,* 12: 21–36.

Condron, L.M., Turner, B.L. and Cade-Menun, B.J. (2005). Chemistry and dynamics of soil organic phosphorus. In J.T. Sims, A.N. Sharpley, eds, Phosphorus:Agriculture and the Environment. American Society of Agronomy, Crop Science Society of America, Soil Science Society of America, Inc., Madison, WI, pp 87–121.

Cordell D., Drangert, J.O., and White S. (2009). The story of phosphorus: global food security and food for thought. *Global Environmental Change,* 19: 292–305.

Devau, N., Le Cadre, E., Hinsinger, P. and Ge´rard F. (2010). A mechanistic model for understanding root-induced chemical changes controlling phosphorus availability. *Annals of Botany,* 105: 1183–1197.

Dou, Z., Toth, J.D., Galligan, D.T., Ramberg, C.F. and Ferguson J.D. (2000). Laboratory procedures for characterizing manure phosphorus. *Journal of Environmental Quality,* 29: 508–514.

Gilbert, N. 2009. Environment: the disappearing nutrient. *Nature,* 461:716–718.

Hansen, J.C., Cade-Menun, B.J. and Strawn D.G. (2004). Phosphorus speciation in manure-amended alkaline soils. *Journal of Environmental Quality,* 33: 1521–1527.

Harrison, A.F. (1987). Soil Organic Phosphorus—A Review of World Literature. CAB International, Wallingford, Oxon, UK, p 257.

Hinsinger, P. 2001 Bioavail ability of soil inorganic P in the rhizosphere as affected by root-induced chemical changes: a review. *Plant and Soil,* 237: 173–195.

Lindsay, W.L., Frazier, A.W. and Stephenson, H.F. (1962). Identification of reaction products from phosphate fertilizers in soils. *Soil Science Society of America Proceedings,* 26: 446–452.

Lindsay, W.L., Vlek, P.L.G. and Chien, S.H. 1989. Phosphate minerals. In JB Dixon, SBWeed, eds, Minerals in Soil Environment, Ed 2. Soil Science Society of America, Madison, WI, pp 1089–1130.

Luengo, C., Brigante, M., Antelo, J. and Avena, M. (2006). Kinetics of phosphate adsorption on goethite: comparing batch adsorption and ATR-IR measurements. *Journal of Colloid Interface Science,* 300: 511–518.

Martins, M.A., Santos, C., Almeida, M.M. and Costa, M.E.V. (2008). Hydroxyapatite micro- and nanoparticles: nucleation and growth mechanisms in the presence of citrate species. *Journal of Colloid Interface Science,* 318: 210–216.

Moody, P.W., Edwards, D.G. and Bell, L.C. (1995). Effect of banded fertilizers on soil solution composition and short-term root growth: II. Monocalcium phosphate with and without gypsum. *Australian Journal of Soil Research,* 33: 899–914.

Oelkers, E.H. and Valsami-Jones, E. (2008). Phosphate mineral reactivity and global sustainability. *Elements*, 4: 83–87.

Parfitt, R.L. (1989). Phosphate reactions with natural allophone, ferrihydrite and goethite. *Journal of Soil Science,* 40: 359–369.

Pierzynski, G.M., McDowell, R.W. and Sims, J.T. (2005). Chemistry, cycling, and potential moment of inorganic phosphorus in soils. In JT Sims, AN Sharpley, eds, Phosphorus: Agriculture and the Environment. American Society of Agronomy, Crop Science Society of America.

Raghothama KG (1999). Phosphate acquisition. *Annual Review of Plant Physiology & Plant Molecular Biology,* 50: 665–693.

Turner, B.L. and Leytem, A.B. (2004). Phosphorus compounds in sequential extracts of animal manures: chemical speciation and a novel fractionation procedure. *Environ Science & Technology,* 38: 6101–6108.

Turner, B.L., Richardson, A.E. and Mullaney, E.J. (2007). Inositol Phosphates: Linking Agriculture and the Environment. CAB International, Wallingford, UK, p 304

Wang, L.J. and Nancollas, G.H. (2008). Calcium orthophosphates: crystallization and dissolution. *Chemical Reviews,* 108: 4628–4669.

Advances in Nutrient Dynamics in Soil - Plant System, pp 21-25

Editors: R. Elanchezhian, A.K. Biswas, K. Ramesh and A.K. Patra

3

Nitrogen Fertilizer Use Efficiency An Overview

A.K. Biswas, K. Ramesh and R. Elanchezhian

ICAR-Indian Institute of Soil Science, Nabi Bagh, Bhopal – 462038, India

To meet the food needs of the burgeoning population, India will need to produce 300 million tonnes of food grains by 2020. At present more than 75% of the total food grains produced in the country are of rice and wheat. Use of nitrogenous fertilizers has contributed much to the remarkable increase in production of rice and wheat in India that has occurred during the past three decades. During the last half-decade or so while fertilizer N consumption is touching new heights, the production of both rice and wheat is showing a trend of plateauing. In fact, fertilizer N efficiency of food grain production expressed as partial factor productivity of N (PFP_N) has been decreasing exponentially since 1965. The PFP_N is an aggregate efficiency index that includes contributions to crop yield derived from uptake of indigenous soil N. fertilizer N uptake efficiency, and the efficiency with which N acquired by the plant is converted to grain yield. A decrease in PFP_N occurs as farmers move yields higher along a fixed N response function, unless other factors shift the response function up. In other words, an initial decline in PFP_N is an expected consequence of the adoption of N fertilizers by farmers and not necessarily bad within a system's context

Reactive N

Reactive nitrogen is defined as all biologically, photochemically, and/or radioactively active forms of N^- in diverse pools of nitrogenous compounds that include organic compounds (eg urea, amines, proteins, amides), mineral N forms, such as NO_3 and NH_4^+ as well as gases that are chemically active in' the troposphere (NOx, NH_3, N_2O) and contribute to air pollution and the greenhouse effect (Galloway *et al.,* 1995). Globally, N fertilizers account for 33% of the total annual creation of reactive nitrogen (Nr) or 63% of all anthropogenic sources of reactive N. Consumption of fertilizer N in developing and developed countries of the world in 1970 was 8.8 and 22.9 Mt, respectively (FAO 2005). Consumption in developing world surpassed that in developed world in 1990; it was only 29.0 Mt in developed countries vis-a-vis 55.7 in the developing countries during 2002. Since ill-effects of excessive reactive nitrogen in the environment are not confined to political boundaries, developing countries like India and China, which need to support huge populations by producing enough food by increasing use of nitrogen fertilizers, will remain in the eyes of the world until and unless high fertilizer N use efficiencies are not achieved at the earliest.

Terminologies

Mosier *el al.* (2004) has described four indices to describe nitrogen use efficiency:

1) Partial factor productivity (PFP_N, kg crop yield per kg N applied),
2) Agronomic efficiency (AE_N, kg crop yield increase per kg N applied),
3) Apparent recovery efficiency (RF_N, kg N taken up per kg N applied) and
4) Physiological efficiency (PE_N, kg yield increase per kg N uptake)

Cassman *et al.* (2002) defined the overall nitrogen use efficiency as the proportion of N inputs that are removed in harvested crop biomass, contained in recycled crop residues and incorporated into soil organic matter and inorganic pools. Applied N not recovered in these sinks is prone to loss from the cropping system via leaching, denitrification and volatilization and contributes to reactive N load that cascades through environment external to the agro-ecosystem.

Typical NUE Values

Nitrogen use efficiency can mean different things to different people and is easily misunderstood. For example a typical irrigated soil under rice-wheat cropping system in the Indo-Gangetic plains contains about 2000 kg N ha^{-1} in the top 30 cm of soil where roots derive majority of N supply from the amount of N derived from indigenous resources during a single cropping cycle typically ranges from 30-100 kg N ha^{-1} that represents onlv 1.5 to 5% of total soil N. Although small in size, the indigenous N supply has a very high N-fertilizer substitution value because of the relatively low RE_N from applied N fertilizer. Further, as C/N ratio of soil organic matter is relatively constant, changes in soil C balance introduced by management practices including fertilizer use affect the soil N balance. The overall fertilizer nitrogen use efficiency can thus be increased by achieving greater RE_N, by reducing the amount of N lost from soil organic and inorganic pools, or both. When soil-N content is increasing, the amount of sequestered N contributes to higher nitrogen use efficiency and the amount of sequestered N derived from applied N contributes to a higher RE_N. Conversely, any decrease in soil N stocks will reduce overall nitrogen use efficiency and RE_N.

A recent review on N use efficiency (Ladha *el al.* 2005) reported average single-year fertilizer N recovery efficiency as 57% for wheat and 46% for rice in the researcher-managed experimental plots. Nitrogen recovery in crops grown by farmers, however, is often much lower. A review of best available information suggests average N recovery efficiency for fields managed by farmers ranges from 20 to 30% under rainfed conditions and 30 to 40% under irrigated conditions. An RE_N exceeding 40% is expected to occur in response to improved N management practices. Cassman *el al.* (2002) found that N recovery from on-farm locations averaged 31% for irrigated rice in Asia and 40% for rice under field-specific management. For wheat grown in India, the recovery averaged 18% under poor weather conditions but 49% when grown under good weather conditions (Table 2). It also highlights the importance of robust crop growth and yields to greater RE_N. Ladha *et al.,* (2005) compiled data on ^{15}N recovery by cereal crops and found that average RE_N ^{15}N was 44% in the first growing season and total recovery of ^{15}N fertilizer in the first and five subsequent crops was only around 50%. Assuming that amount of ^{15}N in the roots becomes negligible in the sixth growing season, the remaining 50% of the ^{15}N fertilizer would have either become part of the soil organic matter pool or was lost from the cropping system (Janssen and Person 1982).

Factors

Fertilizer N use efficiency is controlled by crop demand of N, supply of N from the soil and fertilizers and N losses from soil-plant system. Nitrogen needs of crop plants are met by applying fertilizer N and net N mineralization from soil organic matter. Fertilizer N is applied in forms readily available to plants but mineralization of N is controlled by water, temperature and aeration. Once these factors become optimum, amount of N that is mineralized depends upon the quality and quantity of organic matter in the soil. Also a strong interaction between C and N dynamics is obvious because N transformations are driven largely by biological activity. High rates of net N mineralization can result in dilution of fertilizer N. But if crop demand for N and the amount of fertilizer N remain constant, an increase in net N mineralization will lead to a decrease in observed RE_N. Relative magnitude of different N loss mechanisms will depend upon soil, weather, fertilizer and crop management. On an overall basis, climate exerts the strongest effect on the amount and pathways of N losses.

Challenges

Decisions regarding improvements in fertilizer N use efficiency will begin at the field scale where farmers need to deal with the variability in soils, climates, and cropping patterns. As there exists a large fertilizer-N substitution value of soil N, it is important to know the amount and temporal variations of the indigenous N supply during crop growth for determining the optimal timing and amount of fertilizer N applications.

A great deal of research has been carried out during the past 50 years to improve fertilizer N use efficiency by trying to develop better fertilizers or improved N management practices based mainly on a better synchronization between N uptake by crops and supply though fertilizers. Strategies based on applying N at the right rate, right time, and in the right place have already been developed and are in use. Recent literature .on improving RE_N has emphasized on achieving greater synchrony between crop N demand and the N supply from all sources throughout the growing season (Cassman *el al.*, 2002). This approach explicitly recognizes the need to efficiently utilize both indigenous and applied N because losses of N via different mechanisms increase in proportion to the amount of available N present in the soil at any given time.

Options

Most of the fertilizer-N is lost during the year of application. Consequently, N and crop management must be fine-tuned in the cropping season in which N is applied. Two broad categories of concepts and tools have been developed to increase nitrogen use efficiency. Those in the first category include genetic improvements and management factors that remove restrictions on crop growth and enhance crop N demand and uptake. Management options that influence the availability of soil and fertilizer-N for plant uptake come in the second category. These include site-specific N application rates to account for differences in within-field variation in soil N supply capacity (in large fields), field specific N application rates in small scale production fields, remote sensing or canopy N status sensors to quantify realtime crop N status, better capabilities to predict soil N supply capacity, controlled release fertilizers and fertigation.

Ladha *et al.* (2005) compared different strategies to improve nitrogen use efficiency on the basis of benefit cost ratio and limitations. If a new technology leads to at least a small and consistent increase in crop yield with the same amount or less N applied, the resulting increase in profit is usually attractive enough for a farmer. With very high benefit cost ratio and with no limitation, use of simple and inexpensive leaf colour chart assists farmers in applying N when the plant needs it. As use of leaf colour chart can adequately take care of N supply from all indigenous sources, it ensures significant increase in RE_N and reduced fertilizer N use. This tool is particularly useful for small to medium size farms in developing countries. Similarly, precision farming technologies based on gadgets like optical sensors have demonstrated that variable rate N-fertilizer application has the potential to significantly enhance nitrogen use efficiency by crops like rice and wheat.

Modern N management concepts usually involve a combination of anticipatory (before planting) and responsive (during the growing season) decisions. Improved synchrony, for example, can be achieved by more accurate N prescriptions based on the projected crop N demand and the levels of mineral and organic soil N, but also through improved rules for splitting of N applications according to phenological stages, by using decision aids to diagnose soil and plant N status during the growing season (models, sensors), or by using controlled-release fertilizers or inhibitors. Important prerequisites for the adoption of advanced N management technologies are that they must be simple, provide consistent and large enough gains in fertilizer N use efficiency, involve little extra time and be cost-effective

Conclusion

Nitrogenous fertilizers and their management will remain at forefront of measures to improve the global reactive N balance on both the short- and long-term basis. Fertilizer N use efficiency is governed by N uptake by crops, N supply from soil and fertilizer and N losses from soil-plant system. Innovative fertilizer management has to integrate both preventive and field specific corrective N management strategies to increase the profitability of irrigated rice and wheat and to ensure that there exists synchrony between crop N demand and supply of mineral N from soil reserves and fertilizer inputs. It will lead to maintenance of plant available N pool at the minimum size required to meet crop N requirements at each growth stage with little vulnerability to loss of N to environment.

References

Cassman, K. G, Dobermann, A. and Walters, D. 2002. Agroecosystems, nitrogen-use efficiency, and nitrogen management *Ambio,* 31: 132-140.

FAO 2005. Fertilizer Use by Crop in India. Food and Agriculture Organization of the United Nations Rome.

Fertiliser Statistics 2004-05, Fertiliser Association of India 2005. New Delhi, India

Galloway, J.N., Schlcsinger, W.H., Levy, H., Michaels, A and Schnoor, J.L. 1995. Nitrogen fixation atmospheric enhancement - environmental response. *Global Biogeochemical Cycles,* 9: 235-252.

Jansson, S.L. and Persson, J. 1982. Mineralization and immobilization of soil nitrogen In Nitrogen in Agricultural Soils (F.J Stevenson, Ed.), Agronomy Monograph 22, pp. 229-252 ASA, CSSA, and SSSA, Madison, WI , USA.

Ladha, J. K. , Pathak, H. , Krupnik, T. J. , Six, J. and van Kessel, C. 2005. Efficiency of fertilizer nitrogen in cereal production: *Retrospect and Prospects Advances in Agronomy,* 87: 85-156.

Mosier, A. R., Syers, J. K. and Freney, J.R. 2004. Nitrogen fertilizer: An essential component of increases" food, feed, and fiber production In Agriculture and the Nitrogen Cycle Assessing the Impacts of Fertilizer Use on Food Production and the Environment (A. R. Mosier, J. K. Syers, and J R Freney, Eds), pp 3-15. SCOPE 65, Pans, France.

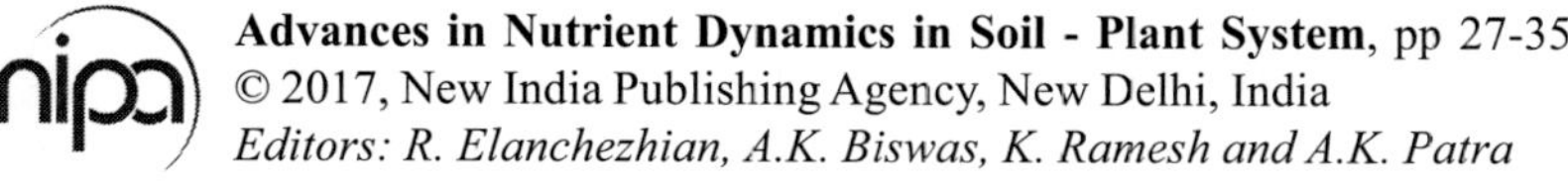
Advances in Nutrient Dynamics in Soil - Plant System, pp 27-35

Editors: R. Elanchezhian, A.K. Biswas, K. Ramesh and A.K. Patra

4

Balanced Nutrition – Key for Improving Nutrient Use Efficiency

Muneshwar Singh and R.H. Wanjari

ICAR-Indian Institute of Soil Science, Nabi Bagh, Bhopal – 462038, India

Introduction

Agricultural lands in India have mostly low native soil fertility. On these soils there must be a regular application of nutrient inputs through chemical fertilizers and/or organic manures for a successful and sustained crop production to replenish soil nutrient reserves depleted by crop removal and other losses. It is essential to recognize that even in agricultural production systems with relatively low productivity level, the quantity of nutrients available for recycling through plant and animal residues is not enough to compensate the amount being removed by crops. Under these circumstances mineral fertilizers play a key role in those areas having low fertile soils by increasing agricultural production to meet out growing food demand. Chemical fertilizers are the major contributors as source of plant nutrients to enhance crop production and also to maintain soil productivity. Fertilizer is one of the key inputs in crop production and India has made rapid growth in fertilizer consumption, particularly after the introduction of high yielding varieties (HYVs) in mid-sixties. Soil fertility depletion due to imbalanced and inefficient use of nutrients has become serious constraint in improving yields. Continuous nutrient mining and non-recycling of crop residues have aggravated the problem of multi-nutrient deficiencies in Indian soils. The nutrient use efficiency is low and declining. The low nutrient use efficiency not only affects crop yields and farmer's profit but also poses a great threat to the environment. Thus, nutrient application plays key role in maintaining productivity and sustainability of the crops/ cropping system.

During the past three decades additional nutrients applied through fertilizers accounted for 50-55% contribution in increasing crop yield in India and other developing countries. In India, in spite of steady increase in the consumption of chemical fertilizers over the years, the nutrient use efficiency with the applied fertilizers continues to be merely at lower range i.e. 30-50% for N, 15-20% for P and 1-2% for Zn, Fe and Cu. It is because of enumerable nutrient losses either from soil as such or due to conversion of applied nutrients into slow cycling/ recalcitrant pools within the soil ecosystem. In this context, National Agricultural Policy in India laid much emphasis on enhancing efficiency of agricultural inputs amongst which fertilizers are found to be important and expensive component. Injudicious and inefficient nutrient use has not only increased the cost of cultivation but also a great threat to biosphere pollution. The economic and ecological considerations have highlighted the compulsive need for more efficient use of nutrients in crop production. Based on the present fertilizer consumption of 15.25 million tones (M t) N and 6.15 M t P_2O_5 in 2013-14at national level and assuming present nutrient use efficiency at 50% for N and 20% for P, if there is increase in efficiency of N and P by merely 1%, it would make a saving of 2.19 lakh tones of N and 1.98 lakh tones of P. Thus, it collectively could save at least Rs. 6250 million annually. Apart from such staggering annual economic benefit, it would additionally reduce the risk of environmental pollution (Singh and Wanjari, 2007). In this context nutrient management have crucial role in reducing the cost of inputs and also environmental hazards. Therefore, one has to be strategic in planning the nutrient application and fertilizer in particular. On the similar option for best fertilizer use one should adopt and follow the principle of "4 R" – as the right fertilizer source, at the right rate, right time and right place to achieve economic, social and environmental goals (Majumdar *et al.,* 2013). This holds true not only for fertilizer but also for nutrient in totality. Thus, balance nutrient management practices act as a key factor to assess progress towards sustainability of soil, crop yield and agricultural system as a whole.

Nutrient Scenario

To meet food grains requirement of 300 Mt by 2025, 45 Mt of N + P_2O_5 + K_2O is estimated to be required per annum (Prasad, 2012). Out of this, 35 Mt is proposed to be met from the chemical fertilizers and the rest from organic manures. A similar estimate of one-and-half times to that in year 2007– 08 consumption of 23 Mt of NPK was made by Tiwari (2007). The present installed capacity of fertilizers is only 12.3 Mt of N and 5.7 Mt of P_2O_5. All potash is imported. During 2007–08, 6.9 Mt of urea, 3.0 Mt of DAP/MAP and 4.4 Mt of muriate of potash (KCl) were imported. Obviously, imports will increase in the coming years to sustain increased food production unless adequate incentives are provided to the fertilizer industry to increase the installed fertilizer production capacity.

Some estimates of N, P and K removed per tonne of food grain produced (Table 1) imply that cereals are heavy feeder of nutrients especially N, P and K. Plant nutrient ($N + P_2O_5 + K_2O$) availability from organic sources such as farmyard manure, compost, vermicompost and green manure is estimated at 13 Mt (9 Mt net) (Table 2). These estimates exclude secondary and micronutrients added, which are sizeable. Organic manures and crop residues can play a major role in recycling K. Concomitant use of organic manure and fertilizer-N can reduce leaching losses of N. Thus, integrated nutrient management is vital for increasing food production.

Table 1: Removal of NPK (kgt^{-1} grain) by rice, wheat and other major food grain crops

Crop	N	P (P_2O_5)	K (K_2O)
Rice	20.4	3.6 (8.2)	20.4 (24.5)
Wheat	22.4	3.8 (8.7)	28.2 (33.8)
Maize	24.3	6.4 (14.6)	18.3 (22.0)
Sorghum	26.1	4.5 (10.3)	21.5 (25.8)
Pearl millet	27.1	8.2 (18.8)	39.7 (47.6)
Chickpea	50.6	8.6 (19.7)	29.7 (35.8)
Pigeonpea	92.1	8.2 (18.8)	30.7 (36.8)

Table 2: Available nutrients from organic manure

Component	Potential availability (Mt)	Actual availability (Mt)	Nutrient value (Mt)
Crop residue	603	201	5
Animal dung	791	287	4
Green manure	4.5*	NA	0.2
Rural compost	184	184	2.6
City compost	12.2	12.2	0.4
Biofertilizer	0.01	Negligible	0.4
Others	96.6	NA	0.9
Total			12.8

Source: Bhattacharya, 2007

Rates of nutrient application optimal for economic yields often minimize nutrient losses (Hong *et al.* 2007). Although it is reasonable to assume that, on a global scale, at least 50% of the fertilizer N applied is lost from agricultural systems and most of these losses occur during the year of fertilizer application. However, these losses could be minimized by adoption of good agronomic practices and best fertilizer management options. P fertilizer applications typically result in cereal yield increases by 20 to more than 50 kg grain/kg P applied. Moreover, due to favorable plant growth condition, most of the agricultural crops recover 20-30% of P applications depending upon the growth stage. A large portion of the unused P accumulates in the soil and is eventually recovered by subsequent crops over, a much smaller fraction of P losses as runoff or through leaching that can be secondary off-site impacts (Dobermann, 2007). On the contrary, it has been reported that in developing countries, K input-output budgets in agriculture are highly negative (Sheldrick, 2002). In countries like India, annual K losses hover around 20-40 kg ha^{-1} and those have been increased steadily during the past 40 years (Majumdar *et al.,* 2013). Therefore, it is expected that there will be high nutrient use efficiency for the K in such areas. Today economic and environmental challenges are demanding more attention to the best nutrient management practices to have high nutrient use efficiency. Moreover, higher prices for both crops and fertilizers have highlighted interest in efficiency improving technologies and practices that also improve productivity. In addition, nutrient losses that harm air and water quality can be reduced by improving use efficiencies of nutrients particularity for nitrogen (N) and phosphorus (P).

Nutrient Use Efficiency

Concept and terminology of nutrient use efficiency

Efficiency is calculated as ratio between input to output in a particular system. A recent scientific review identified about 15 different forms of nutrient use efficiencies. Four of these terms are very commonly used in agriculture.

i. Partial factor productivity (Crop yield per unit of nutrient applied) answers the question: "How productive is this cropping system in comparison to its nutrient input?"
ii. Agronomic efficiency (Yield increase per unit of nutrient applied) answers a more direct question: "How much productivity improvement was gained by the use of this nutrient?"
iii. Physiological efficiency (Yield increase per unit of additional nutrient uptake) answers the question "How much productivity was gained by above-ground crop uptake?"
iv. Recovery efficiency (Increase in above-ground crop uptake per unit of nutrient applied) answers the question: "How much of the nutrient applied did the plant take up?"

In general, we do calculate nutrient use efficiency on the basis of one or two year's experimentation which many a times may not be found realistic. Since NUE is dependent on crop productivity which is further dependent on precipitation and climatic factors. Under such situation, stability of experiment is prerequisite to have real picture of nutrient contribution and long term experiment provides an ideal situation.

Impact of enhanced NUE on input, crop productivity and environment

The enhanced nutrient use efficiency implies the following considerations:

- Lesser nutrients are needed for obtaining a optimum level of production
- More produce per unit of nutrient applied
- Lower cost of production per unit of produce
- Higher returns per rupee invested on nutrient use
- Reduced risk of environmental pollution

Economical significance of enhanced NUE at national level

Since fertilizer nutrients are expensive and used in large quantities at national level, any increase in nutrient use efficiency will lead to a substantial cut in nutrient requirement and huge economic benefit at national level. For example, at present consumption levels with nutrient use efficiency of 50% for N and 20% for P_2O_5, an increase of just 1 unit in the efficiency would reduce costs of nutrients to a very huge amount. Assuming current consumption of 16 million tons (m ton) of N and 5 m tons of P, increase in use efficiency of N and P by one percent would save 0.35 M t of N and 0.26 M t of P_2O_5, respectively, which at present rate would cost Rs. 10056 million annually (Table 3). It is because, P is 100 percent imported, so would save huge expenditure on foreign exchange (Singh and Wanjari, 2013).

Table 3: Dimensions of money saved on increase in use efficiency of N and P by one per cent

Nutrient	Saving of nutrient in million tons	Cost (Rs. million)
N	0.35	3478
P	0.26	6578
Total		10056

Nitrogen use efficiency

Nitrogen is most mobile nutrient and subjected to various kinds of losses like leaching, volatilization and de-nitrification etc. These losses may constitute up to 70 percent of total losses. In Indian agriculture scenario nitrogenous fertilizers are used upto 60 to 70 percent of the total fertilizer consumption. Hence, management of N is most important not only economics point of view but also extent of environmental pollution. Leaching losses of N pollute the water bodies and drinking water and gaseous loss adds to greenhouse gas pool in environment.

In long term fertilizer experiment (LTFE), attempt has been made to calculate nitrogen use efficiency under various nutrient management scenarios. The integrated plant nutrient supply system (IPNS) demands a holistic approach to nutrient management for crop production and it involves judicious combined use of fertilizers, biofertilizers, organic manures (FYM, compost, vermicompost, biogas slurry, green manures, crop residues *etc.*), and growing of legumes in the cropping systems (Prasad, 2009). The IPNS also encompasses balanced fertilization and SSNM. Considerable research on IPNS has been done in India (Rao *et al.*, 2002). Moreover, long-term fertilizer experiments have shown that addition of organic manures in addition to NPK (add-on series) results in high yields over a long period of time as compared to a decline in yield over time when only inorganic fertilizers were applied (Swarup, 2002). In similar context, Sarkar and Singh (2002) reported that for soybean-wheat cropping system in the acidic soils of Ranchi (pH < 5.4), soybean yield (averaged over 28 years) was 0.33 Mg ha^{-1} and wheat yield, 0.43 Mg ha^{-1} for plots receiving N alone as compared to 1.59 Mg ha^{-1} in soybean and 2.65 Mg ha^{-1} in wheat when NPK was applied. Application of FYM with NPK increased the soybean yield to 1.86 Mg ha^{-1} and that of wheat to 3.19 Mg ha^{-1}. Further, the effects of NPK + FYM were at par with NPK + lime, implying that in acid soils continuous application of FYM can also partially offset soil acidity.

The data revealed that integration of nutrients resulted in increased of N use efficiency irrespective of cropping system and soil type (Table 4). On application of N alone, N use efficiency was 16.7 percent which increased to 23.5% and 36.4% on integration, respectively, of nitrogen with P and NP with K doses. Application of FYM further improved the N use efficacy in Inceptisols of Ludhiana in maize-wheat system. A similar trend was noted in Alfisols of Palampur in maize-wheat and Mollisols of Pantnagar in rice-wheat system. Thus data indicated that integration of nutrient and balanced application of nutrient is one of the ways to enhance nutrient use efficiency. Increase in nitrogen use efficiency on integration of nutrient is due to increase in yield due to addition of P and K. In absence of P and K crop is unable to assimilate nitrogen (Singh *et al.*, 2009)

Table 4 Nitrogen use efficiency as affected by balanced use of nutrients under LTFE

Soil Type	Location	Crop	Mean nitrogen use efficiency (%)			
			100% N	100% NP	100% NPK	100% NPK+FYM
Inceptisol	Ludhiana	Maize	16.7	23.5	36.4	40.2
Alfisol	Palampur	Maize	6.4	34.7	52.6	63.7
Mollisol	Pantnagar	Rice	37.5	40.7	44.4	61.7
Inceptisol	Ludhiana	Wheat	32.0	50.6	63.1	67.8
Alfisol	Palampur	Wheat	1.9	35.6	50.6	72.6
Mollisol	Pantnagar	Wheat	42.4	46.1	48.4	47.9

Phosphorous use efficiency

Even though P is most stable plant nutrient in soil and does not move in profile under normal condition. However, one should not worry much about P dynamics in soil but to make farming more profitable by reducing the cost of cultivation by curtailing expenditure on inputs. Increase in P use efficiency will help in reducing the cost of cultivation and environment risk as well. In addition to these benefits, increase in efficiency will be helpful to curtail burden on foreign exchequer of Goverment of India. As India does not have indigenous source of P and 100% P is imported.

Data presented in Table 5, revealed that like nitrogen, integration or balancing of NP with K resulted in increased P use efficiency to a great extent irrespective of soil and cropping system. In Inceptisol of Ludhiana, P use efficiency data reveled that in maize under NP treatment P use efficiency was 10.3 per cent which increased to 21.4 per cent on addition of K in NP treatment and P use efficiency further increased to 26.3 per cent on incorporation of FYM. A similar trend on P use efficiency was also recorded at other sites also.

Table 5: Phosphorus use efficiency as affected by integration of nutrients

Season& Soil Type	Location	Crop	P use efficiency (%)		
			100% NP	100% NPK	100% NPK+FYM
Kharif					
Inceptisol	Ludhiana	Maize	10.3	21.4	26.3
Alfisol	Palampur	Maize	21.8	35.6	41.1
Mollisol	Pantnagar	Rice	18.2	23.3	43.0
Rabi					
Inceptisol	Ludhiana	Wheat	20.6	30.7	34.8
Alfisol	Palampur	Wheat	10.7	15.2	24.6
Mollisol	Pantnagar	Wheat	11.2	10.4	23.3

Potassium use efficiency

Potassium is another primary nutrient which is required in larger quantity by all the crops. Though its consumption is relatively low compared to N and P because it is not frequently used in our country as soils are quite rich in K and supply of K from soil to crop is in sufficient quantity. Even though it has brought importance in recent past as many crops have started showing response to applied K. Data presented in Table 6 clearly demonstrated that integrated nutrient management always has larger K use efficiency like N and P. Irrespective of soil and cropping system, application of organic manure improved K use efficiency in different crop. This is due to increase in productivity level and potential of crops over the years.

Table 6: Potassium use efficiency as influenced by balance and integrated nutrient management

Season& Soil Type	Location	Crop	K use efficiency (%)	
			100% NPK	100% NPK+FYM
Kharif				
Inceptisol	Ludhiana	Maize	43.8	58.2
Alfisol	Palampur	Maize	23.0	38.9
Mollisol	Pantnagar	Rice	34.5	108.3
Rabi				
Inceptisol	Ludhiana	Wheat	88.1	112.8
Alfisol	Palampur	Wheat	22.6	66.8
Mollisol	Pantnagar	Wheat	13.7	35.8

Sulfur use efficiency

Intensification of agriculture and non-recycling of crop residue has compelled the use of sulfur by the farmers recently and it is gaining momentum in agriculture. Data presented in Table 7 and 8 clearly indicated that balanced use of nutrients and incorporation of FYM resulted in increase of use efficiency of N, P, K and S in rice-rice system. However, increase in S use efficiency was not recorded in *Kharif* rice, whereas during *rabi* increase in S use efficiency has been recorded. Increase in use efficiency of S during *rabi* season is due to larger yield of rice in *rabi* season compared to *kharif* because of more sunshine hours during *rabi* season. During *Kharif* season rice yields are low because of cloudy weather and nutrient present in soil are sufficient for that particular level of productivity.

Table 7: Nutrient use efficiency (%) as influenced by INM in rice-rice at Jagtial during *kharif*

Nutrient	Treatment (*Kharif* rice)			
	100% N	100% NP	100% NPK	100% NPK+FYM
N	17.8	41.6	43.2	45.5
P	-	32.4	34.5	47.6
K	-	-	81.0	98.0
S	-	29.0	30.5	32.5

Table 8 Nutrient use efficiency (%) as influenced by INM in rice-rice at Jagtial during *Rabi*

Nutrient	Treatment (*Rabi rice*)			
	100% N	100% NP	100% NPK	100% NPK+FYM
N	14.0	40.3	42.0	58.7
P	-	31.3	31.5	42.0
K	-	-	71.0	77.0
S	-	24.5	38.0	51.0

Conclusion

The nutrient management in crops/cropping system plays key role in enhancing nutrient use efficiency, maintaining productivity and sustainability. It is concluded from the data that integration or balanced use of nutrients is essential to maximize the nutrient use efficiency. The results from long term experiments with different cropping systems also indicated that to sustain the crop productivity and nutrient use efficiency focus should be on balanced and integrated nutrient management. Thus, a strategy on nutrient management should be adopted and follow the right fertilizer source, at the right rate, right time and right place to achieve overall crop productivity and nutrient use efficiency.

References

Bhattacharya, P. (2007). Prospects of organic nutrient resource utilization in India. *Indian Journal of Fertilizer,* 3: 93–107.

Dobermann, A. (2007). IFA International Workshop on Fertilizer Best Management Practices. 7-9 March, Brussels, Belgium.

Hong, N., Scharf, P. C., Davis, J.G., Kitchen, R. and Sudduth, K. A. (2007). Economically optimal nitrogen rate reduces soil residual nitrate. *Journal of Environmental Quality,* 36(2): 354- 362.

Majumdar, Kaushik, Johnston, A.M., Dutt Sudarshan, Satyanarayana, T. and Roberts, T.L. (2013). Fertilizer best management practices concept, global perspectives and application. *Indian Journal of Fertilizer,* 9 (4): 14-31.

Prasad, R. (2012). Fertilizers and manures. Special Section. *Current Science* 102(6):894-898.

Prasad, Rajendra (2009). Efficient fertilizer use: The key to food security and better environment. *Journal of Tropical Agriculture,* 47 (1-2): 1-17.

Rao, A.S., Chand, S. and Srivastava, S. (2002). Opportunities for integrated plant nutrient supply for crops/cropping systems in different agro-ecosystems. *Fertiliser News,* 47(12): 75–90.

Sarkar, A.K. and Singh, R. P. (2002). Importance of long-term fertilizer use for sustainable agriculture in Jharkhand. *Fertilizer News,* 47(11): 107–111.

Sheldrick, .W.F., Syers, J.K. and Lingard, J. (2002). A conceptual model for conducting nutrient audits at national, regional, and global scales. *Nutrient Cycling in Agroecosystem,* 62: 61-72.

Singh, Muneshwar and Wanjari, R.H. (2013). Balanced nutrient management: A key to sustain productivity and soil health on long term basis. *Indian Journal of Fertilisers,* 9 (12):72-81.

Singh, Muneshwar and Wanjari, R.H. (2007). Lessons Learnt from Long-term Fertilizer Experiments and Measures to Sustain Productivity in Alfisols. Research Bulletin. AICRP on Long term Fertilizer Experiments (LTFE) to Study Changes in Soil Quality, Crop Productivity and Sustainability, Indian Institute of Soil Science, Bhopal-462 038

Singh, Muneshwar, Dwivedi, B.S. and Datta, S.P. (2009). Integrated Nutrient Management for Enhancing Productivity, Nutrient Use Efficiency and Environmental Quality pp.55-67. Indian society of Soil Science (Platinum Jubilee Symposium - Proceedings).

Swarup, A. (2002). Lessons from Long term fertilizer experiments in improving fertilizer use efficiency and crop yields. *Fertiliser News,* 47(12): 59–73.

Tiwari, K.N. (2007). Reassessing the role of fertilizers in maintaining food, nutrition and environmental security. *Indian Journal of Fertilizer,* 3: 33–50.

Advances in Nutrient Dynamics in Soil - Plant System, pp 37-50

Editors: R. Elanchezhian, A.K. Biswas, K. Ramesh and A.K. Patra

5

Phosphorus Use Efficiency Through Soil Based Interventions

A. Subba Rao, A.K. Biswas and I. Rashmi

ICAR-Indian Institute of Soil Science, Nabi Bagh, Bhopal – 462038, India

Phosphorus (P) is a critical nutrient in crop production, due to its low bioavailability in soils (Feng *et al.,* 2004). Phosphorus is the vital component of DNA, RNA, ATP and photosynthetic system and catalyses a number of biochemical reactions from the beginning of seedling growth through to the formation of grain and maturity. Many factors influence soil P availability like type of parent material from which the soil is derived, degree of weathering and climatic conditions. In addition to this, erosion, crop removal and phosphorus fertilization and soil phosphorus levels also affect P availability in soil. Rock phosphate is the key raw material used for manufacturing phosphatic fertilizers on which the food production depend. By calculating RP reserve longevity using current reserve and production, Steven *et al.* (2013) predicted P reserves to exhaust over 300 years. With increasing population pressure, global food production will need to increase by 70% by 2050 (Fraiture, 2007). Thus good agronomic management requires the efficient use of fertilizer P for optimum crop production whereas excess soil P can be detrimental for water quality. Phosphorus thus plays a key role in sustainable crop production as well as environmental quality (Subba Rao, 2010).

In 2025 the food grain requirement for India's 1.4 billion people will be about 300 million tonnes (mt) which means nearly about 11 to 13 mt of P_2O_5 will be required for sustaining crop yield (Tiwari, 2001) and the demand for food commodities will further increase in future. Compared to merely 69.8 thousand tonnes of fertilizer consumption in 1950-51, 28.1 mt of fertilizers (N + P_2O_5 + K_2O) was used in 2010-11. At present India is the third largest producer of P fertilizers and second to China in consumption (Prasad, 2012). Currently about 38 per cent of the total fertilizer consumption is fulfilled through imports. The imports of total finished fertilizers have gone up to 21.7 mt in 2010-11 from 3.6 mt only in 2000-

01. Out of 21.7 mt, the import of urea was 6.6 mt, Di Ammonium Phosphate (DAP) 7.4 mt, Murite of Potash (MOP) 6.4 mt and the balance quantity of 1.3 mt comprised of NP/ NPKs, ammonium sulphate, Triple Super Phosphate (TSP) and Sulfate of Potash (SOP). Besides, currently, about 5 mt of rock phosphate, 2mt of phosphoric acid and 1.2 mt of sulphur are imported every year. The low recovery of P by crops, high retention by soil and residual fertilizer P in different P pools necessitates understanding P dynamics in soil and crop management for efficient use of P resources in Indian agriculture. Therefore proper P management is necessary to meet crop demand, to improve P use efficiency and to protect the environment in a given cropping system and landscape.

Phosphorus Status in Indian Soils

The Geographical Information System (GIS) based fertility mapping revealed that 49% of Indian soils are under low category, 45% under medium and 6% under high category of soil P (Muralidharudu *et al.,* 2011). Thus, the data indicated that 94% of Indian soils were classified under low to medium soil P fertility. In the states of Punjab and Kerala, however, no district is placed under low soil P category, apparently due to continuous use of P fertilizers under intensive cropping in these states. The consumption of P fertilizer was the highest in Punjab followed by Andhra Pradesh and Tamil Nadu, whereas Rajasthan and some states in North East were in the low consumption category. There is a problem of overuse of P fertilizers in excess of crop demand and buildup of P in pockets that lead to its inefficient use and consequent environmental pollution. A recent study reported a positive balance of 1.02 t P in Indian soils indicating only 20 per cent of applied P is recovered by crops and the rest is assumed to be locked up in soil (Pathak *et al.,* 2010). Similarly in groundnut-based cropping system with 25 years of annual application of P fertilizer, reported 45-256 kg of residual P accumulation as Olsen P with 43-58% below 60cm depth indicating enormous movement P to deep layers in coarse textured soils of Punjab (Aulakh *et al.,* 2007). The loss of P from high P sorbing soils usually ranges from < 0.4 -5% of applied P whereas in low P sorbing soil the loss can be as high as 40-90% of applied P (Mellan *et al.,* 2008).

Phosphorus Management

Phosphorus management on the basis of cropping sequence instead of individual crop needs to be taken into consideration for improving P use efficiency taking advantage of naturally occurring nutrient mobilization. Knowledge of the rate of increase or decrease of available P and the fate of residual P resulting from fertilization and crop removal is essential for long-term planning of fertilization strategies to sustain crop production (Huang *et al.,* 2011). In semi-arid belts of central India soybean- wheat is the major cropping system, but low native soil available P and high P sorption capacity of the soils result in low P utilization efficiency (Subba Rao and Ganeshamurthy, 1994). Therefore large inputs of P are required to maintain optimum P in soil solution for crop growth. In a five year experiment carried out in soybean- wheat system on Vertisol apparent P recovery (APR) by crops from fertilizer decreased from a mean value of 43.6 to 21% in soybean and 36.2 to 21.7% in wheat with an increase in fertilizer P rate from 11 to 44 kg ha^{-1} (Reddy *et al.,* 1999). The combination of fertilizer with manure showed higher APR compared to those with fertilizer applied alone. The available P status (Olsen P) showed higher build up in treatments with fertilizer and manure applications during five years of cropping. The apparent P balance was positive in higher fertilizer rate (44 kg ha^{-1}) but for control it was negative. The percent of P removed of the total P added was the highest in plots where only manure was added (>100%) indicating that the manure provided a major portion of applied P in plant- available form. In another study on soybean- wheat system illustrated that conjoint use of 16 t ha^{-1} of FYM with 44 kg P ha^{-1} gave the highest yield in both the crops and the APR ranged from 24.9 to 15.1%, the lower APR values with higher fertilizer P application. With FYM applied at 4, 8 and 16 t ha^{-1}, crops recovered 96, 66 and 45% of manure P, respectively. Irrespective of the manure addition, continuous application of fertilizer P resulted in larger positive P balance (Subba Rao *et al.,* 1998). The high P build up in soils of long term fertility experiment of Rothamsted indicated that yield of arable crops increased up to 25 mg P kg^{-1}, and further increase in application led to P concentration increase in drainage water once Olsen P reached 60 mg kg^{-1}. Such critical values should be determined for Indian soil types to determine the threshold values for crop yield and environmental quality. The threshold values for crop yield and environmental quality expressed in terms of degree of P saturation in some Indian soils varied in different agroclimatic zones. These values are dependent on the nature of extractants and soil types (Rashmi, 2013).

The recovery of applied P is only approximately one quarter and the rest remains in the soil and gets fixed which acts as sink of residual P and helps in sustaining productivity of succeeding crops in a system (Raju *et al.,* 2005) and transformation of residual P results in its accumulation which governs the P availability to following crops. Residual effects on yield of subsequent crops of phosphorus applied either to soybean or wheat and on recoveries of the added P and changes in available P were studied in soybean-wheat cropping system in Typic Haplustert (Subba Rao *et al.,* 1996). Phosphorus was applied at rates of 0- 52 kg P ha^{-1} to soybean and 0- 39 kg P ha^{-1} to wheat during the first year and in subsequent years the residual effects were studied in relation to fresh application of 39 kg P ha^{-1} to each crop. Phosphorus applied to soybean showed residual effects in two succeeding crops, whereas P applied to wheat showed residual effect in only one succeeding crop. The total P uptake of both soybean and wheat was higher where P was applied to soybean only as compared to the total uptake in the subsequent year resulting an overall positive P balance. This indicated that recoveries of added P were higher in succeeding two crops or one crop. This study indicated that P recommendation based upon the preceding crop and residue left over for the following crop improved the benefit: cost ratio of the entire system (Raju *et al.,* 2005). In rice wheat cropping system rice is known to utilize residual P more efficiently under submerged condition and therefore wheat crop responded significantly in such soil, while the response reduced with successive rice crops (Goswami and Pasricha, 1992). However, some studies on long term fertility experiment had shown that skipping P doses or blanket reduction of P can adversely affect crop yield in IGP (Jat *et al.,* 2012). Under irrigated condition soybean showed significant response to P application (80 kg P_2O_5 ha^{-1}) where the preceding wheat did not receive any fertilizer P. On the other hand, where wheat received 60 kg P_2O_5 ha^{-1}, the response of succeeding soybean was restricted up to 60 kg P_2O_5 ha^{-1}. It obtained the remaining P from residual fertilizer (Aulakh, 2003). The P solubilises more rapidly under submerged conditions in rice-wheat cropping system. The reduction of Fe and Mn and oxidation of organic matter increases P availability to crops (Stumn and Sulbereger, 1992). This occurs by forming phospho-humate complexes, replacing phosphate ions by humate ions and coating sesquoxides particles by humus to form protective cover which reduces P fixing capacity (Awasthi, 1990).

Influence of Organic Matter and Other Nutrients

Organic matter increases P availability by forming complexes with Al and Fe phosphates by humus coating and reduces P sorption by displacing sorbed phsopahte through organic anions, by complexing with organic phosphate and by mineralization reactions (Haynes and Mokolobate, 2001). The major mechanisms involved in enhancing P availability are orthophosphate incorporations, lower soil pH, increased enzymatic activity, complexing of exchangeable ions like Al, Fe, Ca and Mn. Addition of soybean or wheat residues (SR and WR) to the soil increased labile P fractions and the increase was larger with the SR and SR+FP than the WR and WR+FP (Reddy *et al.,* 2005). The integrated approach of manure and fertilizer was greater than their sole application in improving P fertility status, and efficient utilization of added P by crops (Reddy *et al.,* 1999). Repeated application of P as inorganics viz, fertilizer either alone or in conjunction with organics like cattle manure may affect P balance and its forms and distribution in soil. A long term experiment for six years carried out at IISS, Bhopal under soybean –wheat rotation revealed that conjunctive use of cattle manure with fertilizer P increased P pools ($NaHCO_3$-Pi followed by $NaHCO_3$-Po and NaOH-P) and favoured accumulation of P in organic fractions and encouraged movement of top layer P to subsoil (Ray *et al.,* 2013). Phosphorus uptake and available soil P increased significantly with increasing rates of both FYM (16t) and fertilizer P (44 kg ha^{-1}) in soybean – wheat cropping system in a study carried out on Typic Haplustert of Bhopal (Subba Rao *et al.,* 1998).

In soils, P interacts with other nutrients like N, K, S, Ca, Zn, Mo, Cu, Mn and B. Many studies have shown positive effect on crop yield when N and P are applied together. In studies on interaction between P and K in laterite soils of Kerala, application of 900 and 450 g palm^{-1} year^{-1} of P_2O_5 and K_2O to coconut palm together yielded 233 additional nuts ha^{-1} (Wahid *et al.,* 1988). A lot of literature has shown the beneficial effect of liming on P availability and reduction of Al toxicity. In P deficient acid sulphate soils of lowland paddy, application of P and lime improved available soil P status (Tiwari, 2001). The P and S interaction was synergistic in soybean, mustard, pigeon pea, potato, rice-peanut crops at lower rates of application, but could become antagonistic at higher rates (Rattan and Neelam, 1993). Phosphorus has both synergistic and antagonistic relations with micro nutrients, among which Zn X P interaction is antagonistic, where P induced Zn deficiency or vice versa is reported in many crops. However P interacts synergistically with B and Mn in legume crops.

Management and Efficient Utilization of Applied P

The efficient utilization of P in soil depends upon the P recovery by different crops which ranges from 20 to 30%. Based upon world P literature, recovery due to fertilizer application and residue together can be upto 50-90% when measured by a suitable method over appropriate time scale based upon many cropping systems, across soil types and climatic conditions (Syers *et al.,* 2008). This clearly indicates that application of P into soil can lead to accumulation and enrichment of soil with P if not balanced crop wise or based on cropping systems. The approach "P bank in soil' is more applicable for developed countries with enriched soil P whereas for developing countries like India with low to medium soil P status measure should be taken for judicious use of P fertilizers in conjunction with other resources like organic residues, industrial and municipal solid waste, RP with improved technologies so as to maintain soil fertility levels for sustaining crop productivity.

Inorganic Sources: Phosphorus fertilizers are the best inorganic sources of P applied to soil. Phosphorus fertilizers are divided into three classes based upon their solubility: water soluble (mono (MAP) and di ammonium phosphates (DAP) superphosphate), citrate soluble (Dicalcium phosphate, Thomas slag, basic slag, deflourinated phosphate and fused magnesium phosphate) and acid soluble (phosphate rock and bone meal). The rock phosphate and bone meal are applied in large quantities on acid soils which are sparingly soluble and converted to usable form for plant uptake over a period of time. The water soluble P fertilizer is readily available for crop uptake, but the major problem is that these fertilizers quickly get fixed and become unavailable for crops. The water soluble fertilizers (SSP, DAP, MAP) were found superior over partially water soluble NP and acid soluble RP in term of grain yield and P uptake in wheat (Saha *et al.,* 2014). The use of magnesium ammonium phosphate (MgAP), a highly citrate soluble material prepared out of low grade indigenous phosphate rock was tested for suitability for rice (Misra *et al.,* 2002). Application of high analysis P fertilizers like polyphosphates reduces P fixation and increases P availability in high P fixing soils. The rice crop yield increased by 13% when 80 kg P_2O_5 ha^{-1} was applied through ammonium polyphosphate compared to SSP (Venugopalan and Prasad 1994).

Organic Sources: The organic manure during decomposition forms organic acids, humic acids and chelating substances which help in liberation of insoluble P into soil solution. Organic acids are released by root exudates and microorganisms which are by- products of degradation of complex organic molecules (Yadav and Tarafdar, 2003). The organic sources can even replace 20 to 40% or recommended dose of P fertilizer and can sustain higher productivity. This is supported by the study with the highest productivity in potato –radish system of Himachal Pradesh where 25 to 50% of recommended P and K fertilizer could be replaced with FYM (Jatav *et al.,* 2010). Continuous recycling of green manures with organic amendments not only improve organic carbon but also contribute to P pools in the soils (Balwinder *et al.,* 2008). In a long term study carried out in rice- wheat cropping system in calcareous soils of Bihar addition of green manure (*dhaincha*) and organic manure (5t ha^{-1}) along with 100% NPK increased available soil P and was on par with 100% NPK and green manure. The decomposition of organic matter released CO_2 which enhanced P availability in 6 year study (Kumar and Singh, 2010). Many studies have reported the use of organics with RP is also known to improve P solubility and residual effect on cropping system. Application of lower doses of 30 kg P_2O_5 ha^{-1} with FYM for five years on P deficient soils of Typic Hapludalf of Meghalaya RP improved P use efficiency in soybean crop (Majumdar *et al.,* 2007). Application of P rich residues also improves the mineralizable P fractions in soil. Long-term use of fertilizers and FYM decreased P adsorption even more than a super-optimal application of P fertilizers (Singh *et al.,* 2006).

Best Management Practices for Soluble P: Best management practices (BMP) for P should aim in ensuring P availability in soil solution at appropriate time at a reasonable cost, thus increasing P use efficiency (PUE) in sustaining crop productivity. This can be achieved by using suitable P source which minimize reaction with soil components and makes P pools available to crop, modifying soil component or application method (of P fertilizer) to reduce P fixation. For P '4R nutrient management' i.e., the right fertilizer, the right amount (soil test based), the right time of application (crop growth stage) and the right application method (band placement) and precision application based on management zones is the need of the hour. The best management practices (BMPs) are based upon nutrient requirement of individual crops, the extent of response to crops to P application and the capacity of crops to utilize the residual effect for succeeding crops as shown in the Table 1 for some cropping systems of India. Some specific P BMPs particular to cropping systems are given in Table 2.

Table 1: Some best P management practices (BMPs)

Management Practices	Situation/ Condition
Phosphorus broadcast	Under high speed operations and heavy P application rates
Phosphorus placement	In low soil test P where early season stress is felt
Fertigation	Good under intensive agriculture; increases P fertilizer efficiency; protects environment; sustains irrigated agriculture
Use treated rock phosphate	Incubation with organic matter; addition of P solubilizer, *A. awamori*, during composting
Increasing the effective rooting area	Root symbiosis with arbascular mycorrhizal fungi (AMF)
Increased P availability through rhizosphere modification	Root exudates: phosphatase, oxalates (genotypic difference)
Use of earthworms	Enhanced nutrient availability mainly in tropical soils through casting
Organic residue amendments	A rise in pH in acid soils accompanied by P solubilization; Production and release of organic anions; increased enzymatic activity; complexation of exchangeable ions such as Al^{3+}, Fe^{3+}

Source: Subba Rao and Srivastava, 2012

Table 2: Nutrient BMPs in some cropping systems in India

Cropping sequence	Strategy
Rice-wheat, pearl millet- wheat, soybean-wheat	Apply phosphorus to winter (rabi) wheat and skip P application to kharif crops
Maize- wheat, sorghum- wheat	Prefer to apply P to wheat
Gram- rice	Apply super phosphate to gram and harness the residual effect on rice
Sorghum-castor	Apply P at recommended dose to sorghum and castor crop may be given a reduced dose
Potato based system	P should be applied to potatoes in a potato based cropping system
Groundnut-wheat	Apply recommended dose of P to wheat and skip application to groundnut

Source: Acharya et al., 2003

Some agronomic techniques like placement, time and application of P fertilizers are known to improve PUE and therefore are important aspects in BMP. Phosphatic fertilizer should be placed near root zone for efficient utilization by crops and for improving seedling vigor. Phosphatic fertilizers applied as basal dose after broadcasting should be incorporated in soil during preparation of the field before crop sowing. Among the different methods band placement of P is the best method for many crops which can be practiced in high P fixing soils having low soil P status and during dry season. Liming improves P use efficiency by improving soil pH to neutrality and resulting in more uptake of P by crops. Phosphatic fertilizer is totally imported in India and therefore fertilizer application based on soil testing should be followed for sustaining optimum crop productivity. At the same time this would also improve plant nutrient use efficiency and minimize P accumulation and loss of P from soil subjected to erosion. Selection of crops with high P utilization like cowpea, blackgram and green gram should be encouraged which absorb more P from applied P fertilizers. Among the different crops, P use efficiency is found the highest in pea followed by lentil and chickpea as reported in a study (Joshi *et al.,* 1998). Cereal crops are known to extract more P efficiently from soils because of the fine roots which are highly efficient in drawing P from soil solution. A certain yearly increment in crop yield to some extent can cope up with increasing P fertilizer scenario with better crop management practices.

Nano-Rock Phosphates

Nanotechnology is being visualized as a rapidly evolving field that has potential to revolutionize agriculture. Presently, the application of nanotechnology in soil science research is concentrated on formulation of nano fertilizers, smart delivery systems for nanoscale fertilizers, nanoporous zeolites for slow release and efficient dosage of water and fertilizers for plants, nano sensors for soil quality and plant health monitoring, nano induced polysaccharide powder for moisture retention or soil aggregation carbon build up and nano magnets for removal of contamination from soil and water. The cutting edge research areas are expected to emerge in the coming years (Adhikari *et al.,* 2013). In a study carried out at IISS, Bhopal maize crops treated with Udaipur nano RP (34% P_2O_5) recorded the highest grain and stover yield (5.44 and 7.13 t ha^{-1}) as compared to control. The increase was 44.68 and 13.17% more than control. It was equal to SSP followed by Udaipur nano RP (31%P_2O_5) treated plants. The highest 1000 grain weight and shelling percentage (230.3 and 61.47%) were obtained from Udaipur nano RP (34% P_2O_5) treated plants followed by SSP treatment and control. There was no significant difference in the agronomic parameters of SSP and Udaipur nano RP (34% P_2O_5) treated plant but SSP was consistently superior to RP sources. The P content and its uptake was more in SSP treated plants followed by Udaipur nano RP (34% P_2O_5) which was comparable, whereas less P content and uptake was observed in control. The total P uptake was 40.29, 38.29, 34.42 kg ha^{-1} under SSP, Udaipur nano RP (34% P_2O_5) and Udaipur nano RP (31%P_2O_5) treated plants which was 44.82, 37.63 and 21.24 %, respectively more over control whereas the lowest was recorded by no fertilizer treated plants (17.27 kg ha^{-1}) as shown in IISS Annual Report (2012).

With the introduction of nutrient bound subsidy and opening up of phosphate fertilizer sector, new generation fertilizers which are having high recovery efficiency like polymer coated water soluble fertilizers, rhizosphere- controlled fertilizer (RCF), organic complexed superphosphate, slow release P fertilizers with superabsorbent property and nano phosphate fertilizers gained importance (Ray *et al.,* 2013). In a study on polymer coated MAP, improved plant recovery of fertilizer P was observed and it provided barley grain yield advantage relative to uncoated MAP (Malhi *et al.,* 2002). In rhizosphere controlled fertilizers water soluble fraction acts as starter dose and insoluble P fraction becomes soluble by crop and microorganisms rhizospheric activity (Erro *et al.,* 2011a) and are proved to be more efficient than superphosphate in both alkaline and acid soils. Organic complexed superphosphate is another option for enhancing agronomic efficiency of superphosphate by reducing P fixation by introducing organic chelating agent during superphosphate production (Erro *et al.,* 2011b).

Role of Biological System - Roots and Microbes - in Efficient P Utilization

In developing countries like India P input is a costly affair and therefore farming systems with more efficient utilization of P availability to crops are needed (Lynch 2007). Recovery of P from legumes residues and microbial biomass was 15% to 28%, in contrast to 5% recovery from mineral fertilizer (Bunemann *et al.,* 2004). Efficient utilization of P through crop root system can be achieved by altering root morphology (Johnston *et al.,* 2014). First approach was done by modifying root morphology by increasing root biomass and roots with high specific root length which can cover larger surface area. This can be achieved with arbuscular mycorrhizal fungi (AMF) which increases P absorption. Recently used 'root foraging strategies' can be used in soil with low P and thus will improve yield and will reduce the accumulation of P in moderation to high P fixing soils (Simpson *et al.,* 2011). In wheat high root length density was observed as an important root trait which resulted in improving P uptake and correlated well with P uptake efficiency and P fertilizer applied to soil (Manske *et al.,* 2000). The root growth angle in crops like maize (seminal and crown roots) and soybean (basal roots) which is related with genotype of the crop can increase P acquisition in maize (Tiwari, 2001). Another approach is by introducing P efficient genotypes in cropping systems which can improve P uptake in low P input systems. Phosphorus efficient cultivars of groundnut like M522 due to higher root growth absorbed more P in shoot. Similarly in maize variety like Paras proved more efficient because of more roots and low internal P requirement (Gill and Bhadoria, 2010). Residual P is known to be more efficiently utilized by legumes because of their high cation exchange capacity and organic acid production.

Soil microorganisms are the important contributors to soil P pools which constitute 0.4% to 2.4% of total P in arable soils. They mediate a number of biochemical reactions and thus act as a sink and source of P in soil (Oberson and Joner 2005). They also decompose the organic residue by immobilization and mineralization thus maintaining equilibrium with soil solution P pools. Arbuscular mycorrhizal fungi (AMF) colonise almost all the crop species in agricultural crops and exploit larger volume of soil for P uptake in P deficient soil. Positive response with application of phosphatic biofertilizers like phosphate solubilising microorganism (PSM) and VAM increases the solubility of native P and applied P (Somani and Dadhich, 2005). In general PSM constitutes 0.5 to 1.0% of soil microbial population with bacteria outnumbering fungi (2-150 fold). The crop species, in extremely low available soil P, develops root clusters effective in capturing P by releasing root exudates like organic anions, enzymes, phenolic acids and protons. In five year rotation of soybean-wheat cropping system, farmyard manure and fertilizer P improved rhizosphere activity and AMF colonization of roots (Manna *et al.,* 2006). The treatment where only P fertilizer (44 kg P ha^{-1}) was applied, AMF colonization was 15% lower than in control plots in both soybean and wheat. Co-inoculation of VAM and phosphate solubilising bacteria resulted in high root coloniozation, high VAM spore density and viable counts of *Pseudomonas striata* also significantly increased yield and P uptake in maize- onion cropping sequence under temperate conditions in an inceptisol (Singh *et al.,* 2011). It was observed that selective feeding habit of earthworms and repeated feedings on the casts may be responsible for increased organic C and total N content of casts and increased extractable P and S. The use of inoculants, AMF and plant growth promoting microbes plays significant role in phosphate mineralization from both organic and inorganic sources (Tawaraya *et al.,* 2006).

Future Research Needs

a) Research into reducing phosphate fertilization, specifically on phosphate accumulated soils, must become one main priority. In the future, the P in such enriched soils can possibly be "mined" through biological mobilization.

b) Some research issues that need attention are accounting of residual P in cropping systems especially in long term, developing techniques of the biological mobilization of insoluble/ sparingly soluble P in the rhizosphere.

c) The efficient wastes recycling as integral part of P supply in agricultural production system, developing precision P placement, following strategies of heavy initial application of RP, use of nano RP and state of art controlled release P fertilizers could be an important priority area of research.

d) The use of Indian RP for preparing compost, modified forms of RP by incorporating improved technologies, addition of microorganism, green manures, vermicompost, city refuse modifying the physical properties on large scale can encourage their use at field levels which will not only improve P use efficiency but will reduce the demand of high priced P fertilizers. Encouraging phosphocompost technologies using low grade RP and farm wastes to make composting at field levels could be another important priority area of research.

e) Better soil test P for crop response and environment quality should be designed and used for Indian soils to determine the critical level for crop yield and P loss for each soil type. The development of P index for cultivable fields, watersheds, intensive cropping systems to determine the degree of P saturation of the soils so as to reduce P loss, minimize environmental impact and conserve finite resources and its usage by farmers for enhancing crop production. Source and transport factors should be included to identify the critical sources of P export from watershed.

f) There is need to strengthen the selection criteria for high P efficiency in different crop varieties. Modern breeding tools can be of great help to develop such genotypes, for high P use efficiency. Breeding efforts will definitely lead to identification and development of genotypes efficient in P utilization and P acquisition.

References

Acharya, C.L., Subba Rao, A., Biswas, A.K., Reddy, K.S., Yadav, R.L., Diwedi, B.S., Shukla, A.K., Singh, V.K. and Sharma, S.K. 2003. *Methodologies and packages of practices on improved fertilizer use efficiency under various agro- climatic regions for different crops/ cropping system and soil conditions.* Indian Institute of Soil Science, Bhopal.

Adhikari , T., Kundu, S. and Subba Rao, A. 2013. *Nanotechnology in Soil Science and Plant nutrition* . New India Publishing Agency, New Delhi.

Annual Report 2012-13. Indian Institute of Soil Science, Bhopal (2013).

Aulakh, M.S., Garg, A.K and Kabba, B.S. 2007. Phosphorus accumulation, leaching and residual effects on crop yields from long term applications in the subtropics. *Soil Use Management,* 23: 417-427.

Awasthi, P.K. 1990. Marketing rock phosphate as phosphatic fertilizer. *Fertiliser News,* 35:79-91.

Balwinder, K., Gupta, R.K. and Bhandari, A.L. 2008. Soil fertility changes after long term application of organic manures and crop residues under rice wheat system. *Journal of Indian Society of Soil Science,* 56: 80-85.

Bunemann, E.K., Steinebrunner, F., Smithson, P.C., Frossard, E. and Oberson, A. 2004. Phosphorus dynamics in a highly weathered soil as revealed by isotopic labeling techniques *Soil Science Society of American Journal,* 68:645–1655.

Erro, J., Baigorri, R. Garcia Mina, J.M. and Yuvin, J.C. 2011b.^{31}P NMR characterization and efficiency of new types of water-insoluble phosphate fertilizers to supply plant-available phosphorous in diverse soil types. *Journal of Agriculture Food Chemistry,* 59: 1900-1908.

Erro, J., Baigorri, R. Garcia Mina, J.M. and Yuvin, J.C. 2011a. Patent WO/ 2011/ 080496.

Feng K., Lu, H.M., Sheng, H.J., Wang, X.L. and Mao, J. 2004. Effect of organic ligands on biological availability of inorganic P in soils. *Pedosphere,* 14: 85-92.

Fraiture, C.D. 2007. *Future Water Requirements for Food—Three Scenarios*, International Water Management Institute (IWMI), SIWI Seminar: Water for Food.

Gill, A.A.S. and Bhadoria, P.B.S. 2010. Identification of phosphorus efficient cultivars in maize and groundnut. *Journal of Indian Society of Soil Science,* 58(2), 245-247.

Goswami, N.N. and Pasricha, N.S. 1992. Proceeding of Indian Nutrient Management for sustained productivity, Vol I, pp 30-42, PAU, Ludhiana.

Haynes, R.J. and Mokolobate, M.S. 2001. Amelioration of Al toxicity and P deficiency in acid soils by additions of organic residues: a critical review of the phenomenon and the mechanisms involved. *Nutrient Cycling Agroecosyst*em, 59: 47-63.

Huang, C.Y., Shirley, N., Genc, Y., Shi, B.J. and Langridge, P. 2011. Phosphate utilization efficiency correlates with expression of low-affinity phosphate transporters and non-coding RNA, IPS1, in barley (Hordeum vulgare L.). *Plant Physiology,* 156: 1217–1229.

Jat, M. L., Kumar. D., Majumdar, K., Kumar, A., Shahi, V., Satyanarayana,T., Pampolino, M., Gupta, N., Singh, V., Dwivedi, B. S., Singh, V. K., Singh, V., Kamboj, B. R., Sidhu, H. S. and Johnston, A. 2012. *Indian Journal of Fertilizers,* 8(6): 62-72.

Jatav, M.K., Sud, K.C. and Trehan, S.P. 2010. Effect of organic and inorganic source of phosphorus and potassium on their different fraction under potato-radish cropping sequence in brown hill soil *Journal of Indian Society of Soil Science,* 58(4): 388-393.

Johnston, A.E., Poulton, P.R., Fixwn, P.E. and Curtin, D. 2014. Phosphorus: Its Efficient Use in Agriculture *Advances of Agronomy,* 123: 177- 228.

Joshi, N.L., Singh, D.V., Singh, R.S. and Saxena A. 1998. Fifty years of agronomic research in India. *Indian Society of Agronomy,* New Delhi, 1-32.

Kumar V. and Singh A.P. 2010. Longterm effect of green manuring and farm yard manure on yield and soil fertility status in rice-wheat cropping system. *Journal of Indian Society of Soil Science,* 58 (4): 409-412 .

Lynch, J.P. 2007. Roots of the second green revolution *Australian Journal of Botany,* 55: 1–20.

Majumdar, B., Venkatesh, M.S., Kailash, K. and Patiram. 2007. Effect of rock phosphate, superphosphate and their mixtures with FYM on soybean and soil-P pools in a typic hapludalf of Meghalaya. *Journal of Indian Society of Soil Science,* 55 (2): 167-174.

Malhi, S.S., Haderlain, L.K., Pauly, D.G. and Johnston, A.M. 2002. Improving Fertilizer Phosphorus Use Efficiency. *Better Crop,s* 86: 8-9.

Manna, M.C., Subba Rao, A. and Ganguly, T.K. 2006. Effect of Fertilizer and Farmyard Manure on bioavailable P as influenced by rhizosphere microbial activities in soybean-wheat rotation. *Journal of Sustainable Agriculture,* 29: 149-166.

Manske, G.G.B., Ortiz-Monasterio, J.I., Van Ginkel, M., Gonzalez, R.M., Rajaram, S., Molina, E. and Vlek, P.L.G. 2000. Traits associated with improved P-uptake efficiency in CIMMYT's semidwarf spring bread wheat wheat grown on an acid andisol in Mexico. *Plant and Soil,* 221: 189–204.

Mattingly, G.E.G. 1975. Labile phosphate in soil. *Soil Science,* 119 (5): 369-375.

Melland, A.R., McCaskill, M.R., White, R.E. and Chapman, D.F. 2008. Loss of phosphorus and nitrogen in runoff and subsurface drainage from high and low input pastures grazed by sheep in Australia. *Australian Journal of Soil Research,* 46:161–172.

Misra, U.K., Ghosh, P.C., Das, N. and Pattanayak, S.K. 2002. Evaluation of Magnesium Ammonium Phosphate prepared from Low Grade Phosphate Rock as a Source of Phosphorus for rice. *Journal of Indian Society of Soil Science,* 50: 135-137.

Muralidharudu, Y., Reddy, K.S., Mandal, B.N., Rao, A.S., Singh, K.N. and Sonekar, S. 2011. *GIS based soil fertility maps of different states of India. All Inia Coordinated Research Project on Soil Test Crop Response Correlation.* IISS, Bhopal.

Oberson, A. and Joner, E.J. 2005. Microbial turnover of phosphorus in soil. In: Turner BL, Frossard E, Baldwin DS (eds) *Organic phosphorus in the environment.* CABI,Wallingford, pp 133– 164.

Pathak, H., Mohanty, S., Jain, N. and Bhatia, A. 2010. Nitrogen, phosphorus, and potassium budgets in Indian agriculture. *Nutrient Cycling Agroecosystem,* 86:287–299.

Prasad, R. 2012. Fertilizers and manures. *Current Science,* 102(6): 894-898.

Raju, R.A., Subba Rao, A. and Rupa, T.R. 2005. Strategies for integrated phosphorus management for sustainable crop production *Indian Journal of Fertilizers,* 1(8): 25-28 & 31-36.

Rashmi, I. 2013. Ph.D thesis, Development of phosphorus saturation indices for selected Indian soils. UAS, Bangalore.

Rattan, R.K. and Neelam, S. 1993. Interaction of phosphorus with other macro and micronutrients In: *Phosphorus Researches in India*, (G. Dev, Ed.) PPIC-India Programme, Gurgaon, Haryana, pp.94- 119.

Ray, P., Rakshit, R. and Biswas, D.R. 2013. *Indian Journal of Fertilizers,* 9 (10): 44-53.

Reddy, D.D., Rao, A.S. and Singh, M. 2005. Changes in P fractions and sorption in an Alfisol following crop residues application. *Journal of Plant Nutrition and Soil Science,* 168 (2): 241-247.

Reddy, D.D., Subba Rao, A., Reddy, K.S. and Takkar, P.N. 1999. Effects of repeated manure and fertilizer phosphorus additions on soil phosphorus dynamics under a soybean-wheat rotation. *Field crop Research*, 62: 181-190.

Saha, S., Saha, B., Murmu, S., Patil, S. and Roy, P. D. 2014. Grain yield and phosphorus uptake by wheat as influenced by long-term phosphorus fertilization. *African Journal of Agricultural Research,* 9(6): 607-612.

Simpson R. J., Oberson A., Culvenor R.A., Ryan M. H., Veneklaas E. J. and Lambers H. 2011. Strategies and agronomic interventions to improve the phosphorus-use efficiency of farming systems. *Plant and Soil,* 349(1–2): 89–120.

Singh S.R., Najar, G.R., Singh, U and Singh, J.K. 2011. Phosphorus management in maize-onion cropping sequence under rainfed temperate conditions of inceptisol. *Journal of Indian Society of Soil Science,* 59 (4): 355-361.

Singh, V., Dhillon, N.S, Kumar, R. and Brar, B.S. 2006. Influence of long-term use of fertilizers and farmyard manure on the adsorption–desorption behaviour and bioavailability of phosphorus in soils. *Nutrient Cycling Agroecosystem,* 76: 29-37.

Somani, L.L. and Dadhich, S.K. 2005. Some microbial interventions in phosphorus nutrition of plant. *Indian Journal of Fertilizers,* 1 (2): 21-28.

Steven, J., Van Kauwenbergh, Stewart, M. and Mikkelsen, R. 2013.World reserves of phos phate rock-a dynamic and unfolding story. *Better Crops,* 97: 18-20.

Stumn, W. and Sulbereger, B. 1992. The cycling of iron in natural environments: Consideration based on laboratory studies of heterogeneous redox processes. *Geochimica Cosmochimca Acta,* 56: 3233-3257.

Subba Rao A. and Srivastava S. 2012. *Strategies to Enhance the Efficient use of Low Analysis P Fertilizers and Indigenous P Source that will Reduce DAP Import in India.* In: Souvenir of National Seminar on Strategies to Rationalize and Reduce Consumption of Water Soluble Phosphorus and Potassium in the Country to Minimize Import (Dey *et al.,* Eds) pp 21-25.

Subba Rao, A. and Ganeshamurthy, A.N. 1994. Soybean responses to applied phosphorus and sulphur on vertic ustochrepts in relation to available phosphorus and sulphur. *Journal of Indian Society of Soil Science,* 42: 606-610.

Subba Rao, A., Reddy, K.S. and Takkar, P.N. 1996. Residual effects of phosphorus applied to soybean or wheat in soybean-wheat cropping system on a Typic Haplustert. *Journal of Agriculture Science*, 127: 325-330.

Subba Rao, A., Reddy, D.D., Reddy, K.S. and Takkar, P.N. 1998. Crop yields and phosphorus recovery in soybean-wheat cropping system on a Typic haplustert under integrated use of manure and fertilizer phosphorus *Journal of Indian Society of Soil Science,* 46: 249-253.

Subba Rao, A. ISSS Foundation Lecture In: *75th Annual Convention Ind. Soc. Soil Sci.* 1-20 (2010).

Syers, J.K., Johnston, A.E. and Curtin, D. 2008. Efficiency of soil and fertilizer phosphorus use—reconciling changing concepts of soil phosphorus behaviour with agronomic information. FAO *Fertilizer and Plant Nutrition Bulletin 18*. Food and Agriculture Organisation of the United Nations, Rome.

Tawaraya K, Naito M, Wagatsuma T. 2006. Solubilization of insoluble inorganic phosphate by hyphal exudates of arbuscular mycorrhizal fungi. *Journal of Plant Nutrition,* 29:657–665.

Tiwari K.N. (2001). Phosphorus Needs of Indian Soils and Crops. *Better crops International,* 15(2): 6-10.

Venugopalan, M. V. and Prasad, R. 1994. Nature, behaviour and agronomic value of ammonium polyphosphate as phosphate fertiliser - a review. *Fertiliser News,* 39(11): 27-31 &33.

Vision 2030, Indian Institute of Soil Science, Bhopal. 2011.

Wahid, P.A., Nambiar, P.K.N., Jose, A.I. and Rajaram, R.P. 1988. In *Six decades of Coconut Research.* Kerala Agricultural University, Thrissur, pp. 46-80.

Yadav, R.S. and Tarafdar, J.C. 2003. Phytase and phosphatase producing fungi in arid and semi-arid soils and their efficiency in hydrolyzing different organic P compounds. *Soil Biology and Biochemistry,* 35: 1-7.

Advances in Nutrient Dynamics in Soil - Plant System, pp 51-60

Editors: R. Elanchezhian, A.K. Biswas, K. Ramesh and A.K. Patra

6

Land and Tillage Management Techniques for Enhancing Nutrient Use Efficiency

R.S. Chaudhary and J. Somasundaram

ICAR-Indian Institute of Soil Science, Nabi Bagh, Bhopal – 462038, India

For efficient nutrient use, most important factors are the timely supply/recycling of adequate amount of nutrients to the production system and its retention and release by soil for timely and ready availability in the root zone for its mobility to the plant in the presence of optimum moisture content in the soil. For keeping the applied nutrients retained in the root zone one has to reduce the soil erosion and leaching losses through technologies available in hand which include direct measures of soil and water conservation and maintain the soil aggregation and porosity through suitable tillage and organic residue recycling. For movement of nutrients into plant system we need to ensure optimum soil moisture content through water management.

Recovery of applied inorganic fertilizers by plants is low in many soils. Estimates of overall efficiency of these applied fertilizers have been about 50% or lower for N, less than 10% for P, and close to 40% for K (Baligar and Bennett, 1986, a, and b). These lower efficiencies are due to significant losses of nutrients by leaching, run-off, gaseous emission and fixation by soil. These losses can potentially contribute to degradation of soil, and water quality and eventually lead to overall environmental degradation. These are compelling reasons of the need to increase NUE. In this chapter, we are discussing the role of different methods of land configurations and tillage management practices in nutrient use efficiencies (NUE).

Tillage: Tillage practices mainly influence the physical properties of soil viz., soil moisture content, soil aeration, soil temperature, mechanical impedance, porosity and bulk density of soil and also the biological and chemical properties of soil which in turn influence the edaphic needs of plants viz., seedling emergence and establishment, root development and weed control. Tillage also influences the movement of water and nutrients in soil and hence their uptake by crop plants and their losses from soil-plant system. Tillage affects the WUE by modifying the hydrological properties of the soil and influencing root growth and canopy development of crops. Tillage methods influence wettability, water extraction pattern and transport of water and solutes through its effect on soil structure, aggregation, total porosity and pore size distribution. Tillage system suitable for a soil depends upon soil type, climate and cropping system practiced. Shallow inter-row tillage into growing crops reduces short-term direct evaporation loss from soil even under weed-free condition by breaking the continuity of capillary pores and closing the cracks.

Chaudhary *et al.* (2003) reported that in the six rain fed cotton growing states of India, the reduced tillage (one harrowing and one intercultural operation with pre-emergence herbicide), followed by broad-bed-and-furrow (BBF) planting of rain fed cotton applied with recommended dose of fertilizers, incorporation of green manure and location-specific deficient nutrients gave 32.6% higher cotton-seed yields compared to farmer's practice of rain fed-cotton cultivation on flat-beds with conventional tillage involving one summer ploughing, followed by 2–3 harrowings and with numerous intercultural operations and application of about 70% of the recommended fertilizer dose. The practice also improved soil physical and chemical properties.

Higher contents of available P, Ca, K, organic C and N have been reported for no tillage than for conventional tillage (Lal, 1976; Mahboubi *et al.,* 1993;). Minimum tillage increased root growth in the top 12 cm of soil for barley (*Hordeum vulgare* L) and oat (*Avena sativa* L) cropping systems. Minimum tillage has also been reported to increase root weight, length, and density, increasing the nutrient and water use efficiencies by many workers. Baligar *et al.* (1998b) reported that shoot dry matter yields and root length and density of silage corn in no-till was significantly higher than in conventional tillage. Such improved root parameters contributed to higher yields and uptake efficiencies of N., P, Ca, S, Cu, Fe, and Zn. Improved tillage equipment and practices needs to be developed to increase NUE across different agro ecosystems.

Deep tillage to a depth of 30-45 cm at 60-120 cm intervals helps in breaking subsoil hard pans in alfisols facilitating growth and extension of roots and improving grain yield of crops as well as increasing residual soil moisture. However, the benefit is absent in suboptimal rainfall years and restricted to only deep-rooted crops in high rainfall years. Conservation tillage practice normally stores more plant available moisture than the conventional inversion tillage practices when other factors remain same. The high soil moisture content under conservation tillage is due to both improved soil structure and decrease in the evaporation loss under continuous crop residue/mulch cover. Increase in the available water content under conservation tillage, particularly in the surface horizon, increases the consumptive use of water by crops and hence improves the water use efficiency. Off season tillage or summer ploughing opens the soil and improves infiltration and soil moisture regimes.

Minimum tillage, no tillage, conservation tillage and traditional tillage can bring profound changes in soil quality, SOM and nutrients throughout different soil horizons (Lal, 1976; Mahboubi *et al.,* 1993). Rooting pattern, water holding capacity, water penetration, aeration, soil compaction, and soil temperature are also influenced by type of tillage practices (Arkin and Taylor,1981). Crop rotation and use of cover crops and green manure crops are known to improve soil fertility and physical properties and to minimize pest and weed problems (Delgado, 1998; Delgado *et al.,* 1999; Fageria, 1992; Fageria *et al.,* 1997a).

Land configurations: Land configuration and soil tillage have tremendous potential for its further exploitation and improving NUE. Experiments have been conducted in central India (CRIDA, 1983) on land configuration practices in an effort to conserve more rainfall for stabilizing crop yields and to study NUE pattern. These experiments have documented the advantages of broad bed and furrow systems for soil water conservation (ICRISAT, 1981) and increasing crop yields (Bhatawadekar, 1985). Reddy *et al.* (1992) observed that sowing of crops on a grade and ridging later consistently increased soil water content and crop yield. Similarly compartmental bunding produced a higher grain yield of pearl millet (CRIDA, 1983) compared with that of the flat bed method of sowing. On the other hand, experiments conducted at ICRISAT (1982) showed no distinct yield advantage caused by land configuration adopted during the cropping season. In some Vertisols of south central India, no difference in grain yield of sorghum was observed with compartmental bunding and ridging over traditional flatbed cultivation (CRIDA, 1990). Selvaraju *et al.* (1999) extensively studied the effectiveness of land configuration practices on soil water content, soil fertility, crop establishment under sorghum + pigeon pea and pearl millet+ cowpea intercropping systems in Alfisols and Vertisols of central India (Table 1). They have found that, soil water content in random tie ridging (TR) system was 7% and 14 % higher than the OR (open ridging) and Flat bed (FB) system, respectively. The TR treatment stored 15% more water at 0-15 cm depth and 8%more water at 15-30 cm depth on 30 days after sowing than the FB during 1991-1992, probably because of reduced runoff and greater soil water retention in furrows of TR plots (Hulugalle, 1990). TR stored 20% and 18% higher soil water than did FB at 0-15 and 15-30 cm soil depth, respectively, on 90 days after sowing during 1992-1993. However, soil water stored in the OR did not significantly differ from that in TR except for 0-15 cm depth on 30 days after sowing during 1991-1992. But at 15-30 cm depth, water contents in TR and OR were each significantly different from that in the FB treatment. Further, they have found that broad bed and furrow (BBF) and compartment bunding

(CB) land configurations added more crop residues to the soil at the end of the season and led to higher available nitrogen and organic carbon content (Table 2). In 1992-1993, BBF added 13% and 22% more crop residues under sorghum +pigeon pea and pearl millet +cowpea intercropping, respectively, than did FB. The organic carbon content of the soil under BBF was 11% more than the FB in sorghum + pigeon pea intercropping in 1992-1993. In1991-1992, CB had 21% greater organic carbon than FB. Lal (1995) reported that substantial addition of crop residues enhances soil fertility and increases yield. More soil water storage in BBF, CB and RD practices probably increased the crop residue addition. Sivakumar *et al.* (1992) also reported an increase in soil fertility of the soil due to more water availability.

Table 1: Effect of land configurations on crop residue addition (CRA) and soil fertility.

	Sorghum + Pigeon Pea (1991-1992)		Sorghum + Pigeon Pea (1992-1993)			Pearl millet+ cow pea (1992-1993)		
	Avail. N (kg/ha)	O.C (g/kg)	CRA (kg/ha)	Avail N (kg/ha)	O.C (g/kg)	CRA (kg /ha)	Avail. N (kg/ha)	O.C (g/kg)
CB	149	4.1	628	134	4.0	451	139	4.1
RD	149	3.8	600	129	3.9	427	137	3.8
BBF	144	3.7	642	131	4.1	478	139	4.1
FB	136	3.4	568	125	3.7	391	133	3.7
LSD (0.05)	7.0	0.5	49	NS	0.2	46	NS	NS

Mulching: Mulching influences WUE of crops by affecting the hydrothermal regime of soil, which may enhance root and shoot growth, besides it helps in reducing the evaporation (E) component of the evapotranspiration. Under moisture stress conditions, when moisture can be carried over for a short time or can be conserved for a subsequent crop, mulching can be beneficial in realizing better crop yield.

Conservation of water: *In-situ* conservation of water can be achieved by reduction of runoff loss and enhancement of infiltrated water and reduction of water losses through deep seepage and direct evaporation from soil. Runoff is reduced either by increasing the opportunity time or by infiltrability of soil or both. Opportunity time can be manipulated by land shaping, tillage, mechanical structures and vegetative barriers of water flow and infiltrability can be increased through suitable crop rotations, application of amendments, tillage, mulching etc. Water loss by deep seepage can be reduced by increasing soil-water storage capacity through enlarging the root zone of crops and increasing soil water retentively. Direct evaporation from soil can be controlled with shallow tillage and mulching. Ex-*situ* conservation of water can be achieved by harvesting of excess water in storage ponds for its reuse for irrigation purpose.

Reducing leaching loss: Mobile nutrients (e.g. NO_3^-) are lost from the soil-plant system with the percolating water. Besides reducing the nutrient it may pollute the groundwater. The groundwater having more than 10 mg NO_3^-, N per liter is unfit for drinking purpose (WHO). Leaching loss of NO_3^- can be minimized by balanced fertilization, split application of urea synchronizing with crop demand, manipulation of water application and rooting depth, appropriate crop rotations and use of slow release fertilizers and nitrification inhibitors like N-serve, DCD, AM, CCC and neem-coated urea. Despite the success of synthetic nitrification and urease inhibitors in research farms they have poor acceptability among farmers because of high cost. However, the use of plants products like neem for coating urea can be popularized among the farmers to affect N economy and minimize long-term environmental consequences of denitrification and nitrate leaching.

Dry seeding of rice with subsequent aerobic soil conditions avoids water application for puddling and maintaining submerged soil conditions and thus reduces the overall water demand. The lower water productivity in puddled as well as unpuddled transplanted rice compared to DSR was attributed due to higher water use efficiency during crop season (Table 2). The maximum water saving was with DSR at transplanting time (49.20%) followed by DSR+ Sesbania (43.65%), DSR (39.68%) and raised bed transplanted rice (24.60%).

Table 2. Effect of tillage and crop establishment methods on productivity and water use of rice in Rice- Wheat cropping system.

Tillage/Crop Establishment methods	Grain Yield (t ha^{-1})	Water Used (m)	Irrigation water productivity (kg m^{-3} water)	Rainfall (m)	Gross water productivity (kg m^{-3} water)	Water saving (%)
Conventional transplanting	3.65	1.26	0.29	0.20	0.25	—
Unpuddled transplanting	4.00	1.25	0.32	0.20	0.27	—
Raised bed transplanting	2.95	0.95	0.31	0.20	0.26	24.60
Direct seeded rice (DSR)	3.24	0.76	0.42	0.23	0.32	39.68
DSR +Sesbania	3.65	0.71	0.51	0.23	0.38	43.65
DSR at Transplanting time	2.91	0.64	0.45	0.23	0.43	49.20
CD at 5%	0.32	—	—	—	—	—

Reducing runoff and erosion losses: Many water-soluble nutrients are lost through runoff. This loss can be minimized by proper crops land management and selection of proper crops and cropping systems, tillage and mulching. Nutrients absorbed on the surface of soil particles-clays and silts and soil organic matter are lost when the top soil is eroded by water or wind. Proper soil conservation measures should be adopted to minimize this loss.

Precision farming techniques: Application of N on the basis of soil test value is essential to economize on the cost of fertilizer application. Land leveling and root zone wetting through micro-irrigation systems also lead to efficient use of water and N fertilizer inputs. Employment of drip irrigation and fertigation techniques has gained popularity in recent years, particularly in the widely-spaced high-value crops, precisely in controlled quantity and at appropriate time directly to the root zone as per crop needs at different growth stages. This not only enhances WUE but also enables efficient use of nutrients, particularly N for higher productivity. Using N in accordance with chlorophyll meter has been found to be more efficient than fixed schedule of N fertilizer splits at key growth stages. Precision land leveling has tremendous impact on agronomic efficiency of N, P and K (Jat *et al.,* 2004). Under irrigated agriculture, precision water management has large bearing on the water productivity, higher yield and income. Higher water productivity and NUE was reported under precision drill seeding compared to broadcasting and traditional drill (Pal *et al.,* 2004). In addition to zero-till planting of wheat, raised bed planting and laser land leveling are other technologies being increasingly adopted by farmers. Farmers who have used laser land leveler reported positive effects in improving crop establishment, uniformity of crop maturity, increase in area available for cultivation (2 to 5 percent), improved efficiency of water application and increased water productivity resulting in large savings in irrigation water (upto 35 percent) and in improved use efficiency of applied nutrients. With increased availability of laser land leveling equipment, precision land leveling could become an important basal step for adopting canservation agriculture (CA) systems. The comparison of conventionally leveled and laser leveled fields on the basis of experiments conducted by CSSRI on the farmers' fields is given in Table 3.

Table 3 Wheat yields and water productivity in conventional leveling and laser leveling

Parameters	Conventional levelling	Laser Levelling
Levelling index (cm)	>1.5	<1.5
Irrigation depth (cm)		
Paddy	110-115	90-95
Wheat	30-35	20-25
Pumping requirement per irrigation (hr ha^{-1})		
Paddy	25-27	20-22
Wheat	15-17	9-11
Water productivity (kg m^{-3})		
Paddy	0.37	0.47
Water	1.50	2.44
Profit conventional (Rs ha^{-1})		
1st year	—	1000-1200
2nd year Onwards	—	4000-5000

Source: Ambast, 2006

Interaction with other inputs: The utilization of nutrients can be improved by optimum and synergistic interaction with other inputs viz., water, tillage and mulches. These inputs modify the physical, chemical and biological environment of soil, which influence the nutrient recovery by crop plants. Significant and positive interaction between applied N and water supply was observed on wheat yield and water and nutrient use efficiency by wheat (Bhale *et al.,* 2009). With 80 kg N/ha, N use efficiency increased up to 300 mm water supply in sandy loam soil. Interestingly, with 120 kg/ha, it did not increase when water supply was increased from 50 mm to 125 mm, but increased markedly when water supply was further increased to 300 mm (Table 4). This implied that the balance between these two inputs influenced input use efficiency.

Table 4: Nitrogen and irrigation effects on water use efficiency (kg/grain/mm) and nitrogen use efficiency (kg grain/kg fertilizer N) in sandy loam soil

Irrigation (mm)		WUE N rate (kg/ha)			NUE N rate (kg/ha)		
	0	40	80	120	40	80	120
0	5.3	7.6	8.1	6.0	8.5	5.5	1.5
50	6.3	9.5	11.3	13.3	20.2	18.4	17.8
125	5.7	10.3	11.9	11.8	33.2	25.5	17.0
300	4.6	7.4	9.5	10.2	30.2	30.3	23.7

Source: Bhale *et al.,* (2009)

Application of irrigation and nutrients in conjunction through pressure irrigation system results in efficient utilization of both resources. This will save water as well as reduce nutrient leaching losses and thereby increase WUE as well as NUE. This will increase the yield and quality of crops. There is saving of water and nutrient to the extent of 35 and 22 per cent, respectively. Fertigation is most commonly used for plantation crops like banana, sugarcane and orchards of Maharashtra.

Amelioration of problem soils: Soil related constraints affecting crop production influence the nutrient use efficiency of crops. For example liming of acid soils with calcite, dolomite or paper mill sludge improves the phosphorus use efficiency. Similarly amelioration of alkali and saline-alkali soils with gypsum helps in improving nutrient use efficiency. Any other physical constraint like sub-soil compaction should be ameliorated using appropriate tillage practices to improve the nutrient use efficiency.

Addition of Organic manures and Green manuring: Organic manures are important to enhance use efficiency of fertilizer inputs and also serve as alternative source of nutrients to chemical fertilizers. Combined use of organic manure and N fertilizer maintains a continuous N supply, checks losses and thus helps in more efficient utilization of applied fertilizers. Incorporation and decomposition of organic manures has a solubilising effect on native soil N and other nutrients including micronutrients. Further, such integrated plant nutrient supply (IPNS) systems also help in mitigating the adverse effects of acidity due to chellation of excess Al++ and/or Fe++ by the organic molecules liberated from FYM in the course of mineralization. The effect of FYM was found to be similar to like amendment in these acid soils, which seems mainly due to the formation of Al-organo chellates or complexes, resulting in the reduction of Al++ ion concentration in soil solution to levels beneficial to plant growth. In another study, apparent N recovery was increased when N fertilizer was applied along with organic manures such as FYM and *Eupatorium adenophorum* (Mahajan *et. al.,* 2002).

Effects of soil organic matter (SOM) on physical parameters and nutrient dynamics and how they impact NUE have been reported by several authors (Baligar and Fageria, 1997; Fageria, 1992). The SOM helps to maintain good aggregation and increase water holding capacity and exchangeable K, Ca, and Mg. It also reduces P fixation, leaching of nutrients and decreases toxicities of Al and Mn. Best management practices such as addition of crop residues, green manure, compost, animal manure, use of cover crops, reduced tillage and avoiding burning of crop residues can significantly improve the level of SOM and contribute to the sustainability of the cropping systems and higher NUE.

Inclusion of legumes in cropping systems for green manuring, fodder or grain purposes is an assured agro-technology to improve nutrient-use efficiency, especially that of N. The advantages of green manuring are indicated by increased N availability in soil, higher recovery of green manure N and its greater contribution towards grain production of crop.

Reducing gaseous loss: Part of the applied N is lost from soil by volatilization of ammonia and part of the nitrogenis lost as N_2O and N_2 gas by denitrification. Volatilization loss of ammonia can be minimized by mixing of nitrogen fertilizers in soil rather than broadcasting on soil surface, deep placement of urea super granules (USG) in puddle rice field, using urease inhibitors like thiourea, methyl urea, caprylo hydroxamic acid, phenyl phosphorodiamidate (PPD), ammonium thiosulphate etc. and adding inorganic salts of Ca, Mg or K with urea. Some coated material like sulphur coated urea(SCU), gypsum coated urea (GCU), plastic coated urea (PCU), mud ball urea and synthetic slow release urea based fertilizers viz., iso butylidene diurea (IBDU) and croto bylidene diurea (CDU)etc. may be used to retard the rate of urea hydrolysis and thereby, reducing ammonia volatilization.

Nitrous oxide (N_2O) is mainly produced by denitrification of NO_3^- under anaerobic condition, in lowland rice fields. Nitrous oxide is one of the greenhouse gases that are believed to be forcing global climate change. Dentrification loss can be minimized by avoiding the use of NO_3^- form of nitrogenous fertilizer (e.g. calcium ammonium nitrate, potassium nitrate etc.) in rice and use of nitrification inhibitors viz.,

Dicyandiamide (DCD), N-serve (2-Chloro, 6-Chloro methyl pyridine), AM (2-Amino, 4-Chloro, 6-methyl pyrimidine), coated calcium carbide (CCC), neem coated urea, deep placement of urea sugar granules (USG) in flooded rice field and efficient and efficient water management.

References

Arkin, G. F. and H. M. Taylor (eds.) (1981). Modifying the Root Environment to Reduce Crop Stress. ASAE Monograph No. 4. American Society of Agricultural Engineers, St. Joseph, MI.

Bhale, V. M. and Wanjari, S.S. (2009). Conservation agriculture: A new paradigms to increase resource use efficiency. *Indian Journal of Agronomy*, 54(2):167-177.

Baligar, V. C., and N. K. Fageria (1997). Nutrient use efficiency in acid soils: nutrient management and plant use efficiency pp. 75–93. In: A. C. Monitz,A.M.C. Furlani, N. K. Fageria., C. A. Rosolem, and H. Cantarells. (eds.),Plant-Soil Interactions at Low pH: Sustainable Agriculture and Forestry Production. Brazilian Soil Science Society Compinas, Brazil.

Baligar, V. C. and O. L. Bennett (1986a). Outlook on fertilizer use efficiency in thetropics. *Fertilizer Research*, 10, 83–96.

Baligar, V. C. and O. L. Bennett (1986b). NPK-fertilizer efficiency.Asituation analysis for the tropics. *Fertilizer Research,* 10, 147–164.

Bhatawadekar, P.V. (1985). Industrial raw material crops. In:Balasubramanian, U., Venkateswarlu, J. (Eds.), EfficientManagement of Dryland Crops. CRIDA, Hyderabad, India.pp. 355±372.

CRIDA (1983). All India coordinated research improvement projectfor dryland agriculture. Annual Report. CRIDA, Hyderabad,India.

CRIDA (1990). All India coordinated research improvement projectfor dryland agriculture. Annual Report. CRIDA, Hyderabad,India.

Delgado, J.A. (1998). Sequential NLEAP simulations to examine effect of early andlate planted winter cover crops on nitrogen dynamics. *Journal of Soil and Water Conservation*, 53:338–340.

Delgado, J.A., R.T. Sparks, R.F. Follett, J.L. Sharkoff, and R.R. Riggenbach (1999). Use of winter cover crops to conserve soil and water quality in the San Luis Valley of South Central Colorado. Pp. 125–142.R. Lal (ed.) Soil Quality and Soil Erosion. CRC Press, Boca Raton, FL.

Fageria, N. K. (1992). Maximizing Crop Yields. Marcel Dekker, New York, NY.

Fageria, N. K. V. C. Baligar, and C. A. Jones. (1997)(a). Growth and Mineral Nutrition of Field Crops 2nd edition Marcel Dekker, Inc., New York, NY.

Hulugalle, N.R. (1987). Effect of tied ridges on soil water content, evapotranspiration, root growth and yield of cowpeas in the Sudan Savannah of Burkina Faso. *Field Crops Research*, 17, 219-228.

ICRISAT (1981). Annual Report (1979-80). ICRISAT, Hyderabad, India.

ICRISAT (1982). Annual Report, (1981). ICRISAT, Hyderabad, India.

Jat, M. L., Pal, S.S., Subba Rao, A.V. M., Sirohi, K., Sharma, S.K., and Gupta R.K. (2004). In: Proceedings National Conference on Conservation Agriculture : Conserving resources, enhancing productivity, Sept 22-23, (2004), NASC Complex, Pusa, New Delhi, pp.9-10.

Lal, R. (1976). No tillage effects on soil properties under different crops in Western Nigeria. *Soil Science Socety of America Proceedings*, 40:762–768.

Mahboubi, A. A., R. Lal, N. R. Faussey (1993). Twenty-eight years of tillage effects on two soils in Ohio. *Soil Science Socety of America Journal*. 57:506 –512.

Mahajan, K.K., Kumar, S., Dev, S.P., Bhardwaj, .K.K., and Gupta, S.P. (2002). Evaluation of industrial wastes in wheat (Triticum aestivum)-maize (Zea mays) cropping system in mid hills sub-humid zone of Himachal Pradesh. *Indian Journal of Agricultural Sciences,* 72 (5): 257-259.

Pal, S. S., Subba Rao, A.V.M., Jat, M. L., Singh, J., Chandra, P., Sirohi, K., Chhabra, V., Sharma, G. (2004). In: Proceedings National sysmposium on alternate farming systems: Enhanced income and employment generation options for small and marginal farmers. Sept. 16-18, PDCSR, Modipuram, pp.227-228.

Reddy, K.C., Visser, P., Buckner, P. (1992). Pearl millet and Cowpea yields in sole and intercrop systems, and their effects on soil and crop productivity. *Field Crops Research,* 28: 325-326.

Selvaraju, R., Subbian, P., Balasubramanian, A. and Lal, R.(1999). Land configuration and soil nutrient management optionsfor sustainable crop production on Alfisols andVertisols of southern peninsular India. *Soil & Tillage Research*, 52: 203-216.

Sivakumar, M.V.K., Manu, A., Virmani, S.M., Kanemasu, E.T. (1992). Relation between climate and soil productivity in thetropics. Myths and science of soils of the tropics. *Soil Science Socety of America Journal* (Spl. Publ.) 29: 91-119.

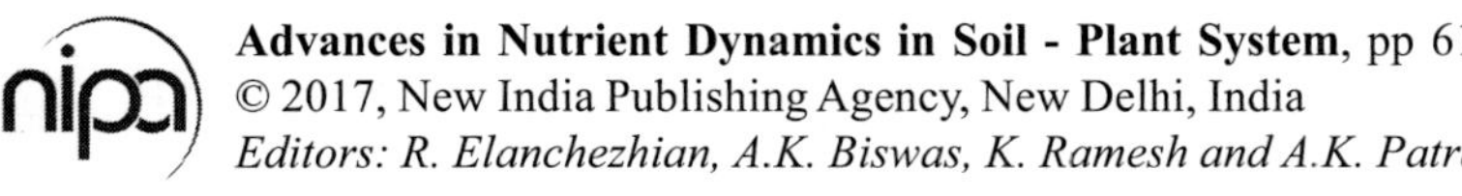
Advances in Nutrient Dynamics in Soil - Plant System, pp 61-74

Editors: R. Elanchezhian, A.K. Biswas, K. Ramesh and A.K. Patra

7

Precision Nutrient Management for Improving Crop Productivity

B.P. Meena, P.K. Tiwari, M.L. Dotaniya, A.O. Shirale and K. Ramesh

ICAR-Indian Institute of Soil Science, Nabi Bagh, Bhopal – 462038, India

Meeting food demand and escalating food prices are serious issues, as current food scenario indicate that crop demand will increase by 100 to 110% from 2005 to 2050 at global level (Tilman *et al.*, 2011). Precision use of fertilizer is the crucial for achieving sufficient food production for growing population. It is costly input to produce and apply, both financially and environmentally. Fertilizer consumption measures the quantity of plant nutrients used per unit of arable land. In India at present the consumption of total nutrient was 25.54 million tonnes, in which the consumption of N, P_2O_5 and K_2O was 16.82, 6.65 and 2.06 million tonnes, respectively during 2012-13. The country has emerged not only as second largest user of fertilizers but also serves as an example for the developing world as to what can be achieved through application of improved technologies. Global consumption of N fertilizers has 107.9 million tonnes, out of the total nitrogen fertilizer consumption 32.7 % was consumed in china and 16.0% in India (FAI, 2012). About of the total increase food grain production has been attributed to the use of fertilizers, and nearly 50% of the current food grains production can be attributed to the fertilizer N application. Despite all this, when food production targets/food demands are to be met in future, the removal gap will further widen. However, indiscriminate use of chemical fertilizer may negatively affect surface water, as well as groundwater, and the atmosphere through leaching and runoff. The usage of fertilizers has been an old-fashioned practice owing to the higher production. Low fertilizer use efficiency means that nutrients otherwise intended for crop uptake are moved from one part of the environment to another. The average fertilizer use efficiency in only about 40-60% for nitrogen (N) and

20-40% for phosphorus (P). The nitrogen use efficiency in main crops different countries are given in Table 1. Raun and Johanson (1999) reported only 33% of nitrogen use efficiency (NUE) for cereal production in global level. Remaining N may remain in the soil, or it will be lost as the gases ammonia, nitrous oxides, and dinitrogen that can be leached down the soil profile. Similarly, phosphorus can be lost as soluble P in water or as P adsorbed on soil particles or as precipitates that move by mass flow with soil erosion resulting in discouragingly low N use efficiency. These nutrients, transfer processes can result in elevated N or P levels in groundwater, which can result in the eutrophication of rivers, streams, and estuaries. The important issue under the current farming practices is the low nutrient use efficiency due to the mis-matched nutrient application particularly nitrogen management.

Therefore, there is an urgent need to enhance the nutrient use efficiency for the higher crop productivity with maintaining environmental quality concerns. To increase the food production in a sustainable manner with precise use of fertilizer with adoption of 4 'R' principles *viz*., right fertilizer, right rate, right time, and right place. The adoption of such as precise application of inputs depending on spatial and temporal variability of crop yield and soil properties to improve resources use efficiency, quality and consistency of agricultural products. It offers variety of potential benefits in crop productivity, profitability, sustainability with enhancing the environmental quality. Precision agriculture is an art and science of using advanced technologies for enhancing crop yield, while minimizing environmental threat that aimed at managing soil spatial variability by applying inputs in accordance with site-specific requirements of a specific soil and crop.

Table 1. Nitrogen use efficiency (NUE) of major cereal crops in the world

Country	NUE (%)			
	Wheat	Maize	Rice	Overall
China	39	33	31	34
India	38	53	30	35
USA	66 (74)*	56 (61)*	32	57
EU	70	45	51	62
World	59	49	39	49

*Based on 2010 grain production and N consumption data (*Source*: FAO, 2011).

Precision nutrient management strategy that employs detailed site specific information too precisely and timely application of N fertilizer to meet plant needs as the across the field variability and micro-climatic condition that occurs within field condition. Therefore, the dimensions of nutrient management recommendation domains manage the temporal as well as spatial variability from large region to farm level or within field. For enhancing the efficiency of farm inputs, increasing productivity and economics of crop production and reducing potential environmental pollution one should ensure to apply farm inputs (i) only when needed, (ii) in specific amounts inputs needed and (iii) in specific locations of the field which can be ensured through the integration of farmers.

Precision Nutrient Management (PNM) tools

1. Geographical information system

Geographical information system (GIS) is computerized data storage and retrieval system which can be used to manage and analysis spatial data relating crop productivity and other agronomic practices. GIS is regarded as brain of precision farming. GIS is a software application that is designed to extract and provide the tools to manipulate and display spatial values. GIS displays the better understanding of the interaction among yield, fertility level, disease and pest etc. and decision making based on such spatial relationship.

2. Global positioning system

Global positioning system (GPS) is a navigation based constellation of satellites, which allow the user to record positional information with accuracy ranging from 100 to 0.01 m. This is introduced by the defence department of the U.S. Army. This can be used every time anywhere on the earth. Precision farming involves positioning information and it can be delivered by the GPS in an efficient manner. GPS provides the accurate positional information, which is useful in locating the spatial variability with accuracy. The GPS can be used in two modes; single receiver mode and differential mode using two receivers. Single receiver collects the timing information and processes it into position. In the differential global positioning system (DGPS), mode one receiver is mounted in a stationary position; usually at the farm office while the other is on the machine/implement. Role of GPS in agriculture and its potential is growing. Farm uses include mapping yields by using combine yield monitor, variable rate planting, variable-rate planter drive, variable rate lime and fertilizer application, variable-rate spreader drive, variable rate pesticide application, variable-rate applicator, field mapping for records and insurance purposes (GPS + mapping software) and parallel swathing (GPS + navigation tool).

3. Remote sensing

Remote Sensing (RS) is the science of obtaining and interpreting information from a distance, using sensors that are not in physical contact with the object being observed (Jensen, 1996). RS helps in detection and measurement of photons of different energies emanating from various objects. It can be also be utilized to detect soil related variables, pest incidence and water stress (Dalal and Henry, 1986). The remote sensing satellites send a known signal towards the earth and a portion of the signal is reflected back. The image data are actively collected by measuring these signals. Data are also collected passively by measuring the sun's energy reflected by an object or electromagnetic energy emanated from an object. It can be of various resolution, spectral coverage, frequency and also use in spatial, temporal, spectral and radiometric resolutions (Venkataratnam, 2001). The inference is that better decision making will provide a wide range of benefits in economic, environmental and social aspects that may or may not be known or measurable at present (Auernhammer, 2001).

4. Sensors

A sensor is a converter that measures a physical quantity and converts it into a signal which can be read by an observer or by an (today mostly electronic) instrument. Soil nutrients are used to detecting the soil fertility status. With the help of these soil nutrient sensors the fertilizer dose is calculated for a place where a nutrient is deficient. Sense soil characteristics: texture, structure, physical character, humidity, nutrient level and presence of clay (Chen *et al.*, 1997). Sense colors to understand conditions relating to: plánt population, water shortage and plant nutrients, monitor crop yield and crop humidity, variable-rate system to monitor the migration of fertilizers and discover weed invasion.

Precision Nutrient Management (PNM) Development Steps

The following steps for the development of precision nutrient management (PNM) technology can be divided into three steps viz., grid sampling and soil fertility mapping assessing field variability and managing field variability.

1. Grid Sampling

Grid sampling was the very first approach used to develop precision application maps where in fields are sampled along a regular grid at sample spacing ranging from 60-150 m depending on the field size and the samples are analyzed for desired properties. Mapping spatial variance related to soils is achieved by grid sampling and geo-statistical methods (Lui, *et al.*, 1996 and Mohamed *et al.*, 1996). However, other rapid and automatic methods are being explored by using mobile systems (soil chemical sensors mounted on smaller vehicles) to measure chemical properties of soils like organic matter, soil moisture, nitrate levels, *etc.* (Borgelt 1992 ; Franzen *et al.*,1996, and Wright, 1996). The results of these analyses are interpolated to unsampled locations by Geo-statistical techniques viz., Kriging and inverse distance weighing (IDW) and the interpolated values are classified using GIS techniques into a limited number of management zones. The delineation of management zones uses three GIS data layers viz., bare soil imagery, topography and farmer's experience. In precision farming, inputs are to be applied precisely in accordance with the existing variable. Therefore, assessing the infield variability soil and crop is very crucial and first step of precision agriculture. Spatial variability of all the determinants of crop yield (topography, soil properties etc.) should be well recognized, adequately quantified and properly located. Construction of condition maps on the basis of the variability is a critical component of precision farming. Condition maps can be generated through (i) Surveys, (ii) Point sampling & interpolation, (iii) Remote sensing (high resolution) and (iv) Modeling.

2. Assessing field variability

The assessing variability is the critical first step in precision farming. In precision farming, inputs are to be applied precisely in accordance with the existing spatial variability of all the determinants of crop yield should be well recognized, adequately quantified and properly located. Construction of condition maps on the basis of the variability is a critical component of precision farming. Use of precision technologies for assessing variability: Faster and in real time assessment of variability is possible only through advanced tools of precision farming.

3. Managing field variability

Once variation is adequately assessed, farmers must match agronomic input to know condition employing management recommendation. In site-specific variability management, we can use GPS instrument, so that the site specificity is pronounced and management will be easy and economical.

Strategies for Precision Nutrient Management (PNM)

The precision nutrient management strategies commonly used in recent years included viz., uniform N rate; site specific nutrient management (SSNM), Grid based SSMZ–based constant yield goal (SSMZ–CYG), and SSMZ-based variable yield goal (SSMZ–VYG). The uniform rate is based on conventional uniform nitrogen application using constant yield target. While grid based strategies uses a variable rate application based upon initial soil fertility status through grid sampling or content target yield, and site specific nutrient management (SSNM) provide need based feeding of crops.

Site specific nutrients management

The site specific nutrient management (SSNM) provides an approach for need based according crop requirement. This SSNM emphasis on managing between-field spatial variability in indigenous nutrient supply capacity of soil, temporal variability in plant nutrient status occurring within one growing season, and temporal variation in soil N, P and K status from season to season (Dobermann and White, 1999). There are three basic approaches of SSNM viz., (1) Site specific N recommendation based on grid soil sampling, residual fertility status, map based fertilizer recommendation, 2) Second approach is to develop specific N response curve and 3) Third approach is monitor to N status by canopy reflectance of light and leaf colour chart (LCC). Even though India has made considerable advance in agricultural research, but still blanket recommendation of fertilizers for adoption over a larger area are in fashion. The blanket uses of fertilizers are no more useful to enhance productivity gains along with certain environmental issues. SSNM practices have to be adopted to enhance the crop productivity and nitrogen use efficiency in maize and rice in different countries is presented in Table 2.

Table 2. Site specific nutrients management in different countries

Crop (country)	N treatment	N applied (kg/ha)	N saved (kg/ha)	Yield (t/ha)	NUE (kg/ha)
Maize (USA)	Conventional	142	-	10.3	73
	SSNM	141	+29	10.4	74
Rice(Philippines)	Conventional	130	+43	7.6	58
	SSNM	87	-	7.5	86
Rice (India)	Conventional	142	-	5.0	35
	SSNM	110	+32	5.0	45

Source: Patil, 2009

Site specific management zone

Precision nutrient management has a prime focus on site specific management zone (SSMZ) as means to generate the maps and improve nutrient (Franzen *et al.*, 1996; Siddiq *et al.*, 2001 and Patil, 2009) management in cropping system. The use of SSMZ for variable rate application (VRA) has proved to be a simple and effective way of enhancing nitrogen use efficiency (Khosla *et al.*, 2002, Hornung *et al.*, 2003). SSNZ means a field can be divided into different homogenous management zone that have similar limiting factors for yield and different methods have been used to delineate management zones including remote sensed imagery, yield data and farmers experience combined with bare soil imagery and topography (Fleming *et al.*, 1999), soil electric conductivity, grid soil sampling, and soil survey information (Franzen *et al.*, 2000). The management zone can be classified by using management zone delineation techniques into three categories viz., high, medium and low productivity potential management zones. Therefore, using management zone approach, all agricultural inputs are described to be applied variably across the field in accordance with the productivity potential of the management zone. When homogenous in a particular area, these attributes should lead to similar results in crop yield potential, input-use efficiency, and environmental impact. Nitrogen application based on variable rate application (VRA) could be reduced by as much as 42 kg N/ha (Mulla and Bhatti, 1997). Khosla and Alley (1999) reported that using VRA reduced total N applied by 22 kg/ha without reduction in grain yield as compared to uniform application of nitrogen. In this view developing accurate variable-rate application maps is a key element to implementing precision nutrient management (Fleming, 2001). There are a variety of variable rate applicator (VRA such as; map-based and sensor based.

Map-based VRA adjust the application rate based on an electronic map, also called a prescription map. The grid –based strategy uses a variable rate N application based upon intensive grid soil sampling with a constant target yield. The grid based strategy increased average N by 1.3 and 30.7%, over the uniform strategy for sites 1 and 2, respectively (Table 3). The SSMZ-CYG strategy increased average N applied over the uniform strategy by 1.3% for site 1, while providing a 37.4% saving in N applied at site 2. The SSMZ-CYG strategy provided a 6.3% and 46.1% saving in average N applied over the uniform strategy for site 1 and 2, respectively. Among the management strategies, SSMZ-VYG was found to be best, which fetched higher net returns at both sites compared to uniform application of N (Koch *et al.*, 2003).

Table 3: Comparison of nitrogen management strategies

N management strategy	Mean N rate ($/acre)		Net returns (kg/acre)	
	Site 1	Site 2	Site 1	Site 2
Uniform (CVG)	158	91	8.16	3.50
Grid based (CVG)	160	119	-0.61	-10.37
SSMZ- (CYG)	160	57	-5.56	16.53
SSMZ- (VYG)	148	49	12.87	14.98

Source: Koch *et al.,* 2003

Real time nutrient management

The leaf color chart (LCC) is an innovative cost effective tool for real-time or crop-need-based N management in rice, maize and wheat. LCC is a visual and subjective indicator of plant nitrogen deficiency and is an inexpensive, easy to use and simple alternative to chlorophyll meter /SPAD meter (soil plant analysis development). It measures leaf color intensity that is related to leaf nitrogen content based on chlorophyll content in the leaves at different growth stages. LCC is an ideal tool to optimize N use in rice/maize at high yield levels, irrespective of the source of N applied, viz., organic manure, biologically fixed N, or chemical fertilizers. Thus, it is an eco-friendly tool in the hands of farmers. Now, it is manufactured with 4 colors called four panel LCC & 6 colors called six panel LCC (Fig. 1). Moreover, LCC is provided with water-proof laminated instruction sticker in the required regional language. A LCC value of 4 indicates that there is 1.4 to 1.5 mg N/g leaf weight. The critical LCC value for rice hybrids and HYVs is 4 and for and basmati rice is 3. These values have to taken from 7-10 DAS (days after sowing) or 20-25 DAT (Days after transplanting) to heading. LCC has shown great response in different countries of Asia (Balasubramanian *et al.*, 1998) and optimizing N use in rice, based on colour of the leaf which in turn reflects total N supply (Balasubramanian *et al.*, 2003; Shukla *et al.*,2004 and 2006; Alam *et al.*, 2005; Thind *et al.*, 2010).

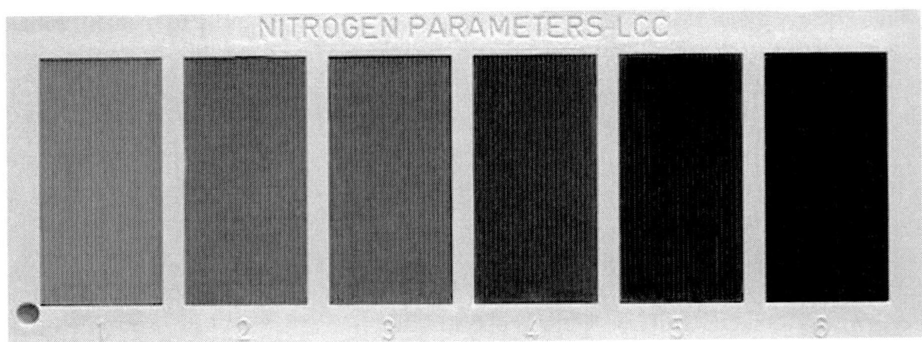

Six Panel Leaf Color Chart (*See colour version on page 377*)

Adhya *et al.* (2010) reported that LCC reduced N requirements from an average of 154 to 122 kg ha^{-1}, a net savings of 32 kg ha^{-1} of applied N, without a loss of rice grain yield. Studies conducted in India (Shukla *et al.*, 2004), Nepal (Regmi and Ladha, 2005), Bangladesh (Alam *et al.*, 2004) showed that LCC-based N management for different rice cultivars produced higher yields as compared to farmers practices. In similar results also observed by Balasubramanian *et al.*, 1998, Shukla *et al.*, 2004 and Thind *et al.*, 2010 based on LCC (real time N application) at different sites of India (Table 4).

Table 4: NUE and grain yield influenced by LCC-based real time N management at different sites

Sites	Treatment	Applied N (kg ha^{-1})	Grain yield (t ha^{-1})	AEn (kg grain.kg N^{-1})	REn (%)
Haryana	LCC 4		124	6.4^{a}	51.6^{a} -
	Farmers' practice	149	6.4^{a}	43.0^{b}	-
Tamil Nadu	LCC 4		108	4.9^{a}	45.4^{a} -
	Farmers' practice	142	5.0^{a}	35.2^{b}	-
Modipuram	LCC 4		142	7.5^{a}	26.0^{a} 53.5^{a}
	Farmers' practice	180	6.9^{b}	17.8^{b}	37.8^{b}
Punjab	LCC 4	115	6.50^{a}	71.6^{a} (PFP)	-
	Farmers' practice	91	6.52^{a}	56.5^{b} (PFP)	-

Same letters (a, and b) between the two treatments indicate a non-significant difference at P d" 0.05 (DMRT-test

Jat and Sharma (2005) also found significant improvement in yield and agronomic efficiency (AE_N) with LCC based N application than the fixed time N application in rice (Table 5). The best way to obtain high NUE is to make crop need-based N application using leaf colour chart (LCC).

Table 5: Precision N management based on leaf colour chart (LCC)

N management practices	N applied (kg ha^{-1})	Grain yield (t ha^{-1})	Agronomic efficiency (kg grain kg^{-1} N)
No-N control	0	2.75	----
Recommended N Mgt	80	3.86	13.9
LCC < 3 (No basal N)	80	4.18	17.9
80 % N basal + LCC < 3	104	3.62	8.4
Farmers' practice (03 splits)	100	3.74	9.9

Optical sensor-based N management

N recommendation based on in-season yield potential and the responsiveness of the crop to additional N, allows variable rate top dressing. Spatially variable nature of the plant available N in agro-ecosystems, crop growth and N requirement vary temporally among and within seasons and spatially among and within fields. Large temporal and field-to-field variability of soil N supply restricts efficient use of N fertilizer when blanket application (Adhikari *et al.*, 1999 and Dobermann *et al.*, 2003). Uniform application of N fertilizer can lead to some areas of a given field receiving excessive N fertilizer while other areas are left under fertilized. Fertilizer N management that does not accommodate temporal and spatial variability may lead to sub-optimal yields and net returns, poor N use efficiency and escape to the environment of excess fertilizer N. Thus, quantifying the optimum in-season N requirement is an important step towards economically and environmentally feasibility in systems. Application of N fertilizer according to spatial variability of the N needs of the crop could lead to enhanced NUE and thus more economically and environmentally sound farming practices (Khosla *et al.*, 1999 and Mulla *et al.*, 1997). Assessing in-field variability of plant N status by collecting in-season biomass samples is cost prohibitive, labour intensive and destructive to the crop. Application of optical sensors in agriculture is increasing rapidly through measurement of visible and near-infrared (NIR) spectral response from plant canopies to detect N deficiencies (Penuelas *et al.*, 1994). Remote sensing is more rapid means to sample multiple crop parameters including photosynthetic capacity, productivity and potential yield. Spectral vegetation indices such as the normalized difference vegetation index (NDVI) have been shown to be useful for indirectly obtaining crop information such as photosynthetic efficiency, productivity potential and potential yield (Raun *et al.*, 2001) and have been found to be sensitive to leaf area index, green biomass.

Spectral reflectance expressed as NDVI was measured using a handheld Green seeker optical sensor unit (N-Tech Industries Incorporation, Ukiah, CA, USA). Green seeker sensors measure crop nitrogen deficiencies in real time, and then predict yield potential for the crop using the agronomic vegetative index NDVI. Green seeker hand held optical sensor is on-the go remote sensing and measures the reflectance of a given crop area over a 0.6 x 0.1 m^2 area when the unit is positioned between 0.6 and 1.0 m^2 above the target area (Bijay Singh *et al.*, 2011). The dimensions of the sensed area remain somewhat consistent over this height range. The sensor utilizes high intensity light emitted diodes (LED) to emit light in the red (650 + 10 nm full width half magnitude) and near infra-red (770 + 15 nm) bands. It records the magnitude of the reflected incident light using a photodiode detector. Green seeker™ sensor embedded software calculates the reflectance in the red or green and near infrared, and then computes the NDVI as follows;

$$\text{NDVI} + \frac{(FNIR - Fred.)}{(FNIR + Fred.)}$$

Where; FNIR and FRED are respectively the fractions of emitted NIR and red radiation reflected back from the sensed area. NDVI is a broadband index that is well correlated to leaf area index and green biomass and is thus sensitive to photosynthetic efficiency. To predict grain yield and make an N fertilizer recommendation with the Green seeker a response index (RI) is utilized. The RI is used to compare the NDVI value of a given area of interest to the NDVI value of an "N rich strip" which has received adequate N fertilizer for crop growth and development. Using the NDVI data collected by the Green seeker, the next step in determining a site-specific N requirement is to estimate the yield potential of the particular area without additional N fertilizer (other factors remaining constant). In-season estimated yield (INSEY) to correct for in-season N deficiency is obtained by dividing the NDVI data by the number of days from planting to remote sensing and it is essentially a measure of the daily accumulated biomass from the time of planting to the day of sensing. The yield potential with no additional fertilization (YP_0) is calculated using an empirically derived function relating INSEY to yield potential. Predicting the yield of an area of interest with additional fertilizer (YP_n) is accomplished using the product of YP_0 and RI. The prediction of wheat response to N applications guided by optical sensor was positively correlated to measured N response; increased NUE and actual grain yield (Raun *et al.*, 2001). Li *et al.* (2009) also conducted a study at china on farm field on optical sensor based N management in wheat and observed that sensor based N management strategy and farmers' practices produced similar yield trend but applied 67 and 372 kg N ha^{-1}, respectively.

Limitations for adoption of precision nutrient management

Precision nutrient management is a proven technology in many developed countries. While in developing countries like India there are some of the obstacles for adoption of precision nutrient management technologies particulars are culture and perceptions of the adaptors or users, heterogeneity of cropping systems with small size land holding and land tenure/ownership restriction, high cost of obtaining site-specific data, complexity of tools and techniques requiring new skills and expertise, resistance to adoption of new technologies and lack of awareness of environmental problems; lack of local technical expertise; high initial investment for buying new machinery and highly costly equipment, uncertainty on returns from investments on new equipment and information management system and knowledge and technological gaps such as inadequate understanding of agronomic factors and their interaction, lack of knowledge of geo-statistics for displaying spatial variability of crops and soils using current mapping software and limited ability to integrate of information from diverse sources.

Opportunities

Despite the many limitations, the opportunities for precision PNM technologies are immense potential in India and other many developing countries. Recently, the governments and non- governments organizations (NGOs) initiated special efforts to promote precision agriculture. Recently, Tata kisan Kendra (TKK) implemented a project in rural India from the bullock-cart age into the new era of satellite and information technology. Tata chemicals limited's extension services, brought to farmers through the TKKs, use remote-sensing technology to analyze soil, inform about crop health, pest attacks and coverage of various crops predicting the final output. This helps farmers adapt quickly to changing conditions. Precision farming models are not complete, unless the parameters related to empowerment of the farmers; especially small and marginal farmers are integrated. Now it is the turn of good news to the Indian farming community. Some of the research institutes such as: Space Applications Centre (SAC, ISRO), Ahmadabad; M.S. Swaminathan Research Foundation, Chennai; Indian Agricultural Research Institute (IARI), New Delhi; Project Directorate of Farming Systems Research (PDFSR), Modipuram; National Remote Sensing Agency (NRSA) in coloration with ICRISAT, CRIDA and ANGRAU, Hyderabad and Central Institute of Agricultural Engineering (CIAE), Bhopal have started working in this direction and in soon it will help the Indian farmers harvest the fruits of frontier technologies without compromising on the quality of land.

It is concluded that precision farming is discussed in developed countries, but it is still at a nascent stage in developing countries, including India. The precision agriculture in developing countries cannot be convincing, if the only environmental benefits are focused. But some technologies are being researched and implemented under various crops and cropping systems. Rice-wheat cropping system is most suitable for adoption of this concept for higher productivity and profitability with maintaining environmental quality. It is evident that, the PNM technologies are successful in their role of enhancing crop production, input use efficiency while minimizing the cost of production and environmental impacts. Opportunities of the PNM technologies including GIS, GPS, remote sensing, yield monitoring system, real time N application using LCC, SSNM, have demonstrated potentialities for improving crop productivity, water productivity and NUE.

References

Adhikari, C., Bronson K.F., Panuallah G.M., Regmi A.P., Saha P.K., Dobermann, A., Olk D.C., Hobbs, P.R. and Pasuquin, E. (1999). On-farm N supply and N nutrition in the rice–wheat system of Nepal and Bangladesh. *Field Crops Research,* 64:273–286.

Adhya, T.K., Shukla A.K. and Panda, D. (2010). Nitrogen losses, N-use efficiency and N management in rice and rice-based cropping systems.51p. In ING Bulletins on Regional Assessment of Reactive Nitrogen, Bulletin No. 10, (Ed. Bijay Singh), SCON-ING, New Delhi, pp i-iv & 1-51. See also FAI (2009) Fertilizer statistics 2008–09. Fertilizer Association of India, New Delhi.

Alam, M.M., Ladha, J.K., Khan, S.R., Foyjunnessa, H.R., Khan, A.H. and Buresh, R.J. (2004). Leaf color chart for managing nitrogen fertilizer in low land rice in Bangladesh. *Agronomy Journal,* 97**:** 949-959.

Auernhammer, H. (2001). Precision farming- the environmental challenge. *Computers and Electronics in Agriculture*, 30: 41-33.

Balasubramanian, V., J.K. Ladha, G.L. Denning. (eds.). (1998). Resource management in rice systems: Nutrients. The Netherlands: Kluwer Academic Publishers. 600 p.

Balasubramanian, V., Ladha, J.K., Gupta, R.K., Naresh, R.K., Mehla, R.S, Bijay-Singh and Yadvinder-Singh. (2003). Technology option for rice in the rice-wheat system in South Asia. In Ladha, JK, Hill JE, Duxbury JM, Gupta RK and Buresh RJ (eds.) Improving the Productivity and Sustainability of Rice-Wheat Systems: Issues and Impact pp. 115-147. American, Madison, Wisconsin, USA-ASA Special Publication Number 65.

Bijay-Singh, Sharma, R.K., Jaspreet-Kaur, Jat, M.L., Martin, K.L., Yadvinder-Singh, Varinder pal-Singh, Parvesh Chandna, Choudhary, O.P., Rajeev K. Gupta and Harmit S. Thind, Jagmohan-Singh, Harminder S. Uppal, Harmandeep S. Khurana , Ajay-Kumar , Rajneet K. Uppal and Monika Vashistha , William R. Raun and Raj Gupta (2011). Assessment of the nitrogen management strategy using an optical sensor for irrigated wheat. *Agronomy for Sustainable Development,* DOI 10.1007/s13593-011-0005-5

Borgelt, S.C. (1992). Sensing and measurement technologies for site specific management", *In:* Roberts P.C. *et al.,* eds., Proceeding of soil specific crop management, A workshop on research and development issues, April 14-16, 1992, *American Society of Agronomy*, Madison, WI, pp.141-157.

Chen, F. Kissel, D. Clark, E.R., West, L.T., Rickman, D., Luval, J. and Adkin, W. (1997). Determining surface soil clay concentration at a field scale for precision agriculture", University of Georgia, GHCC-NASA at Huntsville.

Dalal, R.C. and Henry, R.J. (1986). Simultaneous determination of moisture, organic carbon and total nitrogen by near infrared reference Spectrophotometry. *Soil Science Society of America,* 50:120-123.

Dobermann, A. and White, P.F. (1999). Strategies for nutrient management in irrigated and rainfed lowland rice Systems. *Nutrient Cycling in Agro-ecosystems,* 53: 1–18, 1999.

FAI (2012). Fertilizer Statistics 2012-13, the Fertilizer Association of India, New Delhi.

FAO (2011). FAO Statistics–Agriculture. http://faostat.fao.org

Fleming, K. L., Westfall D.G., Weins, D.W. Rothe, LE., Cipra, J.E. and Heermann, D.F. (1999). Evaluating farmer developed management zone maps for precision farming. P335-343., vol.21, no.1, pp.16-18,

Fleming, K.L. and Westfall D.G. (2001). Fertilizer application by management zone, Colorado State University, *Agronomy Newsletter*, vol.21, no.1, pp.16-18,

Fleming, K.L. Westfall, D.G., Wiens, D.W. and Broadhal, M.C. (2000). Evaluating farmer defined management zone maps for variable rate fertilizer application", *Precision Agriculture,* vol. 2, pp.201–215.

Franzen, D.W. Cihacek, L.J. and Hofman, V.L. (1996). Variability of soil nitrate and phosphate under different landscapes. pp. 521-529, 1996. *In: Proc. of the 3rd International Conference on Precision Agriculture*, June 23-26, 1996, Minneapolis, Minnesota. ASA/CSSA/SSSA.

Frazen, D.W., Halvorson, A.D., and Hofman, V.L. (2000). Management zone for soil N and P levels in the Norhen Great plains. (In) *Precision Agriculture* : P.C. Robort *et al.*, (Ed.) ASA, CSSA and SSSA, Madison, WI.

Hornung, A.R., Khosla, R., Reich, R. and Westfall, D.G. (2003). Evaluation of site specific zone : grain and nitrogen use efficiecny. (In) *Precision Agriculture* J. Stafford and A. Werner (Ed.) In: *Proceedings 4th European Conference*, Berlin, Germany. 15-19 June 2003. Wageningen Academic Publ., Wageningen, The Netherlands.

Jat, M.L. and Sharma, S.K. (2005). Leaf colour chart based nitrogen management in rice. In: *PDCSR Annual Report 2004-05*, Project Directorate for Cropping Systems Research, Modipuram, India.

Jensen, J. R.(1996). Remote sensing of the environment: An Earth Resource Perspective: 3rd Edn., Prentice Hall, USA, pp.1-28.

Khosla, R. and Alley, M.M. (1999). Site specific nitrogen management on Mid Atlantic Coastal Plains soils. *Better Crops*. 83 (3):6-7.

Khosla, R. Fleming, K. Delgado, J.A, Shaver, T. and Westfall, D.G. (2002). Use of site specific management zones to improve the nitrogen management for precision agriculture, *Journal of Soil and Water Conservation* vol.57, pp.515–518, 2002.

Koch, B., Kohosla, R., Frasier, M., and Westfall, D.G., (2003). Economic feasibility of variable rate nitrogen application in site specific management", *Proceedings of Western Nutrient Management Conference,* 5:107-112.

Li, F., Miao, Y., Zhang, F., Cui, Z., Li, R., Chen, X., Zhang, H., Schrodner, J., Raun, W.R., Jia, L. (2009). In: season optical sensing improves nitrogen use efficiency for winter wheat. *Soil Sci Soc Am J.* 73:1566–1574. doi:10.2136/sssaj2008.0150.

Lui, W., Upadhyaya, S.K., Katoaka T. and Shibusawa, S. (1996). Development of a texture soil compaction sensor. *In:* Roberts P. C. *et al.,* Eds, *Proceedings of the 3rd International Conference on Precision Agriculture*, June 23-26, 1996, Minneapolis, Minnesota, ASA/CSSA/SSSA, pp. 617-630, 1996.

Mohamed, S.B. Evans E.J. and. Shiel, R.S. (1996). Mapping techniques and intensity of soil sampling for precision farming", *In*: Roberts P. C. *et al.*, eds. *Proceeding of the 3rd International Conference on Precision Agriculture*, June 23-26, 1996, Minneapolis, Minnesota, ASA/CSSA/SSSA, pp.217-226.

Mulla, D.J. and Bhatti, A.U. (1997). An evaluation of indicator properties affecting spatial pattern in N and P requirements for winter wheat yield. P. 145-153. (In) precision agriculture J.V. Stafford (Ed.) precision Agriculture, 1st European conference precision agriculture, Wazwick Univ., UK. 7-10 September 1997. Bios Scientific Publ., Oxford, Herndon, V.A.

Patil, V.C. (2009). Precision nutrient management: A Review, *Indian Journal of Agronomy,* 54(2): 113-119.

Penuelas, J., Gamon, J.A., Fredeen, A.L., Merino, J., Field, C.B. (1994). Reflectance indices associated with physiological changes in nitrogen- and water-limited sunflower leaves. *Remote Sens Environ* 48:135–146. doi:10.1016/0034-4257(94)90136-8

Raun, W.R., and G.V. Johnson (1999). Improving nitrogen-use efficiency for cereal production. *Agronomy Journal,* 91: 357–363.

Raun, W.R., Johnson, G.V., Stone, M.L., Solie, J.B., Lukina, E.V., Thomason, W.E., Schepers, J.S. (2001). In-season prediction of potential grain yield in winter wheat using canopy reflectance. *Agronomy Journal*, 93:131– 138. doi:10.2134/agronj2001.931131x.

Regmi, A.P. and Ladha, J.K. (2005). Enhancing productivity of rice-wheat cropping system through integrated crop management in the Eastern Gangetic Plains of South Asia. *Journal of Crop Improvement*, 15:1147-1170.

Shukla, A.K., Ladha, J.K., Singh, V.K., Dwivedi, B.S., Gupta, R.K., Sharma, S.K., Balasubramanian, V., Singh, Y., Pathak, H., Pandey, P.S. Padre, A.T. and Yadav, R.L. (2004). Calibrating the leaf color chart for nitrogen management in different genotypes of rice and wheat in a systems perspective. *Agronomy Journal,* 96:1606–1621.

Siddiq, E.A., Rao, K.V. and Prasad, A.S.R. (2001). Yield and factor productivity trends in intensive rice production systems in India, In Preceding, FAO expert consultation on yield gap and productivity decline in rice production, Rome, pp.267-344.

Thind, H.S., Bijay Singh, Pannu, R.P.S., Yadvinder Singh, Varinderpal Singh, Gupta., R.K., Gobinder Singh, Ajay Kumar and Monika Vashistha. (2010). Managing neem (*Azadirachta Indica*)-coated urea and ordinary urea in wheat (*Triticum aestivum*) for improving nitrogen-use efficiency and high yields *Indian Journal of Agricultural Sciences,* 80 (11): 960–964.

Tilman, David, Christian Balzer, Jason Hill, and Belinda L. Befort. (2011). Global food demand and the sustainable intensification of agriculture. *Proceedings of National Academy of Sciences,* 108(50):20260–20264.

Venkataratnam, L. (2001). Remote Sensing and GIS in Agricultural Resources Management", *Proceedings of the 1st National Conference on Agro-informatics*, June 3-4, Dharwad, India, pp: 20-29.

Wright, N.A. (1996). A new mobile soil sampler compared to hand probes and augers for fertility evaluations", In: Roberts P.C. *et al.,* eds, *Proceedings of the 3rd International Conference on Precision Agriculture*, June 23-26, 1996, Minneapolis, Minnesota. ASA/CSSA/SSSA. pp. 631-639.

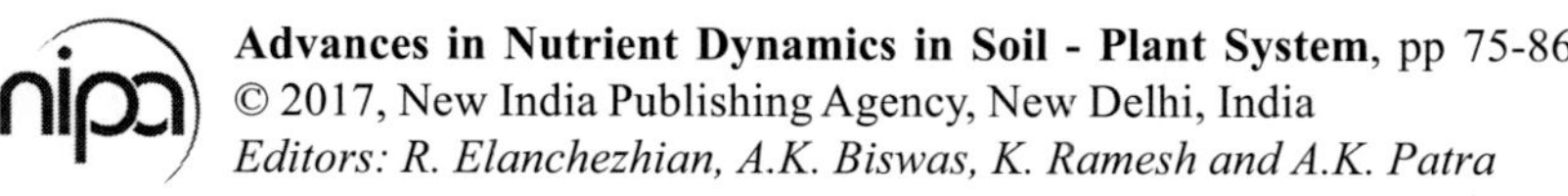
Advances in Nutrient Dynamics in Soil - Plant System, pp 75-86

Editors: R. Elanchezhian, A.K. Biswas, K. Ramesh and A.K. Patra

8

Agronomic Management Options for Improving Nutrient Use Efficiency in Different Soil Systems under Climate Change Scenario

K. Ramesh, A.K.Biswas and R.Elanchezhian

ICAR-Indian Institute of Soil Science, Nabi Bagh, Bhopal-462038, India

Preamble

Crop nutrient management in the climate change scenario is carried out with the basic understanding of nutrient concentrations in soils, plants and their transformations but these strategies might not apply under changed climate conditions where there is an overall change in the environment. It is understood that nutrient acquisition by plants from the soil is closely associated with overall biomass and strongly influenced by root foraging area. It is speculated that change in soil factors due to climate change could restrict root growth and nutrient stress would also occur. Plant size may also change but nutrient concentration may not change; therefore, nutrient removal will scale with growth. The interaction of mineral behavior in the soil with global climate change variables is poorly understood. Studies under forest ecosystem have indicated that terrestrial vegetation is supported by weathered soils with some combination of low amounts of phosphorus, calcium, and high amounts of aluminium and manganese. Some of the major issues outlined in the literature are (1) the effects of transpiration on root acquisition of Ca and Si (2) altered root architecture on the acquisition of P, (3) the effects of altered root exudate production on Al toxicity, and (4) the interaction of photochemical processes with transition metal availability (Lynch and Clair, 2004).

Pedogenic soil barriers can influence P availability in the soils. In the climate change scenario, if high rainfall recurs, iron pans can constrain root access to deeper parts of soil profiles and restrict drainage (McKeague *et al.*, 1983, Kitayama *et al.*, 1997). Further, acidic, high-aluminium sub soils may permit water flow but restrict the growth of roots towards deeper, more P-rich portions of soil profiles (Soethe *et al.*, 2006).

Genetic pool available from different crop varieties may provide varied responses to mineral stress. Root traits are considered as central importance in adaptation to mineral stresses. With a proper understanding of the climate-nutrient interactions, best nutrient management practices for enhancing nutrient use efficiency protocol can very well be applied for various soils.

Understanding the basics behind climate change scenario

There is an upward shift in the atmospheric carbon dioxide concentration from 280 ppm in 1800 to the current value of around 380 ppm, and is expected to approach 700 ppm by the end of 21st century (Houghton *et al.*, 2001). In addition to crop growth, crop tissue quality, specifically tissue C/N will also be influenced by climate change (Kimball *et al.*, 2002). There is confusion in the terminologies while interpreting crop management for the weather fluctuation in an area. There is a clear-cut distinction between climate change and climate variability. Climate change applies to the long-term trends in weather, generally over decades or centuries. This includes long-term changes or trends in the average climate (such as annual average temperature or precipitation) or trends in climate extremes (such as the frequency of intense rainfall events). However, people experience climate as individual weather events, which naturally fluctuate on an annual, seasonal and decadal basis, which is climate variability. In addition to natural variation, climate change implies a shift in the patterns of weather events, over the long term. The magnitude of these climatic changes over the following decades and towards the end of the century will depend on how successful policies are at reducing greenhouse gas emissions and how sensitive the climate system is (IPCC, 2007). In particular reference to agricultural production, frequent drought and floods may hamper crop yields more than other consequences of climate change. Varieties are continuously bred for the new climates and may perform normal even under climate change scenario (Ramesh *et al.*, 2016).

Interactions of elevated CO_2 with soil nutrients

Results conducted elsewhere across the world from 16 free-air CO_2 enrichment (FACE) sites representing four different global vegetation types indicated that only some early predictions of the effects of increasing CO_2 concentration (elevated CO_2) on plant and ecosystem processes are well supported. An increase in productivity with increased N availability is well supported by the FACE results (Nowak *et al.*, 2004). The yield response of a C_3 grass to elevated atmospheric

CO_2 concentration was not significant under low N supply, but the reverse under high supply in a FACE experiment (Schneider *et al.*, 2004). In fertile grasslands, legumes benefit more from elevated atmospheric CO_2concentrations than non-fixing species (Ross *et al.*, 2004; Almeida *et al.*, 2000) since a decline in N availability may be prevented by an increase in biological N_2 fixation under elevated atmospheric CO_2 concentrations.

Results of a Mini-FACE system, when Monoliths of a fertile, N limited, C_3 grassland community were subjected to CO_2 enrichment of 600 µmol mol^{-1} with 3 and 6 cuts per year, species-specific responses were observed. Elevated CO_2 significantly increased the proportion of dicotyledon species to that of the monocotyledon species. However, not all grass species responded negatively to high CO_2 (Teyssonneyre *et al.*, 2002). Notwithstanding to this fact, phosphorus, may also act as a limiting factor for legumes response to elevated atmospheric CO_2 (José *et al.*, 2000).

Soil warming did not significantly change the concentrations of Carbon and Carbon compounds (sugar, starch, hemicellulose, cellulose and lignin) in temperate tree roots. However, an increase in total N concentration in the live fine roots was found in the warmed area to the tune of 10.5%. Zhou *et al.*, (2011) concluded that soil warming increases N mineralization and decreases carbon allocation belowground.

Whether climate change has a role in nutrient availability and acquisition?

Climate change impact on nutrient use efficiency will be primarily affected through direct impacts on root surface area. Kang *et al.*, (2000) opined that a change in precipitation and a rise in air temperature may enhance root zone temperature also due to moisture deficit which may interfere with normal functioning of roots as roots perform the function of acquisition of water and nutrients. Nutrient use efficiency is a function of physical, chemical and biological processes in soils that determine utilization of the form of nutrient ion present in the soil, and available for actual utilization by a plant. Jungk (2002) opined that nutrient acquisition by the plant reflects an array of physiological phenomena that govern nutrient transport in soil as well as into roots and can alter aspects of nutrient availability in the soil. Given that soil moisture and temperature are primary determinants of nutrient availability and root growth and development and that carbon allocation to roots governs nutrient acquisition, it is reasonable to expect that process outcomes will be reflective of the changed climate.

Interference with nutrient mobility in the soil

To accumulate nutrients from the soil solution by plant roots, they should be mobile in the soil. Biological transformation between organic and inorganic pools is also a factor of moisture and temperature, and thus, concentration of nitrogen as well as sulphur might be affected. The soil C pool size would remain unaffected through increased C supply belowground (Kirschbaum, 2000) due to soil warming. Many researchers expect that elevated CO_2 to increase belowground C that will in turn enrich microbial C, but Zak *et al.*, (2000) could not find any consistency in this trend. Pendall *et al.*, (2004) concluded that increased CO_2 may not affect N mineralization *per se* but associated warming, leading to increased solution-phase N.

Temperature and atmospheric N deposition are expected to affect forest growth directly and indirectly by increasing N availability due to higher rates of N mineralization in boreal forest ecosystems. A mature balsam fir stand in Québec, Canada, was subjected during three consecutive growing seasons to (i) increased soil temperature (4 °C) and earlier snow melt (2–3 weeks) as well as (ii) increased inorganic N concentration in artificial precipitation. After three years, there was no significant increase in soil nitrate (NO_3) or ammonium (NH_4) availability either in the organic or in the mineral soil as measured with standard soil extractions (D'Orangeville *et al.,* 2013).

Tissue N concentration under elevated CO_2

Plants grown under elevated atmospheric (CO_2) typically have decreased tissue concentrations of N (Taub and Wang, 2008). Researchers have attributed this phenomenon to

1. Decrease in tissue N concentrations due to dilution of N by increased photosynthetic assimilation of C.
2. A general decrease in the specific uptake rates (uptake per unit mass or length of root) of N by roots under elevated CO_2 probably due to decreased N demand by shoots and ability of the soil-root system to supply N.
3. A decrease in transpiration-driven mass flow of N in soils due to decreased stomatal conductance besides the altered root system architecture.
4. Plants may exhibit increased rates of N loss through volatilization and/or root exudation, further contributing to lowering tissue N concentrations.

Agronomic practices under changing climate conditions

Simple agronomic practices would increase crop yield concomitant with enhanced nutrient use efficiency (NUE). Many of these practices are non-monetary and low-cost suited to varied soil conditions.

Management practices for climate change scenario

1. Crop and/or cultivar selection

Numbers of cultivars with high yielding capacity to accommodate under adverse climate extremes have been developed for different crops which need to be cultivated. The choice of crop and/or variety is one of the prime requirements under the climate change scenario. A change in crop may avert the risk under the new climate and farmers have the choice of selecting improved varieties/hybrids. In most of the field crops only a part of the total dry matter produced is accumulated as economic yield and thus high harvest index is an important criterion for the chosen variety. Specific disease, insect, herbicide resistance would be an added advantage, and above all the efficient utilization of nutrients. A proven variety released at regional level suiting to location specific problems need to be sorted out. The genetic potential of a cultivar can't be fully exploited without adopting proper agronomic management practices.

Bhopal, Madhya Pradesh in the central India is experiencing varied types of weather aberrations over the past six years. As the kharif crop is rainfed in many parts of the state, cultivar selection to the modified rainfall regime has assumed at most significance. A ruling variety of soybean JS 335 (Long duration variety), earlier best suited for the rainfed cultivation is being wiped out due to the weather vagaries. A delay in the onset of monsoon, as experienced in 2014, has forced farmers to adopt JS 9560, a short duration strain. This facilitated the cultivation of timely sown wheat also in the rabi season.

2. Time of sowing

In time seed bed preparation and timely sowings as per the season (Kharif/ Rabi/ summer) for any crop, results in even and better germination, stand establishment and thereby minimize the crop loss. Planting time is the most vital non-monetary input affecting crop yield. A decline in yield due to delayed sowing can't be compensated through excess use of other inputs. Production efficiency of different crop genotypes greatly differs under different planting dates depending upon their foliage characteristics and canopy structure. A genotype, which quickly attains optimum leaf area index and retains it for a longer period is more efficient as it can take better advantage of solar energy available during the growing season. The yield potential of a genotype can be fully exploited through appropriate microclimate-temperature at different growth and development phases. Thus, an agronomist take a lead by providing the planting time of various field crops after studying the relationship between temperature and plant growth and development.

A field experiment with split plot design conducted to study the effect of monetary (three N levels - 125, 150 and 175 kg/ha) and non-monetary inputs (two varieties - BSR 1 and BSR 2) and three planting time (15 May, 15 June and 15 July); monetary inputs *viz.*, three spacing (30 cm x 15 cm, 45 cm x 15 cm, 60 cm x 15 cm) on yield and economics of turmeric formed showed that turmeric variety BSR 2 out yielded BSR 1 in terms yield. Planting the turmeric during middle of May (15 May) was superior compared to 15 June and 15 July plantings. Among spacing, 30 x 15 cm recorded significantly higher growth, nutrient uptake and yield than 45 x 15 cm and 60 x 15 cm. The crop response was better for higher rate of nitrogen (175 kg/ha) than other levels. Economic evaluation indicated that combination of non-monetary inputs *viz.*, planting BSR 2 at 15 May with monetary inputs *viz.*, 30 x 15 cm spacing with 175 kg/ha N would increase the turmeric production and income of the farmers (Kandiannan and Chandragiri, 2008).

Experiments during rabi season of 2006 - 2008 to study the effect of sowing dates and integrated nutrient management on growth, yield and quality of winter maize with three sowing dates (15 October, 25 October and 5 November), three levels of urea (50, 100 and 150 N kg ha^{-1}) and two organic fertilizer (FYM, Azospirillum) have shown that the crop sown on 25 October significantly enhanced the growth and grain yield than early sowing (15 October) and late sowing (5 November) while, 150 kg of N ha^{-1} was better over 100 and 50 kg N ha^{-1}. But, application of 100 kg N ha^{-1} with 7.50 t ha-1 FYM at the sowing of 25 Oct significantly influenced the growth, yield and quality of maize and was recorded 9.35 and 23.07% more grain yield over the other treatment combinations (Verma, 2011).

Soil temperature and moisture influence the sowing time of rapeseed-mustard in various zones of the country. Sowing at optimum time gives higher yields due to suitable environment that prevails at all the growth stages. Delayed sowing resulted in poor yield and oil content and resulted in incidence of insect-pest and disease too. Mustard aphid (*Lipaphiserysimi* (Kaltenbach)) has been reported as one of the most devastating pests in realizing the potential productivity of Indian mustard and the normal sowing (1st week of November) helps in reducing this risk (Shekhawat *et al.*, 2012).

Under Bhopal, Madhya Pradesh in the central India conditions, the normal onset of monsoon falls during the last week of June and sowing are taken up during July first week. Normally sowing is taken up after the soil has just cooled to normal temperature. In case, there is delay in sowing due to unexplained reasons, and situations of continuous rain rendering entry into the field, the soil gets cooled to the extent that the seeds of soybean/ pigeonpea fail to germinate. This is highly expected under the climate scenario. Another study conducted at IISS, Bhopal with two dates of sowing of soybean with a fortnight gap with 10 varieties has indicated that a delay of 15 days could hamper the soybean yield up to an extent of 45%.

3. Plant population

A mild adjustment in the inter and intra row plant population is necessary to minimize the humidity buildup within the crop canopy.The optimum plant geometry or the arrangement of plants in the field has a direct impact on the interception of solar radiation by the crop canopy. Thus row-to-row and plant-to-plant distance should be optimum. It helps to derive the benefit of applied nutrients and enhance NUE. Based on past experiences, the preparation of a database of site specific and crop specific information that are available on plant population, row widths, depths and planting dates etc. is the way to make decisions that optimize performance of the crops without additional input costs.

At the high rate of fertilization, the hybrids effectively use applied P and produce higher yield per unit quantity of P, which is improved even further when closer (more plant population density) per unit area is adopted.

Whenever there appears to be a narrow length of growing season with undue delay in the onset of the monsoon, increasing the population with the same of dose of fertilizers ensures optimum yield with higher nutrient use efficiency. Under delayed monsoon conditions, soybean is sown at a spacing of 30×30 cm instead of 45×30 cm.

4. Crop rotation

It helps as complementary effect and legumes inclusion in system add the nitrogen and residue, thereby increases the water holding capacity of soil as well as reduction in infestation of weeds which are associated with crop and minimize the buildup of pest and diseases. An increase in soil organic matter helps to compensate the warming effect.

5. Strip cropping

Inclusion of erosion permitting and erosion resisting crop help in conservation of moisture and increases in water intake under the change in intensity and amount of precipitation as a result of global warming and climate change.

6. Nutrient management

Proper plant nutrition is considered as one of the best nutrient management strategy to alleviate the temperature stress as nutrients could improve tolerance to this stress.

Recovery from photoinhibition due to exposure to sun light and subsequent acclimation of photosynthesis occurred in *Solanum dulcamara* L. plants when additional nitrogen was supplied (Ferrar and Osmond, 1986). Sinclair and Horie (1987) has noticed that the radiation use efficiency was nearly constant at high leaf CO_2 assimilation rates, but decreased appreciably at low leaf CO_2 assimilation rates in soybean (*Glycine max* [L.] Men.), rice (*Oryza sativa* L.), and maize (*Zea mays* L.). Maize had the greatest biomass accumulation because it had low leaf N contents that allowed the most crop leaf area growth, and it had high radiation use efficiencies. For each rate of N supply to leaves, an optimum leaf N content existed to maximize crop biomass accumulation. Irrespective of carbon pathway,

Sage and Pearcy (1987) concluded that photosynthesis in *Chenopodium album* and *Amaranthus retroflexus* was linearly dependent on leaf N per unit area. The leaf content of RuBPCase followed the same analogy in *Phaseolus vulgaris* and in the shade plant *Alocasia macrorrhiza* (Seemann *et al*., 1987). Hence maintaining optimum nitrogen in the plant is necessary. Although nitrogen deficiency had only a small effect on the quantum yield of CO_2 assimilation but a large effect on the light-saturated rate of photosynthesis (Khamis *et al*., 1990), enhanced the amount of excess light absorption by plants as the stress decreased the plant capacity for photosynthesis (Verhoeven *et al*., 1997). Later Kato *et al*., (2003) has found that higher tolerance to photo inactivation in *Chenopodium album* L. plants grown under high-light and high-nitrogen conditions due to higher photosynthetic capacity. In short, N-adequate plants were able to tolerate excess light by maintaining photosynthesis at high rates and developing protective mechanisms (Waraich *et al*., 2012). Hence maintaining adequate nitrogen levels in plants is necessary for managing the effects of climate change in plants. Notwithstanding to these facts, Terashima and Evans (1998) could not detect any effect of nitrogen under varied irradiance levels in thylakoid membranes. However, the proportion of total leaf nitrogen allocated to soluble protein and RUBP carboxylase increased with an increase nitrogen as well as irradiance levels.

4R nutrient stewardship (right source, right rate, right time and right place) provides a framework to achieve cropping system goals, such as increased production, increased farmer profitability, enhanced environmental protection and improved sustainability. As crop yields increase there is greater demand on our soils to supply both a sufficient and balanced supply of nutrients. The simple agronomic practice of split-applying nutrients does not increase production cost by much, but it significantly improves NUE.

It is now well established that for most crops, N must be applied in 2-3 or more spilt doses coinciding with the crop growth stages when N requirement is high. In case of P, under conditions of low P availability, banded P is usually more effective than broadcast application. Potassium is normally applied at the time of planting, owing to its large availability by diffusion. But in coarse textured soils, leaching of K can occur, and split application may increase K use efficiency plants in such soils.

Increasing NUE is an offspring of balanced fertilization and sound management practices and decisions. Balanced fertilizer use is not only the first requirement but it is a pre-requisite also because agronomic manipulation cannot produce high efficiency out of imbalanced nutrient application.

When balanced fertilization is practiced, one nutrient increases the efficiency of others through a synergistic effect. Traditionally, in India, balanced fertilization indicates the use of N, P and K in a certain ratio (ideally 4:2:1) on gross basis both in respect of areas and crops. This so-called ideal ratio generalizes the nation-wide fertilizer requirement in a cereal-based cropping system.

7. Weed Management

Weeds compete with crop plants for nutrient, soil moisture, sunlight and space. Of course, the intensity of weed competition depends upon the type of weed species, severity of infestation, duration of weed infestation *etc.* Consequently, crop yield is reduced. Reduction in crop yield has a direct correlation with competition. Of the total annual loss of agricultural produce from various pests in India, weeds cause highest loss, about 33%. The most important point is that weeds require the same or greater amount of nutrients and at same time weeds remove plant nutrients more efficiently than crop plants. Therefore, timely and appropriate weed control greatly increases the crop yield and NUE. In the absence of an effective weed management measure, they remove considerable quantity of applied nutrients, resulting in loss of yield.

Climate change may introduce new weed species complexes as Martinez-Ghersa *et al.* (2000) opined that many weed populations arise as a result of the evolution of wild plant colonizers through selection and adaptation to continuous habitat disturbances and pose a multitude of challenges for managing invasive weed species (Kriticos *et al.,* 2003) in the human managed crop ecosystem. However, very little attention has been paid to the imbalance created to biodiversity by those plants in a rapidly changing climate (Crossman *et al.,* 2011). The opinions of Rosenzweig and Hillel (1998) that rising temperature and CO_2 levels could make crop plants less competitive with weeds and a decade later by Wolfe (2008), that weeds would benefit more than cash crops from increasing atmospheric carbon dioxide, was found to be true, as *Amaranthes retroflexus* produced more seeds under Barley cropping, albeit growth of Barley as well as the weed was reduced (Hyvonen 2011) at southern Finland. Certainly *Amaranthus retroflexus* seed production would be proportional to an increase in temperature. Thus under high temperature scenario, the competitive ability of Barley to compete with *Amaranthus retroflexus* was found to be lessened. The reverse may also hold well with a reduction in weed abundance as a result of climatic change.

Implicit in discussion of weed management and climate change is the assumption that we know what to do in relation to soil and crop management, but these strategies might not apply to the unexpected future climate change conditions, particularly weed menace. Soil warming could enhance the availability of certain elements in the soil by faster ion diffusion rate and the soil moisture stress could boom weed proliferation. Judicious agronomic practices would partially help to offset weed pressure, but climate may have over-riding influence on weeds, as they share the same trophic level with crops. Implications of climate change would be identical with crops, aggravating the crop-weed competition. Many of the most troublesome weeds in crop ecosystems follow C_4 pathway. As atmospheric CO_2 increases, it is conceivable that competitive ability of weeds could be similar to C_3crops, such as rice, if there is no dearth of soil moisture and nutrients (Chauhan and Ramesh, 2015).

Conclusion and future outlook

Thus, while summing up, if our goal is to increase the nutrient use efficiency, it is important to know the role of each factor in the uptake and utilization processes of nutrients as well as agronomic management factors. The best management practices, either individually or collectively, has direct bearing on the yield and NUE of crops. Furthermore, NUE greatly depends upon how efficiently nutrient and crop management practices work together in a soil-climate complex in producing higher yield. An integration of quantitative genetics with conceptual models of plant response to mineral stresses is needed if we have to understand plant response to global change in the real world soils, keeping in mind plants do accommodate themselves to the changed climate through plasticity.

References

Almeida, J.P.F., Hartwig, U.A., Hartwig, M., Frehner, M., Nosberger, J. and Luscher, A. (2000). Evidence that P deficiency induces N feedback regulation of symbiotic N_2 fixation in White Clover (Trifoliumrepens L.). *Journal of Experimental Botany*, 51 (348): 1289–1297.

Chauhan, B.S., and Ramesh, K. (2015). Weed regimes in agro-ecosystems in the changing climate scenario–A review. *Indian Journal of Agronomy*, 60 (4): 479-484.

D'Orangeville, L., Houle, D., Côté, B., Duchesne, L. and Morin, H. (2013). Increased soil temperature and atmospheric N deposition have no effect on the N status and growth of a mature balsam fir forest. *Biogeosciences*, 10, 4627-4639.

Ferrar, P.J. and Osmond, C.B. (1986). Nitrogen supply as a factor influencing photoinhibition and photosynthetic acclimation after transfer of shade-grown *Solanum dulcamara* to bright light. *Planta*, 168: 563-570

Houghton, J.T., Ding, Y. and Griggs, D.J. (2001). 'Climate Change 2001: The Scientific Basis. Contribution of Working Group I to the Third Assessment Report of the Intergovernmental Panel on Climate Change' (Cambridge University Press: Cambridge).

Hyvonen, T. (2011). Impact of temperature and germination time on the success of a C_4 weed in a C_3 crop.*Amaranthus retroflexus* and spring Barley. *Agricultural and food science,* 20:183-190.

José, P. F. Almeida, U.A., Hartwig, Marco Frehner, Josef Nösberger, and Andreas Lüscher (2000). Evidence that P deficiency induces N feedback regulation of symbiotic N_2fixation in white clover (*Trifolium repens* L.) *Journal of Experimental Botany,* 51:1289–1297.

Jungk AO (2002) Dynamics of nutrient movement at the soil-root interface. In: Waisel Y, Eshel A, Kafkafi U (eds) Plant Roots: The Hidden Half, 3rd Edn. Marcel Dekker,New York, pp 587–616

Kandiannan, K., and Chandaragiri, K.K. (2008). Monetary and non-monetary inputs on turmeric growth, nutrient uptake, yield and economics under irrigated condition. *Indian Journal of Horticulture*, 65(2): 209-13.

Kang, S., Kim, S., Oh, S. and Lee, D. (2000). Predicting spatial and temporal patterns of soil temperature based on topography, surface cover and air temperature. *Forest Ecology and Management,* 136: 173–18.

Khamis, S., Lamaze, T., Lemoine, Y. and Foyer, C. (1990). Adaptation of the photosynthetic apparatus in maize leaves as a result of nitrogen limitation. Relationships between electron transport and carbon assimilation. *Plant Physiology*, 94: 1436-1443.

Kimball, B.A., Kobayashi, K. and Bindi, M. (2002). Responses of agricultural crops to free-air CO_2 enrichment. *Advances in Agronomy,* 77: 293-368.

Kirschbaum MUF (2000) Will changes in soil organic carbon act as a positive or negative feedback in global warming? *Biogeochemistry,* 48: 21–51.

Kitayama, K., E.A.G. Schuur, D.R. Drake, and D. Mueller-Dombois. (1997). Fate of a wet montane forest during soil aging in Hawaii. *Journal of Ecology,* 85:669–679.

Kriticos, D. J., Sutherst, R. W., Brown, J. R., Adkins, S. W., and Maywald, G. F. (2003). Climate change and the potential distribution of an invasive alien plant: Acacia nilotica ssp. indica in Australia. *Journal of Applied Ecology*, 40: 111–124.

Lynch, J.P. and St.Clair, S.B. (2004). Mineral stress: the missing link in understanding how global climate change will affect plants in real world soils. *Field Crops Research*, 90: 101–115.

Martinez-Ghersa, M.A., Ghersa, C.M., Satorre, E.H. (2000). Coevolution of agricultural systems and their weed companions: implications for research. *Field Crops Research*, 67: 181-190

McKeague, J.A., F. DeConinck, D.P. Franzmeir (1983). Spodosols. In: L.P Wilding, N.E. Smeck, and G.F. Hall, Editors. Pedogenesis and soil taxonomy II.The Soil orders. Elsevier, Amsterdam, pp. 217-252.

Nowak, R.S., David S. Ellsworth and Stanley D. Smith. (2004). Functional responses of plants to elevated atmospheric CO_2– do photosynthetic and productivity data from FACE experiments support early predictions? *New Phytology*, 162:253–280.

Pendall, E., Bridgham ,S., Hanson, P.J., Hungate, B., Kicklighter, D.W., Johnson, D.W., Law, B.E., Luo, Y., Megonigal, J.P., Olsrud, M., Ryan, M.G. and Wan, S. (2004). Below-ground process responses to elevated CO2 and temperature: a discussion of observations, measurement methods, and models. *New Phytology*, 162: 311–322.

Ramesh, K., Biswas, A.K. and Patra, A.K. (2016). Impact of weed management in agriculture. In: Climate change combating through science and technology. Kinhal G.A. et al (Eds), BSMPS Pubishers, p 233.

Rosenzweig, C., and Hillel, D., (1998). Climate Change and the Global Harvest, Oxford University Press, Oxford.

Ross, D.J., Newton, P.C.D. and Tate, K.R. (2004). Elevated [CO_2] effects on herbage production and soil carbon and nitrogen pools and mineralization in a species-rich, grazed pasture on a seasonally dry sand. *Plant and Soil* 260(1-2): 183-196.

Sage, R.F. and Pearcy, R.W. (1987). The nitrogen use efficiency of C3 and C4 plants. 11. Leaf nitrogen effects on the gas exchange characteristics of Chenopodium album (L.) and Amavanthusretrojlexus (L.). *Plant Physiology*, 84: 959-963.

Schneider, M.K., Andreas Lüscher, Michael Richter, Urs Aeschlimann, Ueli A. Hartwig, Herbert Blum, Emmanuel Frossard and Josef Nösberger (2004) Ten years of free-air CO_2 enrichment altered the mobilization of N from soil in Lolium perenne L. swards. *Global Change Biology,* 10:1377–1388.

Seemann, J.R., Sharkey, T.O., Wang, J.L. and Osmond, C.B. (1987). Environmental effects on photosynthesis, nitrogen-use efficiency, and metabolite pools in leaves of sun and shade plants. *Plant Physiology*, 84: 796-802.

Shekhawat, K., S. S. Rathore, O. P. Premi, B. K. Kandpal, and J. S. Chauhan (2012). Advances in Agronomic Management of Indian Mustard (*Brassica juncea* (L.) Czernj. Cosson): An Overview. *International Journal of Agronomy*, Article ID 408284, 14 pages

Sinclair T. R. and Horie T. (1987). Leaf Nitrogen, Photosynthesis, and Crop Radiation Use Efficiency: A Review. *Crop Sci*ence, 29(1): 90-98

Singh G (1993). Integrated weed management in pulses. In: Proc. of International Symposium on Integrated Weed Management for Sustainable Agriculture, Vol. I, *Indian Society of Weed Science*, Hisar, India, pp. 335-342.

Soethe, N., J. Lehmann, and C. Engels. (2006). The vertical pattern of rooting and nutrient uptake at different altitudes of a south Ecuadorian montane rainforest. *Plant and Soil,* 286:287-299.

Taub, D. R. and Wang, X. (2008). Why are Nitrogen Concentrations in Plant Tissues Lower under Elevated CO_2? A Critical Examination of the Hypotheses. *Journal of Integrative Plant Biology*, 50: 1365–1374.

Terashima , I and Evans JR. (1988).Effects of Light and Nitrogen Nutrition on the Organization of the Photosynthetic Apparatus in Spinach Plant. *Cell Physiology,* 29 (1): 143-155.

Teyssonneyre, F., Catherine Picon-cochard, Robert Falcimagne and Jean-François Soussana (2002). Effects of elevated CO_2 and cutting frequency on plant community structure in a temperate grassland. *Global Change Biology*, 8:1034–1046.

Verhoeven, A.S., Demmig-Adams, B. and Adams, W.W. III (1997). Enhanced employment of the xanthophyll cycle and thermal energy dissipation in spinach exposed to high light and N stress. *Plant Physiology,* 113: 817–824.

Verma, N.K. (2011). Integrated nutrient management in winter maize (*Zea mays* L.) sown at different dates. Journal of Plant Breeding and *Crop Sci*ence, 3(8):161-167.

Waraich, E.A., Ahmad, R., Halim, A. and Aziz, T. (2012). Alleviation of temperature stress by nutrient management in crop plants: a review. *Journal of Soil Science and Plant Nutrition* 12 (2): 221-244.

Wolfe, D.W., Ziska, L., Petzoldt, C., Seaman, A., Chase, L., and Hayhoe, K. (2008). Projected change in climate thresholds in the North eastern U.S.: implications for crops, pests, livestock, and farmers. *Mitigation and Adaptation Strategies for Global Change*, 13:555–575.

Zhou, Y., Jianwu Tang, Jerry M. Melillo, Sarah Butler and Jacqueline E. Mohan (2001). Root standing crop and chemistry after six years of soil warming in a temperate forest. *Tree Physiology,* 31: 707–717.

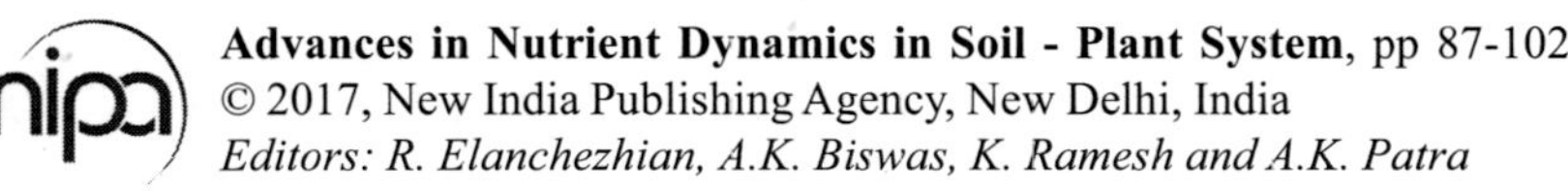
Advances in Nutrient Dynamics in Soil - Plant System, pp 87-102

Editors: R. Elanchezhian, A.K. Biswas, K. Ramesh and A.K. Patra

9

Sustainable Management of Problem Soils for Improving Yield and Nutrient Use Efficiency

P. Dey

ICAR-Indian Institute of Soil Science, Nabi Bagh, Bhopal – 462038, India

The concept of sustainability in agricultural system

All forms of production systems exist today, from nomadic herding to continuous, intensive monocrop systems. Their distribution is determined by soil and climate conditions, and by social and economic factors. In very general terms, as one moves from drier to wetter areas, pastures and animals become less important, and trees more so. Population density is greatest where soils are most fertile, and management systems the most intense. Most production systems have evolved so that they are sustainable in terms of the environmental conditions prevailing at the time—including the level of demographic pressure. Given that demographic pressure has increased dramatically over the past century, and will continue to do so for the next half century, sustainability in relation to agriculture and soil management must be defined to include the need for increases in demand to be met. FAO (1991) has given the following definition: 'A sustainable agricultural system is one which involves the management and conservation of the natural resource base, and the orientation of technological and institutional change in such a manner as to ensure the attainment and continued satisfaction of human needs for present and future generations. Such sustainable development conserves land, water, plant and animal genetic resources, and is economically viable and socially acceptable'. While studying the variation of sugarcane yield under salt affected soil, Dey *et al.* (1996) observed that step down regression analysis revealed that EC and $CaCO_3$ explained 34.7% variation in cane yield; inclusion of available K improved the coefficient of determination significantly (P= 0.05) from 0.387 to 0.712.

Since soil quality is affected by soil genesis and an understanding of soil genesis is essential for sustainable soil management, the genesis of sodic soil and saline soil including salt affected vertisol is provided below:

Genesis of Sodic Soils

Several researches carried out on sodic soils genesis in the Indo-Gangetic plain in India indicated that the salt present in these soils were the result of weathering. The primary minerals comprising of quartz, feldspars, muscovite, biotite, chloritised biotite, tourmaline, zircon and hornblende are similar in the sand fraction of sodic and non-sodic soils. The weathering of alumino-silicate minerals through carbonation yield alkaline bicarbonates and carbonates, apart from silica and alumina. The enormous amount of sodium release from the sodic and non-sodic soils was also investigated to ascertain the source of sodium from mineral lattices through hydrolytic dissolution. Studies on the geochemical source of sodium in sodic soils showed periodic release of sodium from the sand fraction in the Indo-Gangetic Plain. On the other hand relative relief differences between parts of the plain and the outer Himalayas or Siwalik and within the plain, facilitate runoff during the monsoon season carrying part of the products of weathering for deposition in the micro-basin. The process of sodiumisation in these soils therefore begins at the surface. It is low at the deeper horizons where illuviation and sodium saturation occurs with deflocculated clay particles and limited leaching of alkali salt solutions. Repeated cycles of wetting and drying facilitate maximum accumulation of salts on the surface. The most favorable climate is mean annual rainfall between 550-1000 mm with ustic soil moisture regime in Haryana and Punjab and aquic and para-aquic regime in Uttar Pradesh. The soils are characterized by the typical illuviated B horizon enriched with clay and /or sodium, the fine/total clay ratios, and presence of clay cutans on ped faces are prevalent. A shallow underground water table with seasonal fluctuations prevails below some of the sodic soils. The water having a very low degree of mineralization and being non-sodic is extensively used for irrigation and is an active source of soil sodiumisation. A distinct, regular and thick zone of calcium carbonate accumulation, mainly dolomitic exists below the surface in sodic soils. The smooth boundary of the calcic horizon, irregular shape of individual concretions (Kankar), and regularity of occurrence underneath vast areas highlight their pedogenic origin. It originates in presence of fluctuating and shallow water table. On drying and the water table recedes, the precipitation of calcium and magnesium occurs irreversibly to develop carbonate concretions. With every passing hydrological cycle the concretions continue to grow in size and occlude the diverse kinds of soil materials. Sodic conditions bring about degradation of clay minerals leading to accumulation of amorphous oxides of silica, alumina and iron associated with repeated synthesis of clay minerals. The degradation is judged by the presence of amorphous silica and alumina and an equivalent ratio of $SiO_2/Al_2O_3 > 2$. Characteristics of sodic soil is provided in Table 1.

Genesis of saline soils

***Inland saline soils*:** Saline soils occur in the alluvial plain and the shield areas (plains with sand dunes and Aravalli Hills) occupying basins or playas. These soils have commonly cambic and occasionally argillic horizon at a depth below the surface, a calcic horizon at variable depths, coarser soil fabric, high alt content, predominance of chloride and sulfates of sodium, calcium and magnesium, neutral to slightly alkaline pH, usually high SAR, rapid to moderately rapid infiltration rates, saline ground water at varying depths and a petrocalcic horizon occasionally within 1 m depth. The occurrence of saline soils in Rajasthan is related to the climate, topography and practice of irrigated farming. Secondary salinization occurred in areas under canal irrigation, waterlogging and presence of hydrological barrier restricting leaching and drainage of salt within the soil profile. The characteristics of saline soil are provided in Table 1.

***Saline soils of the delta region*:** The process of soil accretion and land subsidence operate in the delta region. The alluvium developed by the rivers in delta region is frequently subjected to inundation during high and low tides with the ingress of sea water through numerous creeks. The problem of soil salinity appeared in the Ganges delta, known as Sunderban, and the Godavari, Krishna and Cauvery deltas of the country. The soil in Sunderban are developed from the micaceous rocks and are characterized by an ocric epipedon, grey to light brownish grey color, presence of yellowish brown mottles, signs of gleying, with grey, black or dark grey colors in the substratum, uniformly fine texture varying from clay loam to clay to silty clay loam, neutral to slightly acidic pH, those occurring in the basins contain highest amount of salts in the epipedon and relatively low EC values below, lower ECe values at the surface that increases with increasing depth in soils of higher elevation, preponderance of chlorides and sulfates of sodium, magnesium and calcium with minor bicarbonates, SAR and ESP varying between 10 to 30, a shallow saline water table, low infiltration rates, absence of calcium carbonate and organic matter content less than one percent. Because of high rainfall in these areas high ECe values are not encountered due to flushing of salt during monsoon. The local relief and aspect however, play the major role in facilitating salt accumulation in this region. The soils in the Godavari, Krishna and Cauvery delta regions differ from the Ganges delta primarily due to their origin from basaltic rocks. The climate is semiarid, the soil exhibit completely different mineralogy and salt regime. The soil properties are similar to vertisols of the delta region. The presence of an ocric epipedon, a uniform fine (clay) texture, angular/ sub-angular blocky structure grading to massive, absence of concretions and calcium carbonate, neutral to slightly alkaline pH, very high ECe, shallow water table, deep crack and the presence of slickensides in subsurface horizon are typical characteristics of these soils. Owing to low rainfall, these soils are higher in salinity than the soils of the Ganges delta.

***Saline soils of the coastal belt*:** The typical soil found in the Guntur and Prakasam districts of Andhra Pradesh show a close relationship between soil characteristics and physiographic settings. The upland soils are free from salinity. The midland soils are characterized by stratified substratum with layers varying in thickness and the texture varying from loamy sand to sandy loam in the surface and sandy clay loam below the surface, In the basin, dark grayish brown to grey with clay textured soils which remain submerged during the monsoon, common presence of marine shells in the soil profile, shallow saline water table in the substratum, high EC throughout the profile, prevalence of chlorides and sulfates of sodium, calcium and magnesium, neutral pH, slow permeability and low to medium organic matter content.

The problem of soil salinity is severe in the coastal belt of Saurashtra in Gujarat located under the arid and semiarid region. Originating in basaltic rocks, these are commonly vertisols or having vertic characteristics. Owing to differences in the relief, there are differences with regards to the menace of soil salinity. These soils are similar to the vertisols of Andhra Pradesh but different in the ESP values ranging from 10 to 60, SAR values from 10 to 25, high hydraulic conductivity but develop sodicity after leaching and poor physical conditions. Almost all the soils are potentially saline with salt reserve in the profile sub-stratum. These are common in the alluvium, and other terraces like flood plain, mud flats along the tidal inlets and mud flats along the coast, which are more saline than the old flood plain, inter-terraces of flood plains and recent flood plains. The Rann of Kachchh constitutes a vast marshy area in the arid coastal region in Gujarat. It is divided into Great Rann and Little Rann having area of 18130 and 5180 sq. km. Rann soils are fine textured, and contain large quantities of chlorides and sulfates of Na, Mg and Ca. Gypsum layers are encountered at varying depths The Rann receives huge discharge of flood water from the Luni, Banas, Saraswati, Rupan, Fulka and Brahmani streams during the monsoon. At the same time strong wind from the south-west force sea water into the area, rendering it as saline. The depth of flood water continues to stand up till December. It dries from January till June. These soils showed high degree of gleying with black matrix, in the epipedon and blue-green to bottle green in the substratum. Reddish brown to brown horizons reported to be the zone of iron pan formation in the substratum. An interesting feature of these soils is the presence of gypsum deposition, iron pan and gleyed horizons within the profile in different sequences. These indicate operation of similar genetic processes over the past geological period as the process of land accretion continued. Saline acid sulfate soils occur along the Malabar Coast in Kerala occupying marshy depressions (lagoons). These have developed on the alluvium derived from laterites under humid and tropical climate. The soils undergo fresh water submergence from May to December and sea water inundation under tidal cycle during the subsequent lean months. The salient features are ocric epipedon, humic horizon in the substratum of the some soil, variety of soil matrix color ranging from pale yellow to very dark grey, grayish brown, and dark yellowish brown, signs of gleying, reduction and bleaching, high EC throughout the profile, prevalence of chlorides and sulfates of sodium, calcium and magnesium, soil pH 3.5 to 7.5, organic matter content varying from 2 to 40%, a shallow saline water table, and in some cases the presence of pyritous clay.

Table 1: Characteristics of Alkali and saline soils

Salt affected soils	pH s	EC (dSm^{-1})	Exchangable Sodium percent (ESP)	Salt in excess
Alkali (Sodic) Soils	> 8.2	Variable	>15	Carbonate and Bicarbonate of sodium salts
Saline	< 8.2	> 4	< 15	Chloride and sulphate calcium, magnesium and sodium salts
Saline-Sodic soils	< 8.2	> 4	> 15	Chloride, sulphate, carbonate and bicarbonate of calcium, magnesium and sodium salts

Saline Vertisols

Among 11 soil orders (Andisols, Alfisols, Aridisols, Entisols, Histosols, Inceptisols, Mollisols, Oxisols, Spodosols, Ultisols, Vertisols), Vertisols form an important soil group. These soils can be defined as clay soils with high shrink-swell potential that has wide, deep cracks when dry. Most of these soils have distinct wet and dry periods throughout the year. Soils with high content of swelling clays, deep, wide cracks develop during dry periods .Soils with 30 % or more clay to a depth of 50 cm and shrinking/swelling properties. Vertisols and associated soils are generally very deep (150-200 cm), fine textured with clay content ranging from 45-68 % and montmorillonite as the dominant clay mineral. The soils exhibit high shrink-swell potential and develop wide cracks of 4-6 cm extending up to 100 cm depth. The water holding capacity is high but permeability is imperfect to poor. These soils are calcareous in nature (2 to 12% $CaCO_3$). The salinity status in the cultivable land varies widely from EC 0.5 dS/m in monsoon to 50 dS/m in summer. The saline Vertisols in Gujarat occurs in Bara tract which experiences a tropical climate. The annual rainfall ranges from 275-1484 mm with an average of 737 mm. The onset of monsoon is erratic which normally affects crop seeding operations, germinations and seedling establishment. Cotton is the dominant crop grown in the kharif followed by sorghum and pearlmillet. Pigeon pea is also grown in some area. Mostly rainfed kharif crops are grown in this area. In the rabi season the land is either kept fallow or some fodder sorghum is grown on the residual moisture.

The Vertisols have low permeability; soils having comparable salinity affect the crop growth in a greater magnitude as compared to the light textured soils. As these soils can sustain the deep rooted crops and are having fine capillary pores, salt concentrations even at a considerable depth affect the crop growth and contribute to surface salinity through capillary rise. The salinity of surface soils varies from 0.46-21 dS/m. The salinity of the sub-soil of Bara tract ranges from 0.4-159 dS/m. This transient salinity fluctuates with depth and also changes with season and rainfall. Even in the absence of contribution of ground water, the excess use of water may also help the sub-soil salinity to come to the surface layer.

Principles of good soil management

Good soil management has always required that the soil be used in such a way that its productivity is maintained or preferably, enhanced. This requires that the chemical and physical condition of the soil does not become less suitable for plant growth than when cultivation commences. Cultivation normally means that the soil will, in fact, deteriorate due both to nutrient removal when harvesting crops, and to physical damage to the soil structure. What is essential is that the deterioration is reversible, by chemical additions to the soil, mechanical manipulation, or natural processes of fertility restoration under pasture or trees. This implies that the soil must be resilient, i.e. after being subjected to the stresses involved in crop production; it must have the ability to return to its former condition, or an improved condition (Greenland and Szabolcs, 1994). The land must produce on a secure basis, the natural resources must be protected, and the management system must be economically viable and socially acceptable. However it must also be recognized that land cannot be managed sustainably unless the soil, which is a component of the land, is properly managed. This requires maintaining and improving soil productivity, avoiding and rectifying soil degradation, and avoiding environmental damage.

Maintaining and improving soil productivity

If a soil is to sustain the production of crops it must:

1. Provide the nutrient requirements of the crop;
2. Provide a physical medium:
    - In which the plant roots can grow adequately so that water and nutrients can be absorbed;
    - Which stores sufficient water for the crop; and
    - Which allows water to enter and move in the soil to maintain the water supply as it is transpired by the crop and evaporates from the soil;

3. Provide a medium in which soil organisms are able to:
   - Decompose organic materials, releasing nutrients to the plants;
   - Assist the transport of nutrients to plant roots;
   - Compete successfully with pathogens which might otherwise infect roots and damage the plants; and
   - Form the soil organic compounds which will have a favourable effect on other soil properties.

Nutrient management in alkali soils

About 3.77 million hectares area is severely affected by sodicity in the Indo-Gangetic plains. Nutrient deficiency and toxicity generally occur in these soils. Fertility of these soils with low nutrient reserves is confounded by the low supply of water and oxygen to roots in profiles with dispersive clays. The main problem is of high pH/ESP, high amount of calcium carbonate, very low amount of organic matter and poor physical conditions limiting nutrient availability and plant growth. Crops grown on these soils invariably suffer nutritional disorders (N, Ca and Zn deficiency and Na toxicity) resulting in low yields (Swarup,1998). Crop production and fertilizer use efficiency in these soils can be increased by following the reclamation technology involving integrated use of amendments preferably gypsum based on gypsum requirement of soil, balanced and integrated use of chemical fertilizers and organic/green manures which help in maximizing and sustaining yields, improving soil health and input use efficiency. Rice based cropping systems like rice-wheat, rice-berseem and rice-mustard are recommended on these soils.

Organic carbon and nitrogen

Alkali soils are highly deficient in organic matter - a storehouse of essential plant nutrients especially available N throughout the soil profile. High exchangeable sodium (ESP >15), high pH (>8.5) and low biological activity, commonly found in these soils, are not conducive for the accumulation of organic matter and its mineralization. Therefore, its efficient management and maintenance assumes greater significance. Results have shown that long-term balanced fertilizer use under rice-wheat system helps in maintaining the organic carbon status of the soil as compared to the control plots. The results further suggest that alkali soils have great potential for carbon sequestration (Lal and Swarup, 2004). Most crops invariably suffer from inadequate N supply. Moreover, nitrogen transformations are adversely affected by high pH and sodicity, thereby affecting the efficiency of applied N.

Numerous experiments have shown that recovery of fertilizer nitrogen normally ranges from 30 to 40 % for rice in alkali soils. Proper management of fertilizer N is thus necessary for better N use efficiency. Because of the adverse physico-chemical conditions, the recovery can be still lower in alkali soils. Under such situations nitrogen use-efficiency can be increased by integrated use of organic and inorganic sources of N.

Phosphorus

Uncultivated barren alkali soils contain high amounts of available (Olsen's extractable) P. This is primarily due to the presence of sodium phosphates, which are water soluble. Water-soluble P increases with soil pH in all the major bench-mark series of alkali soils of the lndo-Gangetic plains, and strongly alkaline calcareous sodic soils have the bulk of soil P as Ca-P (54%) and residual inorganic fractions (28%). When alkali soils were reclaimed by using amendments and growing rice under submerged conditions, Olsen's extractable P of surface soil decreased due to its movement to lower subsoil layers, uptake by the crop and increased immobilization (Swarup, 2004).

The critical values at which crops responds to applied P vary greatly with the nature of the soil (clay content) and stage of its reclamation, initial soil-test value, crop to be grown and the type of amendment used for reclamation. Results of a long-term fertility experiment conducted on a gypsum-amended alkali soil (texture loam, pH 9.2; ESP 32) with rice-wheat and pearl millet-wheat cropping sequence and NPK fertilizer use showed that phosphorus applied at a rate of 22 kg P ha^{-1} to either or both rice and wheat crop in rotation significantly increased the grain yield of rice when Olsen's extractable P (0-15 cm soil) had decreased from the initial level of 33.6 kg ha^{-1} to 12.7 kg P ha^{-1}, which is very close to the widely used critical soil-test value of 11.2 kg P ha^{-1}. Though wheat responded to applied P when available P level decreased close to 8.7 kg ha^{-1}, pearl millet did not respond to applied P at this level of critical soil-test value. The rice and wheat responded to P application in pyrite-amended alkali clay-loam soil (pH 9.3, ECe 3.42 dS m^{-1}, CEC 20.1 meq 100 g^{-1} and ESP 46.7) testing low in available P (4.63 ppm). These studies indicate that recommendations for P fertilization in alkali soils should be based on soil test. Single superphosphate (SSP) is a better source of P than other phosphatic fertilizers because of high Na of alkali soils and as it contains appreciable some amount of calcium sulphate. Recent studies on integrated nutrient management showed that continuous use of fertilizer P, green manuring and FYM to crops significantly enhanced the yield of rice and wheat and improved available P status of the gypsum amended alkali soils .

Potassium

Alkali soils of lndo-Gangetic plains generally contain very high amounts of available K . Studies so far indicate that the crops do not respond to applied K even after 20 years of rice -wheat and pearl millet-wheat cropping systems in alkali soils . Lack of crop response is attributed to the presence of K-bearing minerals and their dissolution and large contribution of non-exchangeable K (> 90%) towards total K uptake by the crops. Potassium application increased the K uptake by plants and reduced the release of K from non-exchangeable reserves from 95 to 70 %. The decrease was about 51 % with the use of K combined with organic manures. The quantity: intensity (Q/I) relationship remained virtually unaltered after continuous cropping . Due to low leaching, a large portion of applied K remained in the top 30 cm soil.

Micronutrients

Alkali soils are sufficient in total zinc but generally deficient in available Zn. Only 3.3 % of the total Zn is attributed to .the exchangeable, complexed, organically bound and occluded forms, which are considered to be available during crop growth. Thus zinc deficiency is very common in rice and its deficiency symptoms appear in the early growth stages (21-25 days), which delay maturity and reduce yields . Therefore significant response to its application is observed. Application of 9 kg Zn ha^{-1} (40 kg zinc sulphate) eliminated Zn deficiency in rice grown on alkali soils treated with gypsum, pyrites, farm-yard manure (FYM) and rice husk and raised the available Zn status of the soil to an adequate level, so as to meet the subsequent requirement of 2-3 crops. With the application of FYM and Sesbania green manure it was possible to prevent the occurrence of Zn deficiency in rice grown on alkali soils. Organic amendments like pressmud, poultry manure and farmyard manure could effectively supply zinc from the native and applied sources to rice crop in a saline sodic soil.

The alkali soils are rich in total Fe and Mn but are generally poor in water-soluble plus exchangeable and reducible forms of Fe and Mn. There exists negative relationship between pH and Fe-Mn availability. Soluble Fe and Mn salts when applied to alkali soils are rendered unavailable because of rapid oxidation and precipitation, and their ecovery by soil-test methods is very low. Thus higher addition of Fe and Mn salts is needed to correct the deficiencies or to have beneficial effect on crop growth. Transformation of Fe and Mn in alkali soils is very strongly influenced by organic matter under submerged conditions; pH per se being relatively less important. This is primarily because of intensely reduced conditions (drop in redox potential) and enhanced PCO_2 created by organic matter under submerged conditions in rice culture Addition of FYM, rice husk and green manures had a marked effect in increasing the extractable Fe and Mn by 10 to 15 times, with corresponding decrease in reducible forms. Available Fe and Mn and rice yield increased significantly when alkali soils were flooded for 15 and 30 days before transplanting rice; the effects being more pronounced at higher levels of ESP. However, benefit of iron application to rice could be realized in sodic soils only when it was applied along with Zn.

Adoption of rice-wheat system for more than two decades on gypsum-amended alkali soils resulted in decline of the DTPA- extractable Mn to a level of 2.7 mg kg^{-1}, where wheat responded to manganese sulphate application at a rate of 50 to 100 kg ha^{-1}. Substantial leaching losses of Mn occur following gypsum application in alkali soils. Foliar application of Mn is better than soil application. Nutrients such as B and Mo are not likely to be limiting factors for plant nutrition in alkali soils, though at higher concentrations they could prove toxic. However, once the alkali soils are amended with gypsum/pyrites and leached, concentrations of these elements in solution drops to within safe limits and remain no longer toxic to plants.

Nutrient management in saline soils

In India about 2.96 million hectares are lying barren due to problems of waterlogging and soil salinity. Out of these 1.146 million ha lie in the various canal commands. Saline soils are those which have excessive amounts of soluble salts, (ECe>4 dSm^{-1}, pHs < 8.5 and ESP <15). These soils predominantly have a high concentration of chloride and sulphate of sodium, calcium and magnesium. Many times these soils have shallow water table representing brackish groundwater, which may be the major cause of salinity due to capillary rise under arid and semiarid climatic conditions. Provision of adequate subsurface drainage to lower the depth of water table and to facilitate leaching of salts has long been recognized as fundamental to the reclamation and management of saline soils. During leaching of these soils release of soil nutrients especially N, P, K, Ca, Mg and Mn and their loss to the ground water have been reported. Moreover, the choice of crops to be grown in saline soils under reclamation is also of paramount importance, since different crops differ widely in their tolerance to salinity.

Nitrogen

Nitrogen is the most limiting nutrient for crop production in saline soils as they are poor in N status and organic matter. Volatilization is a major N loss mechanism that reduce the efficiency of applied N. Volatilization losses increased with increase in salinity. Volatilization losses of N from rice field increased by about 100% when soil salinity (ECe) increased from 4 to 8 dSm^{-1}. Ammonium sulphate showed highest amount of loss being 37.4 per cent at a soil salinity of 8 dSm^{-1}, while fertilizer placed in soil (UPP-urea in paper packet and UB –urea briquette) reduced losses to about 5-6 %. Results also showed that sulphur coated urea followed by urea briquette were more efficient than prilled urea for rice. Poor nitrification rates of NH_4^+ - N at high soil salinity was chiefly responsible for higher volatilization of N from saline soil. Apart from antagonistic effects of high amounts of Cl^- and SO_4^{2-} on the absorption of NO_3, in waterlogged saline soils, poor aeration and anaerobic condition may restrict the availability to and absorption of N by plants leading to low efficiency of applied ammonical fertilizers. Further, high concentration of salts inhibits nitrification and results in ammonical nitrogen accumulation. Due to these reasons, it is better to use NO_3-N fertilizer as compared to NH_4-N in saline soils. High water stress faced by the plants in saline environments further restricts the proper metabolisation of the absorbed nitrogen. These factors along with higher leaching losses of NO_3^- during reclamation of the saline soils results in low availability of N to the plants and therefore nitrogen requirement of crops is higher in saline soils than in normal soils.

Phosphorus

The available P status of saline soils is highly variable. It showed no regular trend in relation to soil salinity probably because of the varied concentration of neutral soluble salts of Ca, Mg and Na in the experimental soil. These may have displaced exchangeable Ca and change the ionic composition of the soil solution thus influencing the extraction of soil phosphorus. Availability of P increases up to a moderate level of salinity but thereafter it decreases.

Application of P enhanced significantly the yield of mustard, wheat and pearlmillet, the effects of being more pronounced at high soil salinity. Increase in salinity decreased P concentration and uptake by the crops. Absence of P in the drain water effluent and available P status of the soil profile after crop harvest indicated very slow movement of P, the large portions being retained by the top soil (30 cm) thereby drastically reducing the chances of ground water pollution through phosphorus fertilization. The availability of fertilizer phosphorus in the soil may be modified by soil salinity due to higher precipitation of added soluble P.

Potassium

Available K status of saline soils is high initially but after continuous leaching and cropping it declines to a level where crops respond to its application. Application of K fertilizer in saline soil increases crop yields in several ways: (i) by directly supplying K, (ii) by improving tolerance of plants to Na uptake (iii) by improving water use efficiency, (iv) by improving N use efficiency. Plants grown under high salinity may show K deficiency due to antagonistic effect of Na and Ca on K absorption and /or disturbed Na/K or Ca/K ratio. Under such condition application of K fertilizer may increase yields. Studies showed that application of K enhanced yield of pearl millet and wheat and also reduced the contribution of non-exchangeable K towards K uptake by plants. The contribution of non – exchangeable K towards total K uptake was 97 per cent in plots receiving no fertilizer K whereas K application at 21 and 42 kg ha-1 reduced it to 83 and 71 per cent respectively. Pearl millet was more exhaustive of K than wheat. This implies that continuous cropping with higher level of K along with N and P would result in rapid depletion of K reserves thereby rendering the soil poor in K fertility.

This suggests that unless K fertility is maintained yield will remain at low levels and will decline. Presence of K in the drain water effluent (3.2 to 8.2 mg K L^{-1}) and higher level of available K into the lower soil depths indicated continuous release of native and applied K from saline soils, thereby contributing towards higher K content of groundwater in the vicinity of saline areas. K concentration and salinity of drainage effluent were lower during rainy season (July-September) than in winter (November–March) and summer season (April–June). Leaching losses of native and applied K were also confirmed in laboratory column experiment when a high saline soil (ECe 43 dS m^{-1}) was leached with good quality water (EC 0.3dS m^{-1}) maintaining a constant water head in the column.

Micronutrients

In a microplot field study (Swarup,1995) effect of micronutrients namely, Fe, Mn and Zn and their combinations was studied on yields of wheat and availability of micronutrients in a reclaimed saline soil with sub-surface drainage system (ECe 5.5 dSm^{-1}, organic carbon 0.36 per cent, DTPA extractable Zn 0.56 mg kg^{-1}, Fe 4.3 mg kg^{-1}, and Mn 2.65 mg kg^{-1}. Results showed significant increase in grain yield following Zn and Mn fertilization. Highest yield was obtained when both Zn and Mn were applied. Application of Fe had no effect on yield (Table 2). After crop harvest recovery of added Fe, Mn and Zn was 25.1, 23.7 and 17.1 per cent, respectively.

Nutrient interactions and balanced fertilization

Nutrients interactions play an important role for sustaining crop production in saline soils. Studies on nutrient interactions showed that N and K interacted significantly on wheat yield, N concentration, uptake and recovery. High dose of N alone had a depressing effect on yield. Application of K had significant effect on yield at all levels of applied N. Increasing rates of N and K enhanced significantly N and K concentration and uptake. In fact, the much higher N and K uptake with the higher K rate indicated that there might be a complementary uptake effect between N and K. It was concluded that K^+ enhanced NH_4^+ assimilation in the plant and that K^+ did not complete with NH_4^+ in the absorption process of the plants. The recovery of N and N-use efficiency increased with K application at all levels of applied N and more so at the highest K rate. These results thus suggest the importance of adequate K for efficient N use. Interactions between nitrogen and phosphorus, and between phosphorus and potassium were significant. However, increasing rates of P and K enhanced significantly N concentration in grain and straw and uptake by pearl millet and wheat, the effect being more pronounced when both P and K were applied together. The highest uptake of N by pearl millet (122 kg ha^{-1}) and wheat 159 kg ha^{-1}) was attained at the highest P and K rate. In fact, the much higher N uptake with the highest P and K rate indicated that there might be a complementary uptake effect between N and P and N and K. This is possibly because of a more balances use of soil nutrients in the presence of adequate phosphorus and potassium for efficient N use by crops in saline soils. Drain water effluent had no NH_4^+ and NO_3^- - N thereby indicating little danger of ground water pollution as a result of leaching of nitrates.

Table 2: Effect of micronutrients on wheat yield and micronutrients availability in soil

Treatments	Yield (t ha^{-1})		DTPA extractable micronutrients (mg kg^{-1})		
	Grain	Straw	Fe	Mn	Zn
Control	5.42	5.85	4.34	2.62	0.58
Fe_{50}	5.45	6.10	9.95	3.65	0.56
Mn_{50}	5.81	6.25	4.40	7.98	0.58
Zn_{11}	5.76	6.18	4.38	2.65	1.42
Fe_{50} + Mn_{50}	5.88	6.30	8.90	9.56	0.57
Fe_{50} + Zn_{11}	5.78	6.21	8.95	4.10	1.32
Mn_{50} + Zn_{11}	6.12	6.43	4.50	8.10	1.45
Fe_{50} + Mn_{50} + Zn_{11}	6.20	6.50	8.56	9.86	1.40
LSD at P=0.05	0.34	0.45	1.18	1.25	0.32

Changes in Soil Quality due to Adoption of Agroforestry in Sodic Soils

Carbon sequestration and its mechanism in sodic soil (Dey, 2009) and organic matter as well as nutrient dynamics (Dey and Singh, 2008) under agroforestry systems in Saraswati forest range Haryana have been studied in details. The soil originally was highly sodic through-out the profile. pH and EC values were highest on the surface (10.7 and 3.3 dS/m) and decreased with depth (Mongia *et al.* 1998). Organic C was very low (0.5 g/kg). A sharp decrease in surface soil pH, EC and ionic concentrations of water extract was observed within three years of growth under all the plantations, the decrease be-ing more under *Prosopis juliflora* followed by *Acacia nilotica, Dalbergia sissoo* and *Casuarina equisetifolia* (Dey *et al.*, 2004a). However, pH and soluble salts increased in the lower depths. The increase in the salts may be due to their translocation through leaching and lowering of pH can be related to the organic matter accumulation because of litter fall and their subsequent decomposition. The lowest pH under *Prosopis juliflora* may be related with the highest amount of organic matter accumulation as evident by organic C content.

Ionic composition of the water extract shows that $CO_3^{=}$, HCO_3^- and Na^+ were the dominant ions in the sodic soils. The ionic concentration as a whole, decreased considerably on the surface following tree plantation. The decrease was highest in case of *Prosopis juliflora* while all the other species were almost at par to each other. A general increase in organic C content was observed throughout the profile under all the plantations, the increase being more in the surface layer and the rate of increase decreased with depth (Dey *et al.*, 1999). The increase in organic C was maximum *Prosopis juliflora* (3.2 g/kg) and the least in *Casuarina equisetijolia* (1.7 g/kg). As regards the available nutrients, available P declined while an increase in available K was observed under all the plantations. The highest value of available K was noticed under *Prosopis juliflora* (Mongia *et al.*, 1998). The higher content of K may be due to release of K from the K-bearing minerals following reclamation and partly due to recycling of K on account of litter decomposition. Calcium carbonate content in the surface and subsurface soil decreased with growing of tree plantations, but it remained more or less constant in the lower horizons of the soil profiles. Tree roots increase the CO_2 level in the soil which helps mobilizing and dissolving in $CaCO_3$ and it results in exchange of Ca^{++} with Na^{+} on the soil ex-change complex, thus resulting in decreased calci-um carbonate content on the surface and subsur-face (Dey *et al.,* 2004b). High variations of Olsen-P in sodic soil can be described by water soluble silicon (Dey *et al.,* 2004c). The Fe and Mn concentrations in the profile increased following plantations. The highest con-centrations of these elements were observed in *Ca-suarina equiselifolia* and the least in *Prosopis juliflora.* Zn and Cu content rather registered a decrease following tree plantations. The variation in the concentrations of these micronutrients in the soil may be due to their differential uptake by the trees and subsequent recycling in the soil through litter decomposition. The afforestation of sodic soil by tree planta-tions helps in reclamation of sodic soil by lowering pH and soluble salts of the soil, creating favourable root environment and building organic matter and fertility status of the soil.

The following package of practices is recommended by ICAR-CSSRI for reclamation of sodic lands, need to be followed for sustaining soil health and crop productivity:

(i) Land leveling and bunding of fields and providing 35-40 cm high bunds to check outflow and entry of outside water from unreclaimed fields. Strong bunding is essential to preserve and utilize rainwater for leaching salts and growing rice crop.

(ii) Suitable surface drains are also required to regulate excessive surface flow of water during heavy rainfall.

(iii) Installation of tubewells to ensure timely irrigation. It is also essential to lower the watertable for attaining permanent reclamation. The shallow cavity tubewells serve as vertical drains.

(iv) Soil sampling and testing to determine gypsum requirement, the quantity of gypsum is decided on the basis of soil pH and texture of the soil and varies from 10 to 15 t/ha.

(v) Apply gypsum powder in the well-ploughed and leveled fields in the month of June or July. It should be mixed in the upper 8-10 cm soils. Gypsum is applied only once in the 1st year of soil reclamation.

(vi) After gypsum application, water is kept standing in the field for 10-15 days before transplanting rice as a first crop.

(vii) Rice is recommended to grow as the 1st crop during soil reclamation. Mainly rice-wheat crop rotation is followed during the reclamation period. In the initial stage CSR10 variety of rice is recommended to grow on newly reclaimed sodic soils. Later on, CSR13, CSR30, PR106, PR107, Basmati (B-370), Pb No.1, Pusa150, Pusa169 etc. can also be grown for economic returns.

(viii) For rice crop, 35-40 days old seedlings grown on the normal soil should be transplanted at a distance of 15 cm keeping 3-4 plants per hill. Seed rate needs to be kept more (40-50 kg/ha) for transplanting one hectare of reclaimed sodic land.

(ix) Apply 25% more nitrogen in rice crop as compared to those of applied in the normal soil. Nitrogenous fertilizers and zinc sulphate are recommended to apply @ 150 kg and 25 kg/ha, respectively. By adopting these practices bumper rice crop can be harvested right from the first year of reclamation.

(x) Phosphorus and potash need not to apply during initial years of the reclamation (5-6 years).

(xi) Wheat crop should be grown during winters. The wheat varieties KRL 1-4 and KRL 19 are the recommended salt tolerant cultivars to grow during initial stage of the reclamation. *Barseem* (*Trifolium alexandrinum*) or *shaftal* (*Trifolium resupinatum*) can also be grown for fodder in subsequent years as per requirement of animals.

(xii) Five to six light irrigations are recommended for the cultivation of wheat crop.

(xiii) *Sesbania* should be grown for green manuring during summers.

(xiv) The field should not allow being fallow during the reclamation period. After few years of continuous cropping, other crops may be introduced to diversify the cropping system.

References

Dey, P. (2009). Transformation and availability of primary nutrients in submerged sodic soils. *In: Improving Sodic Soil Quality, Input Use Efficiency and Crop Productivity through Integrated Nutrient Management* (Eds. Yaduvanshi, N.P.S., Dey, P. and Singh, Gurbachan), Central Soil Salinity Research Institute, Karnal, India, pp. 43-47.

Dey, P. (2009). Carbon sequestration and its mechanism in sodic soils. *In: Improving Sodic Soil Quality, Input Use Efficiency and Crop Productivity through Integrated Nutrient Management* (Eds. Yaduvanshi, N.P.S., Dey, P. and Singh, Gurbachan), Central Soil Salinity Research Institute, Karnal, India, pp. 93-95.

Dey, P., Mongia, A.D and Singh, Gurbachan (1999). Spatial variation of soil properties and tree growth parameters in agroforestry under sodic soil condition. In: *Wasteland Development: Challenges & Opportunities* (A.K. Singh, K.S. Bhatia and J.P. Yadav eds.), C.S.A University of Agri. Technology, Kanpur and State Land Use Board, Dept. of Planning, Yojana Bhawan, Lucknow, pp. 168-172.

Dey, P., Mongia, A.D. and Singh, Gurbachan (2004a). Bio-amelioration of sodic soil. In: *Extended Summaries: International Conference on Susrtainable Management of Sodic Lands*, Lucknow, India, pp. 387388.

Dey, P., Mongia, A.D. and Singh, Gurbachan (2004b). Performance of woody perennials in highly sodic soil of semiarid climatic region under land use pattern of agro-forestry. In: *Proc. 91st Indian Science Congress*, Chandigarh, January 3 to 7, 2004, pp. 49-50.

Dey, P., Mongia, A.D. and Singh, Gurbachan (2004c). Distribution and variation of water soluble silicon in agro-forestry under sodic soil condition. In: *Proc. 91st Indian Science Congress*, Chandigarh, January 3 to 7, 2004, pp. 48-49.

Dey, P. and Singh, Gurbachan (2008). Organic matter and nutrient dynamics in agroforestry system under salt affected soils. *In: Chemical Changes and Nutrient Transformation in Sodic/Poor Quality Water Irrigated soils* (Eds. Yaduvanshi, N.P.S., Yadav, R.K., Bundela, D.S., Kulshreshtha and Singh, Gurbachan), Central Soil Salinity Research Institute, Karnal, India, pp. 224-226.

Dey, P., Yadav, D.V. and Singh, P.N. (1996). Soil factors affecting growth parameters and juice quality in sugarcane grown on salt affected soils. *Indian Journal of Sugarcane Technology,* 10(2): 139-140.

FAO (1991). *FAO/Netherlands Conference on Agriculture and Environment.* S-Hertogenbosch, The Netherlands, 15-19 April 1991.

Greenland, D. J. and Szabolcs, I. (eds) (1994). *Soil Resilience and Sustainable Land Use.* CAB International. Wallingford, U.K.

Lal, K. and Swarup, A. (2004). Effect of afforestation and fertilizer use on functional pools of carbon in alkali soil. In Extended summaries: International Conference on Sustainable Management of Sodic Lands, Lucknow, Feb. 9-14. pp.212-214.

Larson, W. E. and Pierce, F.J. (1994). The dynamics of soil quality as a measure of sustainable management. In: Defining soil quality for a sustainable environment (JW Doran, DC Colemann, DF Bezdicek and BA Stewart Eds.), pp. 3-21. Soil Sci. Soc. Am. Publ. No. 35, Madison, WI.

Mongia, A.D., Dey, P. and Singh, Gurbachan (1998). Ameliorating effect of forest trees on a highly sodic soil in Haryana. *Journal of Indian Society of Soil Science,* 46 (4): 664-668.

Swarup, A. (1995). Management and Balanced Use of Inputs in Achieving Maximum Yield in Salt Affected Soils. *Fertiliser News,* 40(11):39-47.

Swarup, A. (1998). Emerging soil fertility management issues for sustainable crop productivity in irrigated systems. In: Long-Term Soil Fertility Management through Integrated Plant Nutrient Supply. Eds. Swarup et.al pp. 54-68. IISS, Bhopal.

Swarup, A. (2004). Chemistry of sodic soils and fertility management. In Advances in Sodic Land Reclamation. International Conference on Sustainable Management of Sodic Lands, Lucknow, Feb. 9-14. pp. 27- 52.

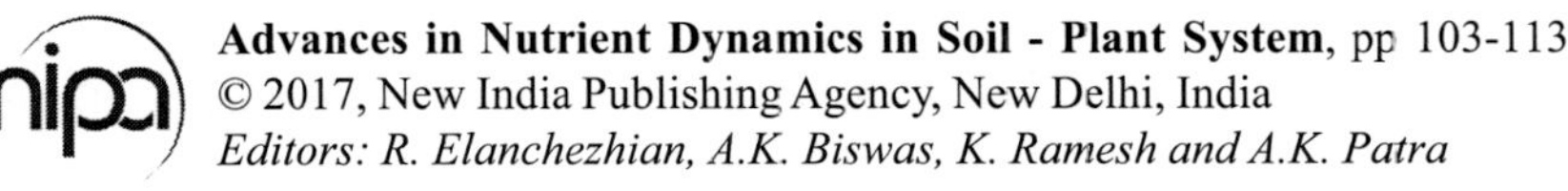
Advances in Nutrient Dynamics in Soil - Plant System, pp 103-113

Editors: R. Elanchezhian, A.K. Biswas, K. Ramesh and A.K. Patra

10

Prospects of Indigenous Sources of Potassium for Crop Production

A.O. Shirale, B.P. Meena, S. Srivastava, A.K. Biswas, A.K. Patra and A. Subba Rao

ICAR-Indian Institute of Soil Science, Nabi Bagh, Bhopal – 462038, India

Since the 1960s, the world population has doubled from three to seven billion, and this trend will persist in the coming decades. Because of this rapid expansion, a massive increase in crop production is required to meet the food and energy demands of future generations. Potassium is the third important plant nutrient after nitrogen and phosphorous. It is seventh most abundant element in the lithosphere. Being chemically active metal, it is never found in its pure elemental state in nature. Potassium (K) plays a particularly crucial role in a number of physiological processes vital to growth, yield, quality, and stress resistance of all crops. The K content of Indian soils varies from less than 0.5–3.0%. The average total potassium content of those soils is 1.52% (Mengel and Kirkby 1987). Potassium content of Indian soils has traditionally been considered as adequate, but in the recent years, the importance of K and the need for its continuous optimal availability for the better crop production has been observed as deficient due to the hidden hunger of K (Leaungvutiviroj *et al.*, 2010; Khawilkar and Ramteke, 1993). Srinivas Rao *et al.* (2011) reported negative balance of K to the extent of 3 million tonnes annually in Indian soils. On average, the surface layer contains 2.6% potassium. The availability of K to the plants depends directly on the concentration of K in soil solution and indirectly on soil (Sparks and Huang1985; Goldstein 1994). The increase in potassium consumption leading to high importing dependence of K fertilizers and the lack of other sources to agriculture have risen interest on the use of indigenous sources of potassium. Considering this, an attempt has been made in this chapter to summarise the possible ways of using indigenous sources of potassium in agriculture. The purpose of this review is first to present the theoretical background that justifies the use of indigenous sources of K as fertilizer and secondly to provide an overview of the results of existing published trials using indigenous rock K as an alternative to conventional fertilizers for sustainable crop production and highlights the need for strategic and consistent future trials.

Role of Potassium in Crop Production

Potassium is an important element for plant and animal on earth's crust. The potassium is involved in more than 60 enzymatic systems in the plant and is required for synthesis of proteins, vitamins, starch and cellulose. It plays important role in photosynthetic process through which plant gets energy for growth and development. Potassium controls opening and closing of leaf stomata and is thus essential for tissue water balance and water use efficiency. Potash increases the yield and quality of agricultural produce, enhances the ability of plants to resist biotic stress (disease, pests and insect attacks) and abiotic stress (cold and drought). Potassium also increases nitrogen fixation in leguminous plants. In carbohydrate rich crops like sugarcane, potato and sugar beet, potassium has special importance owing to its vital role in formation and translocation of starch and sugar. With adequate supply of potassium, cereals produce plump grains and strong stalks. Potassium also enhances the flavor, colour and size of fruits and tubers. In respect of major cereal crops and other commercial crops, the uptake of potassium in kg per tonne of main produce is higher than that of nitrogen.

Forms and Fixation of Potassium in Soils

Potassium is the major nutrient and also a most abundant element in soils but the K content of the soil varies from place to place based on physicochemical properties of soil. Potassium exist in soil in different forms viz., water soluble, exchangeable, non-exchangeable (fixed), mineral K, lattice K and total K. But these forms are not homogeneously distributed in soils. Its amount in soil depends on the parent material, degree of weathering, K gains through manures and fertilizers and losses due to crop removal, erosion and leaching. Usually the amounts of non-exchangeable and total K present in the soil are high compared to water soluble and exchangeable K. The dynamics of potassium in soil depends on the magnitude of equilibrium among various forms and mainly governed by the physicochemical properties of soil. These fractions exist in dynamics equilibrium among themselves and these forms in turn govern the K nutrition in crops. According to increasing order of plant availability, soil K exists in four forms i.e. mineral (5000-25000 ppm), non-exchangeable (50- 750 ppm), exchangeable (40-600 ppm), and solution (1-10 ppm). Soils that are rich in vermiculite and micas can have large amounts of non-exchangeable K, whereas soils containing kaolinite, quartz and other siliceous minerals contain less available and exchangeable K (Martin and Sparks, 1985). The distribution of K forms in the soil and the equilibrium between them determine the K status of the soil and the potential of K supply to plants

(Srinivasarao *et al.*, 2000). There are physical, chemical, biological and climatic factors affecting K forms and equilibrium of K in soil, which could be related to clay mineralogy (Srinivasa *et al.*, 2000), texture (Pal *et al.*, 2001), moisture (Zeng and Brown, 2000), cation exchange capacity (Sardi and Csitari, 1998), pH (Sahu and Gupta, 1987), and concentrations of other ions (Zawartka *et al.*, 1999) of the soil. Besides these soil properties, fertilization and cropping are the most important management factors that influence K equilibrium in soils (Singh *et al.*, 2002).

In addition to releasing K, soil minerals can also fix K, significantly affecting K availability. This involves the adsorption of K ions onto sites in the interlayer's of weathered sheet silicates, such as illite and vermiculite. The degree of K fixation in soils depends on the type of clay mineral and its charge density, moisture content, competing ions, and soil pH. Montmorillonite, vermiculite, and weathered micas are the major clay minerals that tend to fix K (Sparks, 1987). Additionally, soil wetting and drying also significantly affects the K fixation. The fixation process of K is relatively fast, whereas the release of fixed K is very slow due to the strong binding force between K and clay minerals (Oborn *et al.*, 2005). Whether a soil fixes or releases K highly depends on the K concentration in the soil solution (Schneider *et al.*, 2013).

Why Focus on Indigenous Sources

The use of fertilizers in developing countries has exceeded levels of current use in developed countries (FAO, 2011). Because of worldwide concern for food security and the positive impact of fertilizer on crop yields, fertilizer is made readily available in developing countries, and its use increases to compensate for dramatic declines in fertility over time (Arnason *et al.*, 1981). About 90 to 98% of K reserves in soil exists as insoluble K-minerals like feldspar and mica, which constitute a biggest reservoir of soil K. India is totally dependent on foreign countries for potassium supply hence, it becomes costly nutrient in India as it involves cost on import coupled with subsidy to farmers. India ranks 4th in consumption of potassium fertilizers, on an average 2.5 million tonnes of potassium are being imported during 2015 from different countries (FAI, 2015). In India, until the eighties, potassium did not receive much attention because of the general belief that Indian soils were well supplied with potassium. Studies on potassium mining in India revealed the large scale negative K balances in Indian agriculture (Srinivasrao *et al.*, 2011). Bensal and Moiniddin (1999) reported that crop removal of potassium often equals or exceeds that of nitrogen estimated that a total of 13.7 million tons of K_2O year^{-1} are being removed by crops in India against the fertilizer consumption of only 1.57 million tons of K_2O. After considering all the organic and inorganic additions, a net deficit of 7.049 million tonnes K_2O year^{-1} has been estimated which means a depletion of Indian soils.

For attaining and maintaining the self-sufficiency in food grain production, potash fertilizers has played important role in Indian agriculture. One possible alternative could be to fully exploit the reserve of K in the soil. Soil has rich reserves of K, among which only 1–2% can be directly absorbed by plants (Mikhailouskaya and Tcherhysh, 2005). In spite of the fact that a total indigenous resource of potash of the order of 20398 million tonne from various sources viz. evaporate deposit, brines, glauconite etc. and mica constitute of 532237 tonnes exist, but there is no technology available for production of potash fertilizer in the country. In order to reduce the dependence on imported potash, there is need to explore indigenous alternative source of potash. In India, sylvinite deposits do not exist but resources exist (Table 1) as polyhalite (16.2 billion tonnes), sylvite (2.5 billion tonnes) and glauconite (2 billion tonnes). Rajasthan alone contributes 94% to the total resources, followed by Madhya Pradesh (5%) and Uttar Pradesh (1%) (Indian Mineral Year Book, 2013). Finally, the main challenge in the use of rock fertilizers is to increase the solubility of rocks and minerals and enhance nutrient release from fertilizers. This can be done by physically modifying (changing the surface area of the minerals through fine grinding) and/or chemically (modifying the surfaces through acidulation), for enhancing the solubility and nutrient release from minerals. Hence, there is urgent need to divert the attention towards the research on use of indigenous sources as fertilizer in agriculture which in turns helps in reducing the dependence on import of potassic fertilizer import and can saves money on import.

Table 1: Reserves/Resources of Potash (By Grades/States) (In million tonnes)

Grade/State	Resources			Total
	Indicated	Inferred	Reconnaissance	
All India : Total	18142	3652	22	21816
By grades				
Glauconite	878	1068	22	1968
Polyhalite	13985	2179	—	16164
Sylvite	2072	404	—	2476
Unclassified	1206	—	—	1206
By states				
Madhya Pradesh	1206	—	—	1206
Rajasthan	16936	3461	22	20419
Uttar Pradesh	—	190	—	190

Source: Anonymous (2013)

Potential Indigenous Sources of K

Schoenite: Central Salt and Marine Chemical Research, Bhavnagar had developed technology to make Potassium Magnesium Sulphate also known as schoenite, which was found suitable for many crops. However, commercial production of potassium shoenite could not take up due to the uneconomical production costs.

Recently, Rathor *et al.* (2014) revealed that the indigenously developed potassic fertilizer schoenite showed significant effects on yield and yield attributing characteristics of groundnut.

***Feldspar*:**Feldspar is the most abundant rock forming mineral in nature and makes up an estimated 60% of earth crust. The feldspar comprises a group of minerals consisting of silicates of aluminum with varying amounts of potassium (orthoclase and microcline). Among the various feldspar, potassium feldspar is the most common and contains up to 13% potassium as K_2O. Moreover, feldspar can be converted into fertilizer. Also feldspar can be successfully used as fertilizer by giving certain treatments of heat, modifying physical structure or by inoculation with microbial culture.

***Vermiculite*:** The total resources of vermiculites 2.44 million tonnes (UNFC, 2005) of which more than 72% are placed under reserves category. Major resources are located in Tamil Nadu (77%), followed by Madhya Pradesh (11%), Andhra Pradesh (5%), Karnataka (4%), Rajasthan (2%), Jharkhand (1%).

***Sulphate of potash*:** K_2SO_4 is allowed if it is produced from natural sources. Much of the current production of K_2SO_4 comes from the Great Salt Lake. It may not undergo further processing or purification after mining or evaporation other than crushing and sieving. It generally contains 40% K and 17% S.

***Glauconite*:** Greensand is the name commonly applied to a sandy rock or sediment containing a high percentage of the green mineral glauconite. Because of its K content (up to 5% K), greensand has been marketed for over 100 years as a natural fertilizer and soil conditioner. Glauconite occurs in sandstone as a minor component and its K_2O content varies considerably. Because glauconitic sandstone can release potassium slowly, it can be considered as a sustainable source of soil-soluble potassium for long time and reduces the demand for costly K fertilizers. India has glauconite reserves of potash over 2000 million tonnes in states of U.P, M.P, Rajasthan and Gujarat.

***Waste mica*:** It is an important group of rock forming minerals, composed of chiefly orthosilicates of aluminium with potassium, magnesium, iron containing hydrogen and sometime vanadium, lithium chromium, etc. India is fortunate to have the world's largest deposit of micas distributed over a total area of about 3888 sq. km in Munger district of Bihar Koderma and Giridih districts of Jharkhand. The good quality micas are used as electrical insulators. However, waste micas, which are generated in large quantities during the cleaning of raw micas after their mining, are not used in agriculture, though it contains significant amount of K (8 – 12% K_2O) and are dumped near the mica mines. These materials can effectively be used as a source for potassium, if modified or altered by some suitable chemical or biological means.

Polyhalite: Polyhalite occurs in sedimentary marine evaporates, consisting of a hydrated sulfate of K, Ca and Mg with the formula: $K_2Ca_2Mg(SO_4)_4 \cdot 2(H_2O)$. In addition to K it also supplies Ca, Mg and sulphur in sufficient quantities. In general it contains 48% sulphate, 14% K_2O, 6% MgO, and 17% Ca.

Techniques for use of Indigenous Potassium Sources as Fertilizer

Use of K solubilizing microorganism

There are numerous bacteria and fungi species reported by several workers that can solubilize the potassium from K bearing rocks and minerals. The microorganisms like *Pseudomonas* sp., *Burkholderia* sp., *Acidothiobacillicus ferrooxidans*, *Bacillus mucilaginosus*, *Bacillus edaphicus*, *Bacillus megaterium*, *Bacillus circulans*, *Paenibacillus* sp., *Pseudomonas* and *Burkholderia* are able to release K from K-bearing minerals as reported by Sheng *et al* (2002), Lian *et al*. (2002), Rajawat *et al*. (2012), Liu *et al*. (2012), Basak and Biswas (2012) Singh *et al*. (2010). The mechanism involved in K solubilization is organic acids secreted by microbes either direct dissolution of K-containing primary minerals or by chelating the primary mineral's silicon ions which brings the K into solution (Bennett *et al.,* 1998). Hence, inoculation of potassium solubilizing microorganisms has got importance in recent past. The research work done by Sheng *et al*. (2002); Han and Lee (2005); Han and Supanjani Lee (2006); Basak and Biswas (2009), Abou-el-Seoud and Abdel-Megeed (2012) showed beneficial effects of K solubilizers with mica or feldspar on potassium supply. This indicates that these microorganisms can effectively enhance the release of K from clay minerals. At present, there is little information available with regards to efficacy of K solubilizers with field application hence; there is urgent need to conduct research on agronomic field trials for better use of K solubilizers in agriculture.

Use of organic matter

At present huge quantities of crop residues which is generally burnt in the field by the farmers. This residue can effectively be used by converting them into K enriched manure through composting with glauconite, mica, feldspar along with microbial inoculants. Respiration by plant roots and microbial degradation of organic matter can elevate carbonic acid concentrations in the soils and ground water, leading to an increase in the weathering rates of minerals. Badr (2008) prepared K enriched compost with rice straw, silicate dissolving bacteria and feldspar and found that the concentration of available K released from feldspar increased markedly through composting. It was also found that conjunctive use of feldspar compost plus silicate dissolving bacteria recorded maximum potassium use efficiency, total K uptake and yield of tomato (Fig 1). Nishanth and Biswas (2008) also observed similar kind of results with mica enriched compost.

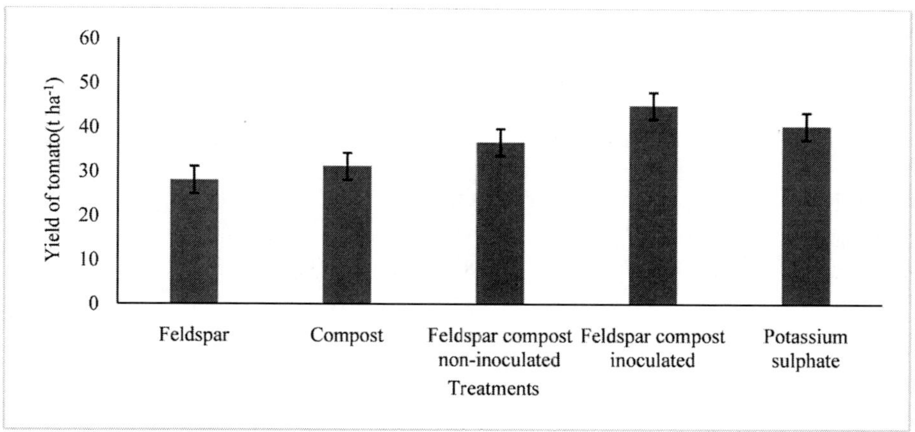

Fig. 1. Yield of tomato as influenced by K enriched compost

Physical modification

Scovino and Rowell (1988), used finely ground feldspar (< 100 mesh, largely sanidine, 7% K) as fertilizers in a pasture experiment, they concluded that the feldspars may be valuable as a slow release fertilizer in low input agricultural system particularly on leached soils of low cation exchange capacity. Also, Priyono and Gilkes (2008) evaluated the effectiveness of intensively milled gneiss and potassium feldspars as K-fertilizers through a glasshouse experiment with ryegrass. They found that the application of K-silicate rock fertilizer will be most advantageous for amending K-deficient soils. Crushed rocks can be considered multinutrient fertilizers carrying silicate minerals containing other macro and micronutrients in variable concentrations. Among these, the main are phosphorus and potassium (Wilpert and Lukes, 2003; Ribeiro *et al*., 2010). In addition to the supply of macronutrients, some crushed rocks promote changes in soil characteristics such as acidity. Srinivas Rao and Subba Rao (1999) reported that addition of different sized particles of glauconite resulted in a significant increase in the dry matter yield as well as K uptake. Dry matter yield response was more in particle-size of 35 ASTM, beyond this size the yield increase was significant but marginal. Potassium concentration and uptake was more under decreasing size of particles (Table 2). Similar kinds of results were also reported by Karimi *et al.* (2011). Hence, glauconitic sandstone can release potassium slowly, it can be considered as a sustainable source of soil-soluble potassium for long time and reduces the demand for costly potassic fertilizers.

Table 2: Effect of particle size fractions of glauconite on dry matter yield, K concentration and K uptake by pearl millet in Neubaur experiment

Particle size of glauconite	Dry matter yield ($g\ kg^{-1}$)	K-content (%)	K-uptake ($mg\ kg^{-1}$)	K release from unit glauconite ($g\ kg^{-1}$)
10 ASTM	5.98 (0.12)	1.29(0.05)	77.3(1.32)	0.101
35 ASTM	7.58(0.12)	1.49(0.05)	113.0(1.49)	0.153
60 ASTM	7.66(0.14)	1.84(0.04)	141.0(1.35)	0.196
100 ASTM	7.91(0.17)	2.37(0.06)	196.0(3.49)	0.266
Control	5.41(0.07)	0.57(0.04)	31.0(1.30)	—
CD (5%)	0.08	0.05	3.6	—

Source: Srinivas Rao and Subba Rao (1999)

Acidulation treatments

Organic acids may affect mineral weathering rates by at least three mechanisms: by changing the dissolution rate through decreasing solution pH or forming complexes with cations at the mineral surface; by affecting the saturation state of the solution with respect to the mineral; and by affecting the speciation in solution of ions such as Al^{3+} that themselves affect mineral dissolution rate. The levels of organic acids are affected by a number of independent variables such as pH and redox potential influence on the dynamics of K release from silicate rocks (Song and Huang 1988). Ugolini and Sletten (1991) added that humic, fulvic and other organic acids have been shown to be aggressive weathering agents in soils, especially with respect to the dissolution of clay minerals. Various organic acids can effectively dissolve minerals and chelate metallic cations. Generally, the great effect of organic acid on dissolution of rocks and minerals is attributed to the presence of hydrogen ions and the formation of cation complexes. The structural cations, released from minerals as a result of the attack of hydrogen ions, tend to form cation-organic complexes with oxalic acid, which has OH^- and $COOH^-$ groups in the ortho position. The chemisorptions of the cation- organic complexes on the mineral surface cause a shift of electron density toward the frame work of the mineral. This charge transfer increases the electron density of the cation oxygen bonds and makes them more susceptible to hydrolysis (Chen, 2000). Acidulation with nitric and sulphuric acid enhancing the release of K for plant uptake through enhancing the solubility and K- release from rocks (Hinsinger *et al.*, 1993). Wafaa *et al.* (2015) reported that acidulation of feldspar with organic acids like humic, fulvic and citric acids is the promising option for supplying potassium through feldspar by changing the dissolution rate through decreasing solution pH or forming complexes with cations at the mineral surface; by affecting the saturation state of the solution with respect to the mineral; and by affecting the speciation in solution of ions such as Al^{3+} that themselves affect mineral dissolution rate.

Conclusion

The studies carried out in past showed that indigenous sources like glauconite, mica and feldspar can release potassium and can be used as alternative to conventional fertilizer. The use of microbial inoculants, organic matter application, acidulation and physical modification can be promising options for efficient use of indigenous sources of potassium. This can save huge amount of revenue on import of conventional fertilizers and subsidy on it. There is need to study the available technologies for use of indigenous sources of potassium in field condition.

References

Abou-el-Seoud, A. Abdel-Megeed. (2012). Impact of rock materials and biofertilizations on P and K availability for maize (*Zea maize*) under calcareous soil conditions. *Saudi Journal of Biological Sciences,* 19: 55–63.

Anonymous (2013). Indian Mineral Year Book. Part- III : Mineral Reviews Part- III : 52th edition. Government of India, Ministry of Mines, Indian Bureau of Mines.

Badr, M.A. (2006). Efficiency of K. feldspar combined with organic materials and silicate dissolving bacteria on tomato yield. *Journal of Applied Science Research*, 2(12):1191-1198.

Basak, B. and Biswas, D.R. (2012). Modification of Waste Mica for Alternative Source of Potassium: Evaluation of potassium release in soil from waste mica treated with potassium solubilizing bacteria (KSB). Academic Publishing, ISBN: 978-3-659-29842-4.

Basak, B.B. and Biswas, D.R. (2009). Influence of potassium solubilizing microorganism (*Bacillus mucilaginosus*) and waste mica on potassium uptake dynamics by sudan grass (*Sorghum vulgare* Pers.) grown under two Alfisols. *Plants and Soil*, 55: 235–55.

Bennett, P.C., Choi, W.J. and Rogers, J.R. (1998). Microbial destruction of feldspars. *Mineral Magazine*, 8: 149–50.

Bensal S.K., Moiniddin P. (1999). Crop responses to balanced nutrition management in India. Proceedings of International Symposium of the soil and Water Research Institute, University of Tehran (Iran), 57-68.

Chen Y.X., Lin Q., Lu F., He Y.F. (2000). Study on detoxication of organic acid to radish under the stress of Pb and Cd. *Acta Scientiae Circumstantiae*, 467–72.

FAI (2015). Fertilizer Statistics 2015–2016. The Fertilizer Association of India, New Delhi

FAO (2011). Current world fertilizer trends and outlook to 2015. Rome: Food and Agriculture Organization of the United Nations.

Goldstein, A.H. (1994). Involvement of the quino protein glucose dehydrogenase in the solubilization of exogenous mineral phosphates by gram-negative bacteria. In: Phosphate in microorganisms: cellular and molecular biology. Eds Torriani Gorni, A., Yagil, E., Silver, S., Washington: ASM Press; ISBN 1555810802, p. 197–203.

Han, H.S. and Lee, K.D. (1964). Phosphate and potassium solubilizing bacteria effect on min-eral uptake, soil availability and growth of eggplant. *Research Journal of Agriculture and Biological Sciences*,1: 176–80.

Han, H.S. and Supanjani Lee, K.D. (2006). Effect of co-inoculation with phosphate and potassium solubilizing bacteria on mineral uptake and growth of pepper and cucumber. *Plant Soil and Environment*, 6: 52-130.

Hinsinger P. and Jaillard, B. (1993). Root-induced release of interlayer potassium and vermiculitization of phlogopite as related to potassium depletion in the rhizosphere of ryegrass. *Journal of Soil Science*, 44: 525–34.

Karimi, E., Abdolzadeh, Hamid Reza Sadeghipour and Arash Aminei (2011). The potential of glauconitic sandstone as a potassium fertilizer for olive plants. *Archives of Agronomy and Soil Science*, 1–11.

Khanwilkar, S.A. and Ramteke, J.R. (1993). Response of applied K in cereals in Maharashtra. *Agriculture*, 84–96.

Leaungvutiviroj C., Ruangphisarn P., Hansanimitkul, .P, Shinkawa, H., Sasaki K. (2010). Development of a new biofertilizer with a high capacity for N_2 fixation, phosphate and potassium solubilization and auxin production. *Biosciences Biotechnology Biochemistry*, 74(5): 1098–101.

Lian, B., Fu, P.Q., Mo, D.M. and Liu, C.Q. (2002). A comprehensive review of the mechanism of potassium release by silicate bacteria. *Acta Mineralogica Sinica*, 22: 179.

Liu, D., Lian, B. and Dong, H. (2012). Isolation of *Paenibacillus sp.* and assessment of its potential for enhancing mineral weathering. *Geomicrobiology Journal*, 29(5): 413–21.

Martins, H.W and Sparks D.L. (1983). Kinetics of non exchangeable K release from two coastal plain soils. *Soil Science Society of America Journal*, 47: 883-887.

Mengel, K. and Kirkby, E.A. (1987) Principles of plant nutrition. International Potash Institute:Worblaufen-Bern.

Mikhailouskaya, N. and Tcherhysh, A. (2005). K mobilizing bacteria and their effect on wheat yield. *Latvian Journal of Agronomy*, 8: 154–157.

Nishanth, D. and Biswas, D.R. (2008). Kinetics of phosphorus and potassium release from rock phosphate and waste mica enriched compost and their effect on yield and nutrient uptake by wheat (*Triticum aestivum*). *Bioresource Technology*, 99: 3342–3353.

Oborn, I., Andrist Rangel, Y., Askekaard, M., Grant, C.A., Watson, C.A. and Edwards, A.C. (2005). Critical aspects of potassium management in agricultural systems. *Soil Use and Management*, 21: 102–112.

Pal, D.K., Srivastava, P., Durge S.L. and Bhattacharyya, T. (2001). Role of weathering of fine-grained micas in potassium management of Indian soils. *Applied Clay Science*, 20: 39–52.

Priyono, J. and Gilkes, R.J. (2008). High-Energy Milling Improves the Effectiveness of Silicate Rock Fertilizers: A Glasshouse Assessment. *Communication in Soil Science and Plant Analysis*, 39 (3&4): 358-369.

Rajawat, M.V.S., Singh, S., Singh, G. and Saxena A.K. (2012). Isolation and characterization of K-solubilizing bacteria isolated from different rhizospheric soil. In: Proceeding of 53rd Annual Conference of Association of Microbiologists of India, p. 124.

Rathore, S.S., Chauchary, D.R., Vaisya, L.K., Kalpana Shekhawat and Bhatt, B.P. (2014). Schoenite and potassium sulphate: Indigenous potash fertilizer for groundnut (*Arachis hypogaea* L.). *Indian Journal of Traditional Knowlwdge*, 13(1): 222-226.

Ribeiro, L.S., Santos, A.R., Souza, L.F.S., Souza, J.L. (2010). Rochassilicaticasportadoras de potássiocomofontes do nutriente para as plantas. *Revista Brasileira de Ciência do Solo,* 34: 891-897.

Sahu, S. and Gupta, S.K. (1987). Fixation and release of potassium in some alluvial soils. *Journal of Indian Society of Soil Science*, 35: 35–40.

Sardi, K. and Csitari, G. (1998). Potassium fixation of different soil types and nutrient levels. *Communication in Soil Science and Plant Analysis*, 29: 1843–1850.

Schneider, A., Tesileanu, R., Charles, R. and Sinaj, S. (2013) Kinetics of soil potassium sorption-desorption and fixation. *Communication in Soil Science and Plant Analysis,* 44: 837–49.

Sheng, X.F., He, L.Y. and Huang, W.Y. (2002). The conditions of releasing potassium by a silicate dissolving bacterial strain NBT. *Agriculture Sciences in China*, 1: 662–669.

Singh, G., Biswas, D.R. and Marwah, T.S. (2010). Mobilization of potassium from waste mica by plant growth promoting rhizobacteria and its assimilation by maize (*Zea mays*) and wheat (*Triticum aestivum* L.). *Journal Plant Nutrition*, 33: 1236–51.

Singh M., Tripathi, A.K. and Reddy, D.D. (2002). Potassium balance and release kinetics of non-exchangeable K in a Typic Haplustert as influenced by cattle manure application under a soybean-wheat system. *Australian Journal of Soil Research*, 40: 533-541.

Song, S.K. and Huang, P.M. (1988). Dynamics of potassium release from potassium-bearing minerals as influenced by oxalic and citric acids. *Soil Science Society of America Journal*, 52: 383-390.

Sparks, D.L. and Huang, P.M. (1985). Physical chemistry of soil potassium. In: Munson RD, *et al.*, editors. Potassium in agriculture. Madison, WI: ASA, p. 201–76.

Sparks, D.L. (1987). Potassium dynamics in soils. *Advances in Soil Science*, 6: 1–63.

Srinivas Rao, CH. Satyanarayana, T. and Venkateswarlu, B. (2011). Potassium mining in Indian agriculture: Input and output balance. *Karnataka Journal of Agricultural Sciences*, 24 (1): 20-28.

Srinivasa Rao, C. and Subba Rao, A. (1999). Characterization of indigenous glauconitic sandstone for its potassium supplying potential by chemical, biological, and electroultrafiltration methods. *Communication Soil Science Plant Analysis*, 30(7&8): 1105-1117.

Srinivasa Rao, Ch., Subba Rao, A. and Rupa, T. (2000). Plant mobilization of reserve potassium from fifteen smectitic soils in relation to mineralogy and soil test potassium. *Soil Science,* 165: 578–586.

Wafaa, S.M.A. and Mona, A.O. (2015). Impact of Feldspar Acidulation on Potassium Dissolution and Pea Production. *International Journal of ChemTech Research*, 8(11): 1-10.

Wilpert, K. and Lukes, M. (2003). Ecochemical effects of phonolite rock powder, dolomite and potassium sulfate in a spruce stand on an acidified glacial loam. *Nutrient Cycling in Agroecosystem*, 65:115-127.

Zawartka, L., Huszcza-Ciolkowska, G. and Szumska, E. (1999). Effects of Poly and Orthophosphates on the Dynamics of some Macro and Micronutrient Elements in Soil Material of Varied pH III. Potassium. *Communications in Soil Science and Plant Analysis*, 30: 655-661.

Zeng, Q. and Brown, P. H. (2000). Soil potassium mobility and uptake by corn under differential soil moisture regimes. *Plant and Soil,* 221: 121-134.

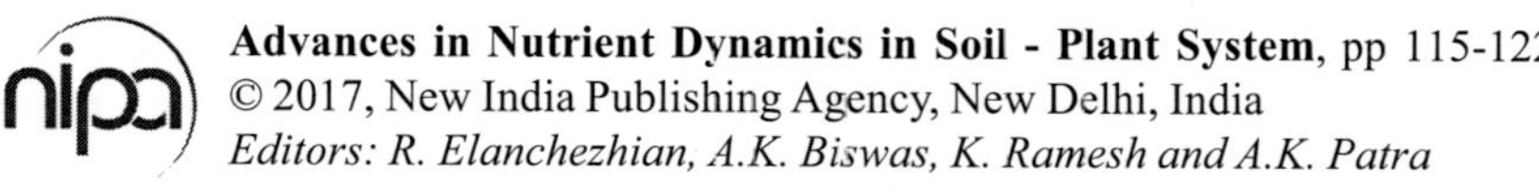
Advances in Nutrient Dynamics in Soil - Plant System, pp 115-122

Editors: R. Elanchezhian, A.K. Biswas, K. Ramesh and A.K. Patra

11

Nanotechnology Applications in Soil and Plant Nutrition Research

S.Kundu, K. Ramesh and R. Elanchezhian

ICAR-Indian Institute of Soil Science, Nabi Bagh, Bhopal – 462038, India

Nanotechnology is a field of study which deals with materials at nano-scale like nanometer which is one billionth of a meter (1 nm = 10^{-9} meter). At nano-scale, physics, chemistry and biology converge towards the same principles and tools. Nanotechnology is domain in science and technology, which is multidisciplinary in nature and any scientist, engineer, medical doctor can do research on nanotechnology. Presently, it is more of a science and less of a technology because of a lack of proper tools for studying the properties of nano-material. Nano-particles (NPs) are materials that are small enough to fall within the nanometric range, with at least one of their dimensions being less than a hundred nanometres. This reduction in size brings about significant changes in their physical properties with respect to those observed in bulk materials. The presence of naturally occurring NPs is implicated in interplanetary and interstellar space (Hochella Jr, 2002) as well as our earth in which life from the beginning has evolved in their presence (Becker and Reitmeijer, 2006). Currently it is observed that NPs can be synthesized from many biological organisms including plants.

Nano particles (NPs) are generally defined as materials that are <100 nm size in at least one dimension and these particles can be of three dimensional (Spherical, Cuboids etc), or two dimensional (Ultra thin film) or one dimensional (fine rods) in nature. The properties of NPs viz. chemical (reactivity, solubility etc), mechanical (elasticity, hardness etc), electronic (conductivity, redox behavior) and nuclear (magnetic) properties often change as a function of size. These changes can be, and often are dramatic, used in various novel applications. Such properties make them exceptional in scientific and commercial value and their anticipated enigmatic behaviour in already extraordinarily complex earth environment.

The unique physical properties of nanomaterials can be successfully exploited for novel applications. As per the British Standard Institution (BSI, 2005) and American Society for Testing and Materials (ASTM, 2006), nanotechnology is defined as "Design, characterization, production and application of structure, devices and systems controlling shape, size and composition at the nanoscale". Nanoparticles can also be defined based on the size at which fundamental properties differ from those of the corresponding bulk material (Banfield and Zhang, 2001). NPs have the property of colloidal solutions since their size range overlap with the size of colloidal fractions between 1 nm and 1 mm (Buffle, 2006). Nanomaterials can be made from a range of materials like magnetic materials, metal oxide ceramics and silicates, emulsions, dendrimers etc. Nanomaterials exist since the evolution of the earth. Some of the naturally occurring materials having the dimension in nano-scale are viz. atom/molecules (< 1nm), colloids (1-1000 nm), DNA double helix (~2 nm diameter), virus (10-100 nm), red blood cell (2000-5000 nm), bacteria (250 nm-1000 nm) tissue/cell (10000 nm) and width of human hair (80000 nm).

Owing to the smaller size of nanoparticles, the physical, chemical, electronic properties of nano-structures changes and are very different from that of their bulk counterparts. There are more atoms on the surface of nano-particles compared to the interior of the particle which lead to large surface to volume ratio of particular atoms/ molecules, which in turn leads to higher reactivity of nano-particles. This can be best illustrated in a cube of 1 mm size which if broken into cubes of 1 nm size, the total volume remains the same, but the surface area is increased by 10^6 times. On the contrary bulk or large particle has less proportion of the atoms that are associated with its surface in comparison to nanoparticles. Thus the behavior of the atoms on the surface of the nanoparticle becomes more potent as compared to those atoms that are inside the particle as surface area increases in comparison to the volume. Moreover, these particles exhibit quantum mechanical behavior once they become small enough and do not obstruct light since their size is smaller than the order of wavelength.

The chemical properties of NPs is attributed mainly due to more interaction between atoms in intermixed materials as a consequence of large surface to volume ratio, which may lead to increased strength, increased heat resistance etc. It is observed that the melting point of gold in bulk is 1337 °K whereas melting point of gold nano-particle (~2 nm) is 650 °K. This indicated that the magnetic properties of nano-clusters are very different from that of the corresponding bulk material. For example nanoclusters of certain materials like Pd, Na, K and Rh are ferromagnetic, where as in bulk form, these elements are paramagnetic. Super-paramagnetism is a phenomenon that arises from the small size of nanoclusters. Discretezation of energy level is an important property of nano-particles. In bulk, the overlapping of the molecular orbitals of a large number of atoms results in a continuum of energy levels or energy band. But in nano-particles, due to fewer atoms the overlapping of their orbiltals is not much and thereby exhibit discrete energy levels. Thus discretization of the electronic energy levels takes place in nanoparticles along with an increase in the electronic band gape energy, which in turn, results in interesting optical and electronic properties in nano-particles.

Surface plasmon resonance (SPR) is the resonant oscillation of conduction electrons at the interface between a negative and positive electric charge of the material stimulated by incident light. SPR in subwavelength scale nanostructures can be polaritonic or plasmonic in nature. SPR is the basis of many standard tools for measuring adsorption of material onto planar metal (typically gold or silver) surfaces or onto the surface of metal nanoparticles. It is the fundamental principle behind many color-based biosensor applications. SPR is a collective excitation of the electrons in the conduction band near the surface of the nano-particles. The surface of the nano-particles is like plasma having free electrons in the conduction band with positively charged nuclei. The position of the specific Plasmon absorbance band indicates the presence of specific size of nano-particles. For example, a single absorbance band at 529 nm confirms the presence of gold nano-particles (7-20 nm) and specific plasmon absorbance at 412 nm (γ_{max}) indicates the presence of silver nano-particles.

The decrease in size of particle reduced to the nano level, increases the the ratio of surface to the bulk atoms and thereby increases energy of the system as a whole. However, it leads to a decrease in the system stability. Numerous studies indicated that nano-particles will aggregate after their entry into water. The reported size of selected engineered nano materials (ENMs) in water indicated that their size increases after entry into aqueous system (Table 1). Hence, the stabilization of ultra fine or nano-particles in suspension is very important for both controlling the particle size and for developing process based application of these suspensions to achieve a desired result. The formation, stabilization and sedimentation of nano-particles depend upon the discreet steps of nucleation, condensation and coagulation into larger particles. Therefore, stabilization requires the optimization of these competing factors.

Table 1: Hydrodynamic particle size of ENMs

ENMs	Individual particle size (nm)	Hydrodynamic particle size (nm)
Ag	26.6 ± 8.8	216
Cu	26.7 ± 7.1	94.5-447.1
Al	41.7 ± 8.1	4442
Co	10.5 ± 2.3	224-742
Ni	6.1 ± 1.4	44.9-446.1
TiO_2	20.5 ± 6.7	220.8-687.5
ZnO	50-70	320 ± 20
SiO_2	10	740 ± 40
Fe_2O_3	5-25	200 ± 10
Fe_2O_3	9.2	46.2
Fe_3O_4	<10	120
CeO_2	8	323-2610
Al_2O_3	60	763
CdSe/ZnS	2.1	12.5

(Adapted from Adhikari *et al.*, 2013)

Stability of NPs in aqueous environments is a key factor controlling their transport and ultimate fate in such environments. Engineered and natural NPs, after entering the aqueous environment, will interact with the ubiquitous natural aquatic colloids which affect the stability and subsequent environmental behavior of both NPs and aquatic colloids. Larger aggregates of NPs will quickly precipitate out and their transport and bioavailability will be greatly restricted. However, well-dispersed NPs will be widely transported and have higher chances to interact with and cause potential harm to organisms.

Another important parameter that has been extensively investigated for stability of nano-particles is the zeta potential of the nano-particles. It is observed that nano-particles with low zeta potentials tend to coagulate or flocculate (Table 2), whereas particles with high zeta potential (negative or positive) will be stable in a suspension. Under such situtation, pH is a major factor determining the zeta potential of nano-particles. The nano-particles exhibit minimum stability (i.e., exhibit maximum coagulation/flocculation) when pH is at point of zero charge (*pzc*) or isoelectric point. When the pH is lower than the *pzc* value, the nano-particle surface is positively charged and the zeta potential will increase with decreasing pH below the *pzc*. Conversely, at pH above *pzc*, the surface is negatively charged and the zeta potential will be more negative with increasing pH (Adhikari *et al.,* 2013).

Table 2. Effect of zeta potential on colloidal stability

Zeta potential (mV)	Stability behaviour of colloids
From 0 to ± 5	rapid coagulation or flocculation
From ± 10 to ± 30	incipient instability
From ± 30 to ± 40	moderate stability
From ± 40 to ± 60	good stability
More than ± 61	excellent stability

(Adapted from American Society for Testing and Materials, 1985)

Natural and incidental NPs dominate the number distribution of particles in the atmosphere in polluted urban, rural, remote continental and marine atmosphere, typical concentration ranges are 10^5 to $4x10^6$, $2x10^3$ to 10^4, 50 to 10^4 and 100 to $400/cm^3$, respectively. (Seinfeld and Pandis, 2006). Most of the NPs in the atmospjere are either as ultrafine particles, a category consisting of particles smaller than 10 nm (referred to as nucleation mode) or those from 10 to 100 nm (referred to as Aitken mode). Particles in this size range originate mainly from combustion and photochemical reactions. For NPs, coagulation is dominated by Brownian (thermal) motion that leads particles to collide with each other, grow in size, and shrink in number. Coagulation rates are low when particles are of same size but increases by orders of magnitude as the differences in size grows. Such nanoparticles are removed from atmosphere mainly by brownian diffusion, gravitational settling (dry deposition) and wet deposition in precipitation to the earth's surface. The wet deposition is highly dependent on the frequency of precipitation events in which the atmospheric NPs may be removed in a matter of minutes to hours (Tiwari and Marr, 2010). The most common types of naturally occurring particles in the atmosphere consists of black carbon or soot, a combustion byproduct which is thought to very harmful because it contains fullerenic compounds and carbon nanostructures.

NPs may be retained by soil matrix or break through soil matrix and reach ground water after entering into soil environment, which is also determined by the properties of NPs and soil. Moreover, there are significant physical and chemical similarities between the most widely manufactured ENPs and naturally occurring nanoparticles, although in a number of cases the exact size, shape, and coatings/surface functional groups may be quite different from ENPs. Earth scientists have been studying at least some major classes of natural nanoparticles for many decades and now nanoscience has become an integral part of many area of science including soil science that goes far beyond the study of clay minerals (Maurice and Hochella, 2008; Theng and Yuan, 2008). Presently this field of geology and nano science is called as nanogeoscience, and have been extensively developed in the last decade (Banfield and Navrotsky, 2001; Hochella, 2002; Hochella, 2008). This is because an exceptionally wide variety of nanoparticle exist on earth, and are in fact ubiquitous in both the biotic and abiotic compartments of earth (Gilbert and Banfield, 2005; Hochella,2008).

Biological synthesis of nanoparticles is an emerging research area, many organisms synthesize nanomaterials. Bacteria in sediments may synthesize electrically conductive pilli, called nanowires, for sensing neighbors or for transferring electrons and energy (Blango and Mulvey, 2009; Gorby *et al.,* 2006). Bacterial reduction of urantyl, U^{6+}(aq), to U(IV) oxide (uraninite) is an important bioremediation strategy (Bargar *et al.,* 2008). Manceau *et al.,* (2008) found that wetland plants, or their symbionts, synthesize copper (Cu) nanoparticles in their rooting zone when grown in contaminated soils, thereby reducing Cu uptake. Such strategy can be effectively used in phytoremediation of contaminated soils.

Studies on interaction of NP with plants is an emerging area of research where limitted information is available. The uptake of many types of NPs in the bacterial cell (prokaryotes) is very much limited as they do not have mechanisms for transport of NPs across the cell wall but in eukaryotes, cellular internalization of NPs occurs through the process of endocytosis and phagocytosis (Moore, 2006). Plants are an important component of the ecological system and may serve as a potential pathway for NPs transport and a route for bioaccumulation into the food chain. Plant cell wall acts as a barrier for easy entry of any external agent including nano-particles into plants cells. The sieving properties are determined by pore diameter of cell wall ranging from 5 to 20 nm (Fleischer *et al.*, 1999). Hence, only nano-particles or nanoparticle aggregates with diameter less than the pore diameter of the cell wall could easily pass through and reach the plasma membrane (Navarro, *et al.,* 2008, Moore, 2006).

NPs may also increase the permeability of plant cell walls under stress and then permeate into the cells (Lin and Xing, 2008). There is also a chance for enlargement of pores or induction of new cell wall pores upon interaction with engineered nano-particles which in turn enhance nano-particles uptake. Further internationalization occurs during endocytosis with the help of a cavity like structure that form around the nano-particles by plasma membrane. They may also cross the membrane using embedded transport carrier proteins or through ion channels. When nano-particles are applied on leaf surface, they enter through the stometal openings or through the bases of trichomes and then translocated to various tissues (Uzu *et al.,* 2010, Eichert *et al.,* 2008).

After entering the cells, NPs may be able to transport between cells via plasmodesmata, which are microscopic channels of plants traversing the cell walls and enabling transport and communication between cells. Plasmodesmata or intercellular bridges were reported to be cylindrical channels about 40 nm in diameter (Tilney *et al.,* 1991). Thus, NPs with diameter less than 40 nm may enter and transport in the plant cells through the plasmodesmata once they are in the plant cells. In the cytoplasm, the nano-particles may bind with different cytoplasmic organelles and interfere with the metabolic processes at that site (Jia. *et al*., 2005). Plant uptake can be a critical transport and exposure pathway of NPs in the environment. Further research is required to understand the interactions between plants and NPs, such as the mechanism of uptake and their translocation at the molecular and cellular level. Such studies will help us understand the plant uptake of ENMs as a potential transport and exposure route and its role in bioaccumulation through the food chain.

There are many unknown areas regarding the environmental fate, transport and ecotoxicity of NPs and therefore, we need strong research support for the development and implementation of environmentally safe nanotechnology. The basic information regarding the health and environmental risks of NPs is still lacking despite several works. Inhalation of NPs has been associated with oxidative stress, inflammation, fibrosis, multifocal granulomatous inflammation and pneumonia, free radical production, reduced cell viability, induction of apoptosis, and subpleural fibrosis, some of which exhibit dose-dependent behavior. NPs can enter any cells by diffusion through cell wall membrane as well as through endocytosis and adhesion. The size distribution of NPs is very important in terms of toxicity in biological system. For example, 20 nm NPs were found to deposit mostly in the alveolar region, whereas, 5 to 10 nm particles deposited in tracheobronchial region, and particles smaller that 10 nm accumulate mostly in the upper respiratory tract. Hence, more research efforts are needed to focus on understanding (1) the form, route, and quantity of NPs entering in the human body, (2) the transformation and ultimate fate of NPs inside human body, (3) the transport, distribution and bio-availability, (4) biochemical response to NPs and their expression and (5) chronic exposure to low concentration of nanoparticles in animal system including human system.

References

Adhikari, T., Kundu, S. and Rao, A.S. (2013). Nanotechnology in soil science and plant nutrition research, NIPA, New Delhi.

American Society of Testing and Materials (1985). Zeta Potential of Colloids in Water and Waste Water, ASTM Standard D 4187–82 ASTM, West Conshohocken, PA.

Banfield, J. F., and Zhang, H. (2001). Nanoparticles in the Environment, In "Nanoparticles and the Environment." (Banfield, J. F., & Navrotsky, A. Eds), 1–58, Mineralogical Society of America, Washington DC.

Banfield, J., and Navrotsky, A. (2001). Nanoparticles and Environment.Reviews in Mineralogy and Geochemistry, 44, Mineral. Soc. of Am. Washington, DC.

Bargar, J. R., Bernier-Latmani, R., Giammar, D. E., and Tebo, B. M. Biogenic Uraninite Nanoparticles and their importance for Uranium Remediation. *Elements,* 4: 407–412.

Becker L and Rietmeijer, F. J. M. (2006) Natural Fullerenes and Related Structure of Elemental Carbon, Springer, Dordrecst, Netherland, 6, 95–121.

Blango and Mulvey, M. A. (2009). Bacterial landlines Contact-dependent Signaling in Bacterial Populations, *Current Opinion in Microbiology*, 12: 177–181.

BSI (2005) British Standard Institute. Vocabulary-Nanoparticles, Publicly available Specification (PAS) 71, 25. Department of Trade Industries and British Standard Institution London.

Buffe J (2006). The Key Role of Environmental Colloids/Nanoparticles for them Sustainability of life, *Environmental Chemistry*, 3: 155-158.

Eichert T, Kurtz A., Steiner, U., and Goldbach, H. E. Size Exclusion Limits and Lateral heterogeneity of the Stomatal Folier Uptake Pathway for Aqueous Solutes and Water Suspended Nano-particles, *Physiologia Plantarum*, 134: 151–160.

Fleischer, M. A., Neill, O., and Ehwald, R. (1999). The Pore Size of Non-graminaceous Plant Cell Wall is rapidly decreased by Borate Ester Cross-Linking of the Pectic Polysaccharide Rhamnogalacturon II, *Plant physiology*, 121: 829–838.

Gilbert, B., and Banfield, J. F. (2005). Molecular-scale Processes Involving Nanoparticulate minerals in Biogeochemical Systems. Rev. *Mineral Geochemistry*, 59: 109–155.

Gorby, Y. A., Yanina, S., McLean, J. S., Rosso, K. M., Moyles, D., Dohnalkova, A., Beveridge, T. J., Chang, I. S., Kim, B. H., Culley, D. E., Reed, S. B., Romine, M. F., Saffarini, D. A., Hill, E. A., Shi. L., Elias, D. A., Kenned, D. W., Pinchuk, G., watanabe, K., Ishii, S., Logan, B., Nealson, K. H., and Fredrickson, J. K. (2006). Electrically Conductive Bacterial Nanowires Produced by Sbewanella oneidensis Strain MR-1 and other Microorganisms. *Proceedings of National Academy of Sciences* USA, 103:11358–11363.

Hochella, Jr. M.F. (2002). Nanoscience and Technology: the Next Revolution in the Earth Sciences. *Earth and Planetary Science Letters,* 203: 593–605.

Hochella Jr. M.F. (2008). Nanogeoscience from Origins to Cutting Edge Applications, *Elements* (Chantilly, VA, U.S.), 4: 373–379.

Jia G *et al.,* (2005). Cytotoxicity of Carbon Nano-Materials, Single-wall Nano-tube, Multi-Wall Nano-tube and Fullerene, *Environmental Science & Technology*, 44: 1036–1042.

Lin, D. H., and Xing, B. S. (2008). Root Uptake and Phytotoxicity of ZnO Nanoparticles, *Environmental Science & Technology*, 42: 5580–5585.

Manceau, A., Nagy, K. L., Marcus, M. A., Lanson, M., Geoffroy, N., Jacquet, T., and Kirpichtchikova, T. (2008). Formation of Metallic Copper Nanoparticles at the Soil Root Interface, *Environmental Science & Technology*, 42: 1766–1772.

Moore, M. N. (2006). Do Nano-particles Present Ecotoxicological Risks for the Health of the Aquatic Environment, *Environment International*, 32: 968–976.

Navarro, E., Baun, A., Behra, R., Hartmann, N. B., Filser, J., Miao, A. I., Quigg, A., Santschi, P. H., and Sigg, R. (2008). Environmental Behaviour and Ecotoxicity of Engineered Nanoparticles to Algae, Plants and Fungi, *Ecotoxicology,* 17: 372–386.

Seinfeld JH and Pandis, S. N. (2006). Atmospheric chemistry and physics, From Air Pollution to Climate Change, John Wiley & Sons, Hoboken, N. J.

Theng, B. K. G., and Yuan, G. (2008). Nanoparticles in the Soil Environment. *Elements*, 4: 395–400.

Tilney, L. G., Cooke, T. J., Connelly, P. S., and Tilney, M. S. (1991). The structure of plasmodesmata as revealed by plasmolysis, detergent extraction, and protease digestion. *Journal of Cell Biology*, 112: 739–747.

Tiwari, A. J., and Marr, L. C. (2010). The Role of Atmospheric Transformations in Determining Environmental Impacts of Carbonaceous Nanoparticles. *Journal of Environmental Quality*, 39: 1883–1895.

Uzu, G., Sobanska, S., Sarret, G., Munoz, M., and Dumat, C. Foliar lead uptake by lettuce exposed to atmospheric pollution, *Environmental Science & Technology,* 44:1036–1042.

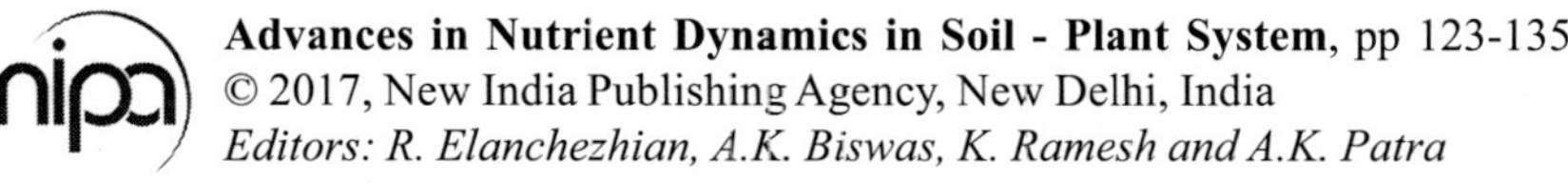

Advances in Nutrient Dynamics in Soil - Plant System, pp 123-135

Editors: R. Elanchezhian, A.K. Biswas, K. Ramesh and A.K. Patra

12

Physiological and Molecular Basis for Improving Nutrient Use Efficiency of Crops

R. Elanchezhian[1] and Renu Pandey[2]

[1]ICAR-Indian Institute of Soil Science, Nabi Bagh, Bhopal – 462038, India
[2]Division of Plant Physiology, IARI, New Delhi – 110012, India

Through green revolution, we have developed high yielding varieties albeit with high nutrient requirements putting much more reliance on chemical fertilizers. The major challenge facing plant biology right now is to improve crop production and to feed an expanding world population. This is against a background of pressure on agricultural land use and climate change having negative impacts on growing conditions. The adverse effects of agriculture, and specifically fertilizer use, include damage to the environment, a large carbon footprint for the manufacture and use of agrochemicals, and the utilization of non-renewable resources. One solution is to increase the area of land for agriculture, as well as increasing production while maintaining the current rate of inputs; however, this is predicted to have substantial negative impacts on the environment (Tilman *et al.,* 2002), is unsustainable in terms of phosphate use, and would have a huge economic footprint in terms of energy demands for nitrogenous fertilizer production. The challenge is to increase yield, decrease inputs, and improve resistance to abiotic and biotic stresses. Improving crop nutrient use efficiency ideally requires an understanding of the whole system, from the macro (agro-ecosystem) to the molecular level. While acknowledging the critical contribution of agronomy to improving efficient nutrient use, particularly in classically inefficient systems, there is a point at which crop genetic improvement becomes essential for further improvement. This may be achieved by either conventional breeding or by marker-assisted breeding utilizing genetic information derived from basic plant science, and by the utilization of this same information to produce transgenic crops.

Nutrients, along with light, temperature, and water, are critical determinants of crop production. However, fertilizers are costly inputs and inappropriate, imbalanced and overuse can have many ecologically damaging effects, making efficient use of fertilizers a major issue for agriculture. For example, excessive nitrogen use results in a major fraction of anthropogenic nitrous oxide and methane emissions, which contribute substantially to climate change, and inefficient nutrient uptake may result in pollution of inland and coastal waters by leaching and runoff. Worldwide, it has been estimated that nitrogen use efficiency (NUE) for cereal production is only 33% (nitrogen removed in grain as a percentage of that applied). Both agronomic practice and plant breeding have a responsibility to optimize efficient nutrient use, particularly nitrogen, in crop systems. Furthermore, crop improvements to anticipate changing patterns of rainfall and temperature must include an anticipation of nutritional demands influenced by changing cropping systems and crop ideotypes.

Optimal plant growth demands a balanced nutrient supply, with a deficiency of any individual essential nutrient having a detrimental effect on production as determined by the law of the minimum. Some nutrients are required at high levels (the macronutrients nitrogen, phosphorus, potassium, and secondary nutrients like calcium, sulfur, magnesium), while some are only required at low levels (the micronutrients: iron, zinc, manganese, copper, boron, nickel and molybdenum). In some cases, excess or luxury accumulation of nutrients in plant organs is an issue, negatively impacting on crop growth or quality for the consumer.

Agricultural production systems have a range of demands for nutrients; low input as compared with intensive highly managed systems, will have different issues and the solutions will be specific for each system. Solutions for efficient fertilizer capture and conversion to biomass or yield in high-input agriculture will be quite different from targets in extensive, organic, or low-input agriculture. In recent years, the emphasis has alternated from a primary objective of improving yields to minimizing impacts on the environment, and back to yield in order to achieve global food security. In the 21st century, sustainable food production has become a major issue with a growing world population, negative impacts of climate change, and demands on land use.

To optimize progress on crop improvement, an understanding of nutrient use efficiency from the agronomic or agro-ecosystem level down to the molecular level is required. Substantial progress is being made on the functions and regulation of genes and proteins; molecular data are usually interpreted at the cellular level; however, it is essential that this understanding is placed at the organ, plant, and whole-crop levels. Targets for improvement need to take into account the different agricultural systems, crop physiology and yield components, and the demands of the consumer. Safe, sustainable, and secure food, feed, fiber, and fuel production will demand optimized genetic material including the trait of nutrient use efficiency.

Crop Improvement for Better Crop Yield and Fertilizer Use Efficiency

Crop improvement through breeding has resulted in considerable increases in yield for many crops like maize, wheat, sugar cane, with the principal target being economic yield and resistance to biotic and abiotic stresses. Economic yield is a good measure of nutrient use efficiency, especially as related to nitrogen (Barraclough *et al.,* 2010). NUE is defined as quantum of grain or total biomass (depending on crop) yield divided by available nutrient. As field conditions are seldom as ideal as research breeding plots, which have optimum inputs and agronomy, potential or best yields are seldom achieved in practice. This discrepancy is inevitable as the "yield gap" is usually reported on a national level and will encompass growth of varieties in a range of conditions and environments (Fischer and Edmeades, 2010). A greater issue is the observed plateau of yield improvements, probably due to abiotic and biotic stress. The large improvements in cereal yields were brought about by introduction of dwarfing genes and the consequent improvement in harvest index (HI). However, in recent years, there have been incrementally smaller improvements in theoretically achievable yields of many crops. In some cases, theoretical yields may not be achievable due to limiting fertilizer application.

In general, plant breeding is performed under "ideal" conditions, which usually include high fertilizer inputs. In those breeding experiments, nutrient use efficiency has seldom been a key target; even though yield and NUE are closely related at a given fertilizer input. Nutrient use efficiency is the product of both uptake and utilization efficiencies, and therefore selecting for yield effectively selects for the combination of these two very separate traits. Small improvements or even negative trends in acquisition or uptake may be hidden by gains in the other utilization efficiency. Therefore, selecting for yield alone may not be sufficient for optimal nutrient acquisition characteristics, especially at reduced inputs or in environments with specific nutrient deficiencies. It is evident that maximum acquisition is determined by availability; however, the efficiency of scavenging mechanisms will have a substantial impact on acquisition, although this may not be sustainable in the long term in any single location if nutrients are being mined and not replaced. Moreover, there are immense practical difficulties for assessment of root-associated properties. Under such scenario, it is better to determine overall nutrient uptake, but this is not ideal as selection for the individual component traits is not achieved. Moreover, there is little data to indicate whether best uptake performance as selected at high inputs equates to best performance at low inputs.

Under ideal conditions, increasing inputs of nitrogen fertilizer will result in increasing yield. Similarly, fertilization with any limiting nutrient will improve yield. The yield plateau and declining response has been observed for nitrogen fertilizers in recent decades which are indicative of limiting due to other nutrients or may be due to limiting photosynthesis. As a result of yield plateauing, there is decreased NUE, specifically attributable to the nitrogen utilization efficiency component.

Yield Relations with Nutrient Content

Yield is determined by photosynthetic carbohydrate production and storage in many crops, including grain crops. As yield is increased, other nutritional components like protein or micronutrients in grain are often diluted (Monaghan *et al.,* 2001). Moreover, it has been suggested that post-anthesis nitrogen uptake is an important contributory mechanism (Bogard *et al.,* 2010) to arrest such nutrient dilution. The deep rooting systems, under such conditions, will have access to untapped nutrient reserves at deeper depth in the soil profile and such nutrients that are taken up are preferentially allocated to the economic produce.

NUE under Diverse Agricultural Systems

Agricultural production systems comprised of wide range of input systems including organic systems, low input systems, and intensively farmed high-input systems. Each of these situations requires specific crop or species. Hence, one needs to take into account these contrasting agro-ecosystems, the associated agronomic practices, and end-product requirements to consumers for targeting improvement of nutrient use efficiency. It is apparent that substantial improvements in NUE can be achieved through improved agronomic practice alone in certain cases. On the contrary, genetic improvements are likely to be small and incremental when compared to agronomic practices. With ever more nutrient capture or efficient varieties, there is a danger of mining reserves, leaving land completely unproductive in long run under low and no input systems. Here the target for improvement needs to be low-nutrient-requiring genotypes combined with improved agronomy to supply minimal nutrition. In low input systems, where fertilizers are applied at seed sowing, improved early capture is a critical phenotype. However, in intensive, high-input systems, conversion to biomass is the principal concern, along with minimization of losses from the system from imbalanced or over-fertilization.

Nutrient Use Efficiency: Targets for Improvement

Efficient use of any nutrient comprises two fundamental aspects: acquisition efficiency and utilization efficiency. It is important to separate the individual processes and identify the respective genes involved, monitoring improvements with the appropriate physiological measures for improving nutrient use efficiency of crops. Hence, resolving NUE into two component traits, nitrogen uptake efficiency (NUpE) and nitrogen utilization efficiency (NUtE), is a first step to resolving the complexity, and subsequently each of these traits can be sub-divided into many specific physiology traits, each of which are complex traits in itself, the result of networks of biochemical pathways, encoded by multiple genes and subject to complex regulatory processes (Gojon *et al.,* 2009).

Nutrient Acquisition Efficiency and Root Architecture

Nutrient capture (NUpE) is essentially a root trait, although to be fully expressed, it also requires adequate sinks for temporary storage or final deposition of the nutrients. Efficient acquisition will depend first on root architecture, root functions in terms of transporters and exudates, and often the presence of symbiotic associations such as mycorrhiza. Early root establishment is essential for scavenging soil nutrients prior to the application of fertilizer, or alternatively to capture fertilizer applied at the time of sowing. Nutrients will be immediately available in the soil solution, and further availability will be depend on mineralization of organic matter and release from sparingly soluble soil minerals (oxides, clays, etc.). High activity of the high-affinity transporter systems required for uptake into root cells, expressed in the plasma membranes of cells of roots, root tips, root hairs, or in associated organisms (mycorrhiza), will be important in this situation as diffusion of nutrients through soil is the rate-limiting factor. In addition, a well-developed shallow root system will be ideal for intercepting further applications of fertilizer. Deeper roots assume importance with the depletion of surface nutrients, as water near the surface becomes limiting and restricts uptake of nutrients, or in the case of high water supply, for the interception of nutrients that would be potentially leached from the soil profile. In some cases, local proliferation of roots in response to nutrient supply is observed which is controlled by specific transcription factors (Forde, 2002).

An alternative approach to enhancing capture mechanisms (root architecture and function) to improve acquisition is to enhance mechanisms for increasing bio-availability of nutrients (e.g., for phosphorus by acid secretions or to inhibit nitrification losses by the secretion of bioinhibitors of this process (Subbarao *et al.,* 2007).

Nutrient Acquisition Efficiency and Transporter Systems

A key step in mineral nutrient acquisition is the initial trans-membrane transport step. In many cases, for any individual nutrient, there are gene families encoding multiple homologs. In *Arabidopsis*, for example, there are two gene families for nitrate trans-porters, *NRT1* and *NRT2*, with 53 and 7 members, respectively, a gene family of 14 sulfate transporters (Hawkesford, 2003) and 9 members of the phosphate transporter family pht1 (Smith *et al.,* 2003). While in most cases there are families specific for a single nutrient, there are instances of nonspecificity: Sulfate transporters effectively transport selenate and molybdate (Shinmachi *et al.,* 2010).

While there is some potential redundancy of function with these large gene families, it has become apparent that there is tissue, developmental, and even membrane specificity with regard to expression patterns. Functionally, there are usually both high and low affinities for the substrate ions, depending on functional requirements: In relation to primary uptake into root cells, the most common functionality is for high-affinity uptake, as required for effective acquisition from soil solutions with low concentrations of ions. Patterns of expression within the root are often complex to effectively transfer the respective ions from the soil solution to the vasculature for transfer to the shoot material. In some instances, vacuolar storage may also play an important part (Kataoka *et al.,* 2004). Many studies have focused on the impacts of nutrient limitation on patterns of transporter expression and the contribution to overall nutrient use efficiency strategies of plants in limiting nutrient availability (Buchner *et al.,* 2010). For phosphate and sulfate, there is an apparent de-repression system controlling gene expression, facilitating increased expression when nutrient demand exceeds availability (Hawkesford and De Kok, 2006). For nitrate, the pattern is more complex, with some transporters induced and others repressed, depending on the presence of nitrate and the nutritional status of the plant.

The transporters play essential roles, contributing to nutrient use efficiency, for the most part extremely effectively scavenging nutrients from the soil (potentially present at low concentrations), and particularly in con-junction with effective root proliferation. As targets for improvement of NUE, sophisticated strategies are likely to be important. Modifications to the selectivity (Rogers *et al.,* 2000) may enhance preferential uptake of beneficial ions and exclude toxic ions. Overriding negative feedback mechanisms may facilitate luxury uptake, but appropriate sinks or temporary storage would also be required. One approach that apparently overrides limits on nitrogen uptake is the overexpression of alanine amino transferase in root exodermal tissues, thus channeling nitrogen away from metabolites involved in negative feedback. In some instances, enhancing remobilization and optimizing partitioning to harvested organs may require optimization of transporter expression.

Metabolic Responses to Nutrient Availability

Plant responses to nutrient availability are complex and involve changes in pathway fluxes, in activity of pathway enzymes mediated by post-translation modifications and/or changes in substrate/inhibitor ratios (allosteric effects), as well as changes in expression of genes encoding the pathway enzymes and many additional proteins. The challenge for the plant is to optimize growth and development given the available nutrient inputs. Matching availability to demand may entail many regulatory steps and sensory mechanisms. It is essential to understand these networks before intervention through trans-genesis or molecular breeding. For the most part, our knowledge of these regulatory loops is restricted in plants (Gojon *et al.,* 2009).

Nutrient use efficiency, although simply divided into uptake and utilization, encompasses all processes of plant growth and development, and all aspects of metabolism. Potential targets for nutrient use efficiency improvement are therefore diverse. Obvious targets in, for example, nitrogen metabolism include genes of the assimilatory pathway. Glutamine synthetase has been a specific target for transgenic approaches, as it is not only involved in primary assimilation but also has a role in efficient recycling of ammonia during senescence processes (Kichey *et al.,* 2006). Generally, results of single-gene manipulation have been disappointing, in part because metabolic pathways form networks that have a great plasticity in responding to perturbations, whether due to gene targeting or environmental fluctuations, for example, in nutrient supply. Typically, nutrient uptake is balanced by nutritional requirement for growth, and a coordination of pathway expression and activity is seen (Hawkesford and De Kok, 2006; Gojon *et al.,* 2009) and excess uptake of nutrient is avoided. Excess accumulation of some ions does occur but only to the point at which available storage pools are saturated (for example, nitrate accumulation in vacuoles); this is a strategy to aid with fluctuating supplies of nutrients but is not helpful when one nutrient becomes permanently limiting.

Utilization Efficiency

Efficiency of utilization may be defined as biomass production (predominantly fixed carbon) as a function of nutrient taken up. This is most often applied for nitrogen, as total canopy nitrogen content reflects the extent of photosynthetically active biomass, as the greatest proportion of the total nitrogen content in this tissue is a major component of proteins involved in photosynthesis. The effectiveness of this capacity in producing harvestable biomass is defined by NUtE. The key attributes that will enhance NUtE are photosynthetic activity, canopy size, longevity, and sink organ capacity.

Photosynthetic activity includes the ability to intercept light, which is clearly linked to canopy architecture and the light harvesting complex density, as well as the biochemistry of the carbon fixation processes, particularly Rubisco, for efficient fixation of carbon dioxide (Parry *et al.,* 2003). An alternative and radical solution is to engineer C_4 photosynthesis, which is up to 50% more efficient than C3 photosynthesis, into C_3 plants such as rice (Hibberd *et al.,* 2008). Attributes of canopy development and architecture include rapid establishment, followed by proliferation and eventual canopy closure (full coverage of the ground), and then effective architecture to intercept radiant light. Depending on the harvestable product, which may be the canopy itself, or it may be biomass derived from this, for example, woody stem or generative material such as seed, the canopy must be photosynthetically active for as long as possible. Delaying senescence and prolonging the period of photosynthesis results in increased carbon fixation. However, the complexity of processes involved in leaf senescence is highlighted by transcriptome analysis, emphasizing the difficulty in manipulating this process to enhance yield (Gregersen and Holm, 2007). As a target, this process has huge potential for crop improvement, as by definition for a fixed amount of nutrient (nitrogen) taken up, the more carbon that is fixed, the better the NUE.

NUE and Sink Demand

An important attribute for uptake efficiency is having adequate sinks to store acquired nutrients, whether nitrogen or minor but important nutritional components including Fe, Zn, and Se. Adequate sinks will prevent negative feedback regulation on the initial acquisition/assimilatory processes and should provide important remobilizable storage that can be accessed should supply be limiting as well as during production of harvested organs such as seed. Sinks may be subcellular, for example, vacuoles, may be chemical such as nitrogen stores in protein, or may be defined at the organ level, for example, stems . Attempts have been made to engineer both metabolism and protein sinks to enhance nutritional quality with high methionine, cysteine, or lysine content (Tabe and Higgins, 1998; Nikiforova *et al.,* 2002). As already indicated, one explanation for the remarkable improvement in NUE seen by the overexpression of alanine aminotransferase is that alanine is a local metabolic sink for nitrogen that does not have negative feedback effects on uptake, unlike glutamate.

HI and Partitioning of Nutrients

Nutrient use efficiency will be optimum if HI is high and nutrients are partitioned to the harvested material. This ignores the impact at the whole ecosystem level, and there may be merit in not harvesting some nutrients but allowing them to be recycled within the field; this might particularly apply to phosphorus; it does, however, assume that leaching losses will be minimal between crops. In many instances, partitioning to the cropped organ is preferable, for example, nitrogen in the case of grain protein and minerals for human nutrition. As indicated, a major improvement in yields and NUE was obtained with the introduction of dwarfing genes (into wheat and rice), minimizing the non-harvested fraction of wheat and rice. Although there are efforts to extend the repertoire of dwarfing genes, which may have additional benefits (Ellis *et al.,* 2005), overall, as the HI for many crops has already been optimized, there is likely little benefit from further manipulation of the HI. For those crops for which this is not the case, improving the HI is a high priority. While nutrient harvest index (NHI) for nitrogen is usually high in cereals, this is not the case for all minerals in all crops. Even in wheat, different minerals are partitioned with varied efficiency: selenium and molybdenum were shown to be differentially partitioned to grain (Shinmachi *et al.,* 2010). In *Brassica napus* (oil seed rape, Colza), sulfur is very inefficiently partitioned to the seed (Blake-Kalff *et al.,* 1998).

Manipulating senescence to produce stay-green phenotypes, thus enhancing yield may have a detrimental effect on the HI and NHI. In this case, additional selection for late but rapid nutrient remobilization would be required. Conversely, enhancing the rate of senescence by the introduction of a NAC transcription factor increased remobilization and enhanced grain protein as well as zinc and iron content (Uauy *et al.,* 2006; Waters *et al.,* 2009). However, increasing the onset and rate of senescence may have a negative effect on yield.

Many nutrient deficiencies lead to changes in biomass allocation between roots and shoots, generally increasing the root:shoot ratio (Hermans *et al.,* 2006). This is an adaptive strategy, facilitating the ability to scavenge for nutrients. In nutrient-poor environments, this would be an advantageous trait; however, this does need to be balanced with production of harvestable material. Little is known about the signaling pathways involved; however, the signals may be linked to imbalances in nutrient accumulation in shoot tissues. Signals from the shoots to the roots (Forde, 2002) may be hormonal (Signora *et al.,* 2001) or may be metabolites, for example, carbohydrates (Hermans *et al.,* 2006). Ultimately, partitioning to the cropped organ is of most importance in agricultural production.

Strategies for the Genetic Improvement of NUE Traits

There is considerable pressure to improve fertilizer use efficiency, and this has been traditionally achieved by agronomic practice and breeding for yield in specific environments and agronomic systems. In the case of nitrogen, breeding for yield is equivalent to breeding for nitrogen utilization efficiency at any given nitrogen input, and there has been considerable progress in improving yields and, therefore, the NUtE component of NUE. Generally, wheat varieties responding well at high inputs also respond well at low inputs (Ortiz-Monasterio *et al.,* 1997; Barraclough *et al.,* 2010). However, there is a strong case for selection at varied inputs and for seeking new and untested germplasm to find new alleles for greater NUtE efficiency.

However, acquisition efficiency has been much less specifically selected for, in part due to the difficulties of phenotyping roots. Here, selection at low inputs is vital. Additionally, it will certainly be necessary to introduce wider germplasm pools (land-races, wild relatives) into screening pro-grams, as alleles for high efficiency of acquisition will have almost certainly been lost from the gene pool without the selection pressure for high acquisition efficiency.

Many new technologies for gene discovery (microarrays, deep sequencing tilling transformation SNP detection) are now available and may be combined with established breeding approaches (breeding, quantitative trait loci, germplasm screening). In combination with the identification of new traits, technologies for introducing these into modern breeding lines are required (synthetic polyploids, alien introgression, gene transformation; Able *et al.,* 2007). As already mentioned and cautioned by others, single-gene introductions, perhaps through crop transgenesis, are often not successful, particularly when gene selection is from only preexisting bio-chemical knowledge (Sinclair *et al.,* 2004), almost certainly due to a lack of appreciation of the complexity of the systems being manipulated. Ideally, in order to effectively target the trait, this needs to be resolved to the smallest subcomponent, encoded by just a few genes. After trait prioritization, assessment of variation is required. This may be either natural variation, or variation induced by mutation or by crossing an examination of mapping populations, and will provide material directly for commercial breeding. Variation may also be used to aid in the identification of the target genes. Transcriptome approaches will indicate genes co-expressed with traits of interest; however, these candidates are usually very numerous (Wang *et al.,* 2003, 2004; Lu *et al.,* 2005; Gregersen and Holm, 2007). Examination of occurrence across diverse germplasm and expression patterns under multiple conditions will narrow these candidate lists to a few key genes worth further investigation. Definitive implication in crop improvement with respect to NUE may require transgenesis in the crop of interest. Such genetically modified crops may be the end product, or the genes may be used as "perfect" markers for screening other natural populations, avoiding the need for transgenesis. Critically, genes identified by such a route may be more robust than selection based on bio-chemical pathways alone.

Future Prospects

The major targets are improving nutrient capture and interception to avoid losses, modifying requirements (reducing if possible) and enhancing utilization efficiency by generally improving carbon fixation and yield. An ideotype of an idealized set of traits for nutrient use efficiency can be defined and will be crop specific (Foulkes *et al.,* 2009). Furthermore, such ideotypes will be specific to different environments and cropping systems.

Nutrient use efficiency in its broadest sense indicates how effectively a plant is able to capture and utilize nutrients to produce biomass. It is most usually specified for nitrogen as this is a main driver for production. However, healthy and productive crop growth requires a balanced nutrition including several macronutrients and many micronutrients. Irrespective of the quantity needed, all are essential and any limitation will impact on plant growth and crop yields.

In almost all cases, the nutrient in question must be obtained from the pedosphere and therefore uptake processes dependent on architecture and functioning of the roots are critical. Subsequent to this, partitioning within the plant is a vital prerequisite to efficient utilization of the element as part of the plant's growth and developmental cycle. Independent but simultaneous selection for both of these traits must be performed. A radical and alternative solution to providing nitrogen fertilizer would be the transfer of nitrogen fixation capacity, or the ability to form the required symbioses, to non-legume crops.

NUE is an essential component of crop production, and irrespective of the agronomic system, low-input or intense, efficient utilization of valuable resources will be essential for future sustainable food production. NUE is a complex trait that can be broken down into subtraits, all of which are also complex in nature. Few instances can be expected where single genes or a single locus will have a huge benefit; dwarfing genes were an exception. Modern tools and resources available to plant scientists and the agronomy and breeding communities should aid further improvements in NUE and hence crop production. Great variability exists in the extent to which individual crops have been optimized in relation to NUE, and while large improvements may be anticipated for some crops, for the major world grain crops such as wheat maize and rice, smaller incremental improvements are likely. The prospect of step changes in primary production by engineering the photosynthetic process itself will require additional concomitant improvements in nutrient acquisition efficiency.

References

Able, J.A., Langridge, P., & Milligan, A.S. (2007). Capturing diversity in the cereals: many options but little promiscuity. *Trends in Plant Science,* 12: 71–79.

Barraclough, P.B., Howarth, J.R., Jones, J., *et al.* (2010). Nitrogen efficiency of wheat: genotypic and environmental variation and prospects for improvement. *European Journal of Agronomy,* 33: 1–11.

Baxter, I., Muthukumar, B., Park, H.C., *et al.* (2008). Variation in molybdenum content across broadly distributed populations of *Arabidopsis thaliana* is controlled by a mitochondrial molybdenum transporter (MOT1). *PLoS Genetics,* 4: e1000004.

Blake-Kalff, M.M.A., Harrison, K.R., Hawkesford, M.J., *et al.* (1998). Distribution of sulfur within oilseed rape leaves in response to sulfur deficiency during vegetative growth. *Plant Physiology,* 118: 1337–1344.

Bogard, M., Allard, V., Brancourt-Hulmel, M., *et al.* (2010). Deviation from the grain protein concentration–grain yield negative relationship is highly correlated to post-anthesis N uptake in winter wheat. *Journal of Experimental Botany,* 61: 4303–4312.

Buchner, P., Parmar, S., Kriegel, A., *et al.* (2010). The sulfate transporter family in wheat: tissue-specific gene expression in relation to nutrition. *Molecular Plant,* 3: 374–389.

Ellis, M.H., Rebetzke, G.J., Azanza, F., *et al.* (2005). Molecular mapping of gibberellin-responsive dwarfing genes in bread wheat. *Theoretical and Applied Genetics,* 111: 423–430.

Fischer, R.A.T. & Edmeades, G.O. (2010). Breeding and cereal yield progress. *Crop Science,* 50: S85–S98.

Fitzpatrick, K.L., Tyerman, S.D., & Kaiser, B.N. (2008). Molybdate transport through the plant sulfate trans-porter SHST1. *FEBS Letters,* 582: 1508–1513.

Forde, B.G (2002). Local and long-range signaling pathways regulating plant responses to nitrate. *Annual Review of Plant Biology,* 53: 203–224.

Foulkes, M.J., Hawkesford, M.J., Barraclough, P.B., *et al.* (2009). Identifying traits to improve the nitrogen economy of wheat: recent advances and future pros-pects. *Field Crops Research,* 114: 329–342.

Gojon, A., Nacry, P., & Davidian, J.C. (2009). Root uptake regulation: a central process for NPS homeo-stasis in plants. *Current Opinion in Plant Biology,* 12, 328–338.

Gregersen, P.L. & Holm, P.B. (2007). Transcriptome analysis of senescence in the flag leaf of wheat (*Triticum aestivum* L. *Plant Biotechnology Journal,* 5: 192–206.

Hawkesford, M.J. (2003). Transporter gene families in plants: the sulphate transporter gene family— redundancy or specialization? *Physiologia Plantarum,* 117: 155–163.

Hawkesford, M.J. & De Kok, L.J. (2006). Managing sulphur metabolism in plants. *Plant, Cell & Environment,* 29: 382–395.

Hermans, C., Hammond, J.P., White, P.J., *et al.* (2006). How do plants respond to nutrient shortage by biomass allocation? *Trends in Plant Science,* 11, 610–617.

Hibberd, J.M., Sheehy, J.E., & Langdale, J.A. (2008). Using C-4 photosynthesis to increase the yield of rice - rationale and feasibility. *Current Opinion in Plant Biology,* 11: 228–231.

Kataoka, T., Watanabe-Takahashi, A., Hayashi, N., *et al.* (2004). Vacuolar sulfate transporters are essen-tial determinants controlling internal distribution of sulfate in Arabidopsis. *The Plant Cell,* 16: 2693–2704.

Kichey, T., Heumez, E., Pocholle, D., *et al.* (2006). Combined agronomic and physiological aspects of nitrogen management in wheat highlight a central role for glutamine synthetase. *The New Phytologist,* 169, 265–278.

Lu, C.G., Hawkesford, M.J., Barraclough, P.B., *et al.* (2005). Markedly different gene expression in wheat grown with organic or inorganic fertilizer. *Proceedings of the Royal Society B-Biological Sciences,* 272: 1901–1908.

Monaghan, J.M., Snape, J.W., Chojecki, A.J.S., *et al.* (2001). The use of grain protein deviation for identifying wheat cultivars with high grain protein con-centration and yield. *Euphytica,* 122: 309–317.

Nikiforova, V., Kempa, S., Zeh, M., *et al.* (2002). Engineering of cysteine and methionine biosynthe-sis in potato. *Amino Acids,* 22: 259–278.

Nikiforova, V.J., Kopka, J., Tolstikov, V., *et al.* (2005). Systems rebalancing of metabolism in response to sulfur deprivation, as revealed by metabolome analysis of Arabidopsis plants. *Plant Physiology,* 138, 304–318.

Parry, M.A.J., Andralojc, P.J., Mitchell, R.A.C., *et al.* (2003). Manipulation of Rubisco: the amount, activity, function and regulation. *Journal of Experimental Botany,* 54: 1321–1333.

Rogers, E.E., Eide, D.J. and Guerinot, M.L. (2000). Altered selectivity in an Arabidopsis metal trans-porter. *Proceedings of the National Academy of Sciences of the United States of America,* 97: 12356–12360.

Shinmachi, F., Buchner, P., Stroud, J.L., *et al.* (2010). Influence of sulfur deficiency on the expression of specific sulfate transporters and the distribution of sulfur, selenium, and molybdenum in wheat. *Plant Physiology,* 153: 327–336.

Signora, L., De Smet, I., Foyer, C.H., *et al.* (2001). ABA plays a central role in mediating the regulatory effects of nitrate on root branching in Arabidopsis. *The Plant Journal,* 28: 655–662.

Sinclair, T.R., Purcell, L.C., & Sneller, C.H. (2004). Crop transformation and the challenge to increase yield potential. *Trends in Plant Science,* 9: 70–75.

Smith, F.W., Mudge, S.R. & Rae, A.L., *et al.* (2003). Phosphate transport in plants. *Plant and Soil,* 248: 71–83.

Subbarao, G.V., Tomohiro, B. & Masahiro, K., *et al.* (2007). Can biological nitrification inhibition (BNI) genes from perennial *Leymus racemosus* (Triticeae) combat nitrification in wheat farming? *Plant and Soil,* 299: 55–64.

Tabe, L. & Higgins, T.J.V. (1998). Engineering plant protein composition for improved nutrition. *Trends in Plant Science,* 3: 282–286.

Tilman, D., Cassman, K.G. & Matson, P.A., *et al.* (2002) Agricultural sustainability and intensive production practices. *Nature,* 418: 671–677.

Tomatsu, H., Takano, J. & Takahashi, H., *et al.* (2007). An *Arabidopsis thaliana* high-affinity molybdate trans-porter required for efficient uptake of molybdate from soil. *Proceedings of the National Academy of Sciences of the United States of America,* 104: 18807–18812.

Uauy, C., Distelfeld, A. and Fahima, T., *et al.* (2006). A NAC gene regulating senescence improves grain protein, zinc, and iron content in wheat. *Science,* 314, 1298–1301.

Wang, R.C., Okamoto, M., Xing, X.J., *et al.* (2003). Microarray analysis of the nitrate response in Arabidopsis roots and shoots reveals over 1000 rapidly responding genes and new linkages to glucose, trehalose-6-phosphate, iron, and sulfate metabolism. *Plant Physiology,* 132: 556–567.

Wang, R.C., Tischner, R., Gutierrez, R.A., *et al.* (2004). Genomic analysis of the nitrate response using a nitrate reductase null mutant of Arabidopsis. *Plant Physiology,* 136: 2512–2522.

Waters, B.M., Uauy, C., Dubcovsky, J., *et al.* (2009). Wheat (*Triticum aestivum*) NAM proteins regulate the translocation of iron, zinc, and nitrogen compounds from vegetative tissues to grain. *Journal of Experimental Botany,* 60: 4263–4274.

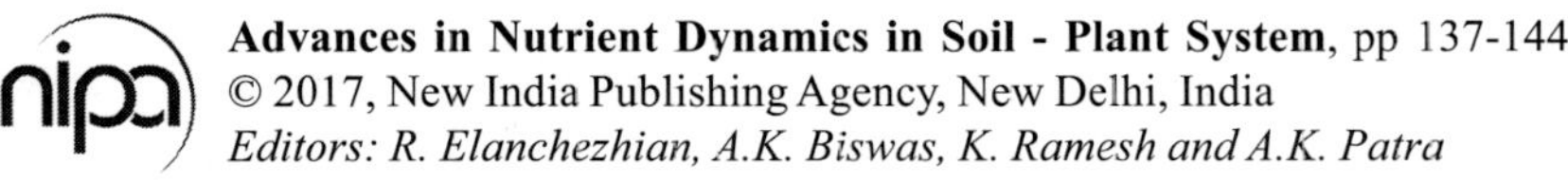
Advances in Nutrient Dynamics in Soil - Plant System, pp 137-144

Editors: R. Elanchezhian, A.K. Biswas, K. Ramesh and A.K. Patra

13

Entry and Internalization of Nanoparticle into Plant Systems

R. Elanchezhian, K. Ramesh and A.K. Biswas

ICAR-Indian Institute of Soil Science, Nabi Bagh, Bhopal – 462038, India

---

Nutrient supply is one of the major inputs for agricultural crop production in a sustainable manner. The factor productivity of inputs especially the fertilizers in agriculture was found to be in a declining phase during the last four decades. This had a detrimental effect on annual compound growth rate of major crops as wells on soil quality. The current status of nutrient use efficiency in different cropping system is low in case of major macro nutrients likes Nitrogen (30-50%) and Phosphorus (15-20%). Because 50-70% of the nitrogen applied using conventional fertilizers, with particle size dimensions greater than 100 nm, is lost to the soil due to leaching, nutrient utilization efficiency (NUE) by plants is low. Attempts to increase the NUE in conventional fertilizer formulations have thus far resulted in little success. On the other hand, the emerging nano strategies indicate that, due to the high surface area to volume ratio, nano fertilizers are expected to be far more effective than even polymer-coated conventional slow-release fertilizers. In this context, nanotechnology offers an important role in improving the nutrient use efficiency of agricultural systems.

Plants have evolved in the presence of natural nanomaterials (NMs). Nanotechnology holds a greater promise of controlled release of chemicals and fertilizers for improved nutrient utilization and enhanced plant growth (Nair *et al.*, 2010). However, the uptake, bioaccumulation and biotransformation of NMs in the agricultural crops are still not well understood. Very few NMs and plant species have been studied, mainly at the very early growth stages of the plants (Rico *et al.*, 2011). Most of the studies, except one with multi walled carbon nanotubes performed on the model plant *Arabidopsis thaliana* and another with ZnO nanoparticles (NPs) on ryegrass (Lin and Xing, 2008), reported the effect of NMs on seed germination or 15 day old seedlings. However, this provided limited information because of the incomplete plant root and vascular system development by the germination stage. The effect of slow / controlled release fertilizers coated with NMs were studied with wheat and found that germination and emergence and growth of seedlings were not affected (Liu *et al.,* 2006; Zhang *et al.,* 2006). They also showed that the leaching loss of nitrogen is lowest with these fertilizers coated by NMs.

Natural NMs like zeolites when mixed with nitrogen, phosphorus, and potassium compounds, enhances the action of such compounds as slow release fertilizers. Furthermore, these mixtures give the plant access to water and nutrients for longer, which results in a significant saving in water resources and reducing the amount of fertilizer to be applied, thus helping to decrease the amount of water used per crop and the contamination of aquifers resulting from the overuse of fertilizers (Perez *et al*., 2008). Similarly a slow and sustained release of nitrogen over 60 days was observed in the urea-modified hydroxyl apatite nanoparticle-encapsulated *Glyricidia sepium* nanocomposite (Kottegoda *et al.,* 2011). Hoeung *et al.* (2011) developed a granular urea-zeolite slow release fertilizer with a granule size of 3-4 nm using inclined pan granulator and showed that release of nutrient is very slow spanning 3 months which will be enough for rice crop.

Lopez-Moreno (2010) investigated the uptake and accumulation of ZnO NPs (8 nm) by soybean (*Glycine max*) seedlings. They treated the soybean seeds with ZnO NPs in the range of 500–4000 mg L^{-1}. The Zn uptake by the seedlings was significantly higher at 500 mg L^{-1}, perhaps because at this concentration the NPs have lesser aggregation. However at higher concentration aggregation was proposed to occur which makes passage through the cell pore walls difficult; thereby, reducing uptake and accumulation. Zhu *et al.* (2008) studied the uptake of magnetite (Fe_3O_4 NP, 20 nm diameter) by pumpkin seedlings in hydroponic conditions. It was reported that the signal for magnetic NPs were detected in roots, stems, and leaves of pumpkin plants. However, the uptake of the NPs was also seen to depend on the growth medium, because no uptake was observed when grown in soils and reduced uptake when grown on sands. It seems that the uptake also depends on the species of plant because no uptake of Fe_3O_4 NPs was found to occur in NP treated lima bean plants (*Phaseolus limensis*). On the other hand, Wang *et al.* (2010) did not notice any uptake of 25 nm Fe_3O_4 NPs by the pumpkin plants. It has been hypothesized that it is difficult for the large size NPs to penetrate through the cell walls and transport across the plasma membranes. The cell wall pore sizes vary from 2–20 nm, while the size of ions and water molecules are about 0.28 nm respectively (Carpita *et al.,* 1979). Thus ions and water find their ways through ion channels and aquaporins.

Nanoparticle Uptake, Entry into Plants and their Accumulation

Plant cell wall acts as a barrier for easy entry of any external agent including nanoparticles into plant cells. The sieving properties are determined by pore diameter of cell wall ranging from 5 to 20 nm. Hence, only nanoparticles or nanoparticle aggregates with diameter less than the pore diameter of the cell wall could easily pass through and reach the plasma membrane. There is also a chance for enlargement of pores or induction of new cell wall pores upon interaction with engineered nanoparticles which in turn enhance nanoparticle uptake. Further internalization occurs during endocytosis with the help of a cavity like structure that form around the nanoparticles by plasma membrane. They may also cross the membrane using embedded transport carrier proteins or through ion channels. In the cytoplasm, the nanoparticles may bind with different cytoplasmic organelles and interfere with the metabolic processes at that site. When nanoparticles are applied on leaf surfaces, they enter through the stomatal openings or through the bases of trichomes and then translocated to various tissues. However, accumulation of nanoparticles on photosynthetic surface could cause foliar heating which results in alterations to gas exchange due to stomatal obstruction that produce changes in various physiological and cellular functions of plants.

Nanoparticles may be able to travel between cells via plasmodesmata, which are cylindrical channels about 40 nm in diameter (Tilney *et al.,* 1991). Hence, NPs with diameter less than 40 nm may enter and transport in the plant cells through the plasmodesmata once they are in the plant cells. In the cytoplasm, the NPs may bind with different cytoplasmic organelles and interfere with the metabolic processes at that site (Jia. *et al*., 2005). Studies on the mechanism of uptake and formation of nanoparticles within plants have also led to more investigations on the use of plants as source for nanoparticle synthesis.

Khodakovskaya *et al.* (2011) reported that multiwall carbon nanotubes (CNTs) induce few changes in gene expression in tomato leaves and roots, particularly, up-regulation of the stress-related genes, including those of pathogens and the water-channel gene. They have also demonstrated the detection of such multiwall CNTs in roots, leaves, and fruits down to the single nanoparticle and cell level with higher sensitivity and specificity, compared to existing assays. They found that tomato plants grown in soil containing CNTs are able to uptake nanoparticles from soil and bio-distribute them in various tissues, including fruits. Thus, the documented presence of nanoparticles in certain plant tissues or individual plant cells can shed light on the possible mechanisms of the positive effects of nanoparticles on plant growth and development.

It is observed that single-walled carbon nanotubes (SWCNTs) is passively transported and irreversibly localized within the lipid envelope of extracted plant chloroplasts (Giraldo *et al.,* 2014). These SWCNTs were found to promote over three times higher photosynthetic activity than that of controls, and enhance electron transport rates. The SWCNT–chloroplast assemblies also enabled higher rates of leaf electron transport *in vivo* through a mechanism consistent with augmented photoabsorption. Concentrations of reactive oxygen species inside extracted chloroplasts are significantly suppressed by SWCNT–nanoceria complexes. Moreover, they showed that SWCNTs enabled near-infrared fluorescence monitoring of nitric oxide both *ex vivo* and *in vivo*, thus demonstrating that a plant can be augmented to function as a photonic chemical sensor. This indicated that the interface between plant organelles and non-biological nanostructures has the potential to impart organelles with new and enhanced functions.

The fate of Nanoparticles (NPs) in soils and their interaction with naturally occurring inorganic and organic substances in soils has been studied in maize (Zhao *et al.,* 2013). They have cultivated maize (Zea mays) plants for one month in soil amended with 10 nm sized ZnO NPs (0–800 mg kg^{-1}) and sodium alginate (0-100 mg kg^{-1}). The results indicated that ZnO NPs coexisting with Zn ions were continuously released to the soil solution to replenish the Zn ions or ZnO NPs scavenged by roots. At 400 and 800 mg kg^{-1}, without alginate, ZnO NPs significantly reduced the root and shoot biomass production; however, plants treated with these NP concentrations, plus alginate, had significantly more Zn in tissues with no reduction in biomass production. It was observed that alginate significantly reduced the activity of stress enzymes catalase and peroxidase, which could indicate damage in the defense system.

In transgenic Bt cotton crop, SiO_2 nanoparticles decreased significantly the plant height, shoot and root biomasses (Le *et al.,* 2014). These SiO_2 nanoparticles affected the contents of Cu, Mg in shoots and Na in roots of transgenic cotton; and SOD activity and IAA concentration were significantly influenced by SiO_2 nanoparticles. In addition, as examined by TEM analysis, SiO_2 nanoparticles were found to be present in the xylem sap and roots showing that the SiO_2 nanoparticles were transported from roots to shoots via xylem sap. These NPs were found to promote the transport of Mg in Bt-transgenic xylem sap and Fe in both the xylem sap of non-transgenic and Bt-transgenic cotton.This study provides evidence for the bioaccumulation of SiO_2 nanoparticles in plants, which shows the potential risks, if any, of SiO_2 nanoparticles impact on food crops and human health.

The impact of ZnO nanoparticles on chickpea (*Cicer arietinum*) seedlings were compared with zinc sulphate and ZnO of normal size (Burman *et al.,* 2013). Maximum positive response with respect to shoot dry weight was observed in seedlings treated with 1.5 ppm ZnO nanoparticles while at 10 ppm the nanoparticles exerted adverse effects on root growth. However, overall biomass accumulation improved in the ZnO nanoparticle treated seedlings. Similar responses were observed in maize, soybean and wheat treated with Fe, Cu and Zn micronutrients (Elanchezhian *et al.,* 2015, 2016). Such responses were attributed to low reactive oxygen species (ROS) generation in cells, which have resulted in less lipid peroxidation. This was also associated with lower activity of antioxidant enzymes viz. superoxide dismutase (SOD), and peroxidise in Zn, Fe, Cu nanoparticle treated seedling compared to control. The study indicated the importance of precise application of nano sized micronutrients, where plant response varies with concentration and is important in understanding the mechanism of action of specific nanomaterials.

Recently, Mehrian *et al.,* (2015) conducted an experiment to study the effects of different concentration of silver nanoparticles (AgNPs) on the contents of free amino acids, protein, lipid peroxidation (MDA) and antioxidant enzymes activity, viz., SOD, CAT and POX in tomato plants. Amino acid analysis revealed that all the amino acids except methionine and tryptophan exhibited a linear increase with the increase in concentration of silver nano-particles. The greater increases in amino acids content were observed at 75 and 100 mg l^{-1} concentrations. More than 6 fold increases in glutamine and asparagine were observed in nano-particle treated plants than the control. A remarkable decrease in total soluble protein was observed with increase in concentration of AgNPs. Activities of SOD, CAT and POX increased significantly in both shoots and roots of treated tomato plants, though SOD activity declined significantly in the roots at 100 mg l^{-1} concentration. Also enhanced malondialdehyde content indicated the oxidative stress induced by AgNPs. It seems that the increase in protein degradation, resulting in amino acid contents, and antioxidant enzymes activity are strategies to modulate oxidative stress induced by AgNPs in tomato plants.

The interaction and stability of chitosan nanoparticles suspensions containing N, P, and K fertilizers were evaluated for agricultural applications (Corradini *et al.,* 2010). Similarly, for gradual release of nitrogen with the crop growth, urea modified hydroxyapatite nanoparticles were synthesized (Kottegoda *et al.,* 2011). These nano-fertilizers showed slow release of nitrogen up to 60 days of plant growth compared to commercial fertilizer which shows release only up to 30 days. The huge surface area of HA facilitates the greater amount of urea attachment on the nanoparticle surface and strong interaction between nanoparticles and urea contributes to the slow and controlled release of urea. Wanyika *et al.,* 2012 reported that polymer-based mesoporous nanoparticles (silica nanoparticles with 150 nm size) can provide efficient carrier system to agrochemical compounds which improves the efficiency and economical utilization. It has been observed that 15.5 % of urea was loaded inside the nanoparticles pores and with fivefold increase in controlled urea release profile in soil and water. Zinc solubility and dissolution kinetics of ZnO nanoparticles and bulk ZnO particles coated on macronutrient fertilizers (urea and monoammonium phosphate) have been compared by Milani *et al.* (2012) in which they reported that coated monoammonium phosphate granules show faster dissolution rate.

Plant nutrients have a crucial role in suppression of plant diseases. Micronutrients, especially, are critical in the defense against crop diseases, with tissue infection inducing a cascade of reactions commonly resulting in the production of inhibitory secondary metabolites. Notably, these metabolites are often generated by enzymes that require activation by micronutrient cofactors. For example, Mn, Cu, and Zn enhance disease resistance by activating the host defense enzymes phenylalanine ammonia lyase and polyphenol oxidases. Works in this direction has suggested significant potential for nanoscale micronutrients, either by foliar or root application, to suppress disease and increase crop yield (Servin *et al.,* 2015). The antifungal activity of CuO NP on *Phytophthora infestans* with tomato using foliar application resulted in significantly greater protection (73.5 %) from the pathogen, compared to the bulk amendment of CuO (Giannousi *et al.,* 2013). Similarly the inhibitory effects of Ag NP foliar application on powdery mildew was investigated in field-cultivated cucumber and pumpkin and 25% effective fungal control was reported (Lamsal *et al.,* 2011).

Genetic Transformation of Plants using Nanoparticles

Nanobiotechnology offers a new set of tools to manipulate the genes using nanoparticles, nanofibres and nanocapsules. Properly functionalized nanomaterials serve as a platform to transport large number of genes as well as chemicals that trigger gene expression in plants. Nanofibre arrays which can deliver genetic material to cells quickly and efficiently have potential applications in drug delivery, crop engineering and environmental monitoring. Reports came on the integration of carbon nanofibres which are surface modified with plasmid DNA with viable cells for controlled biochemical manipulations in cells. The successful delivery and integration of plasmid DNA was confirmed from the gene expression. This process has similarity with microinjection method of gene delivery and hence possible with the plant cells in which the treated cells could be regenerated into whole plant that would express the introduced trait. It is possible to make DNA tethered on carbon nanofibers without allowing them to integrate into host genome but still allowing some transcriptions of the tethered genes and hence this technique does not pass modified traits to further generations. This differ it from the existing genetic engineering methods and a onetime modification of the cells is possible. The application of fluorescent labeled starch-nanoparticles as plant transgenic vehicle was reported in which the nanoparticle biomaterial was designed in such a way that it bind and transport genes across the cell wall of plant cells by inducing instantaneous pore channels in cell wall, cell membrane and nuclear membrane with the help of ultrasound. It is possible to integrate different genes on the nanoparticle at the same time and the imaging of fluorescent nanoparticle is possible with fluorescence microscope thus understanding the movement of exterior genes along with the expression of transferred genes. Hence successful generation of pores on cell wall and cell membrane by suitable agents help in nanoparticle mediated DNA transfer that might be more successful in regenerative calli and soft tissues.

Nowadays gene gun or particle bombardment is one of the popular tools to deliver DNA into intact plant cells. Particles used for bombardment are typically made of gold since they readily adsorb DNA and are non-toxic to cells. Since MSNs are too light, it is difficult for delivering foreign DNA attached on MSNs by gene gun method. This problem was solved by capping MSNs with gold nanoparticles which increased their momentum after acceleration by the gene gun. Experiments showed that the plasmid DNA transferred by gene gun method using gold-capped MSNs was successfully expressed in intact tobacco and maize tissues. The major advantage is the simultaneous delivery of both DNA and effector molecules to the specific sites that results in site targeted delivery and expression of chemicals and genes respectively. This makes the nanoparticle mediated plant transformation better than the conventional genetic engineering methods like electroporation, microinjection, etc.

The interaction of plant cell with the nanoparticles results in modification of plant gene expression and associated biological pathways which ultimately affect plant growth and development. Hence there is a need to clarify the nanotoxicity to plants, possible uptake and translocation of nanoparticles by plants, and physical and chemical properties of nanoparticles in rhizosphere and on root surfaces. Moreover the biotransformation of NMs in food crops is still unclear and the possible transmission of the NMs to the next generation of plants exposed to NMs is still unknown. The mechanism of formation of NPs outside the plant system and their translocation into and within plant system needs clarification and further research.

References

Corradini, E., De. Moura, M.R., and Mattoso, L.H.C. (2010). A preliminary study of the incorporation of NPK fertilizer into chitosan nanoparticles. *Express Polymer Letters,* 4:509–515.

Elanchezhian R, Dameshwar Kumar, K. Ramesh, A.K. Biswas and A.K. Patra (2015). Morphological, physiological and biochemical responses of maize plants towards nano-micronutrient fertilization of Fe, Cu and Zn. In 3rd International Plant Physiology Congress held at JNU, New Delhi during 11-14 Dec 2015.

Elanchezhian R, M.R. Sahu, K. Ramesh, A.K. Biswas, A. Guhey and A.K. Patra (2016). Nano-micronutrient fertilization: impact on physiological and biochemical traits of soybean and wheat. In 9th NABS National Conference held at MKU Madurai during 11-12 Aug 2016.

Giannousi, K., Avramidis, I., and Dendrinou-Samara, C. (2013). Synthesis, characterization and evaluation of copper based nanoparticles as agrochemicals against Phytophthora infestans. *RSC Advances,* 3(44):21743–21752.

Jia, G., Wang, H., Yan, L., Wang, X., Pei, R., Yan, T., Zhao, Y. and Guo, X. (2005). Cytotoxicity of carbon nanomaterials: single-wall nanotube, multi-wall nanotube, and fullerene. *Environmental Science and Technology*, 39: 1378–1383.

Kottegoda, N., Munaweera, I., Madusanka, N., Karunaratne, V. (2011). A green slow-release fertilizer composition based on urea-modified hydroxyapatite nanoparticles encapsulated wood. *Current Science*, 101:73–78.

Lamsal, K., Kim, S.W., Jung, J.H., Kim, Y.S., Kim, K.S., and Lee, Y.S. (2011). Inhibition effects of silver nanoparticles against powdery mildews on cucumber and pumpkin. *Mycobiology,* 39(1): 26–32.

Lopez-Moreno, M.L., Rosa, G., Hernández-Viezcas, J.A., Castillo-Michel, H., Botez, C.E., Peralta-Videa, J.R., and Gardea-Torresdey, J.L. (2010). Evidence of differential biotransformation and genotoxicity of ZnO and CeO2 NPs on soybean plants. *Environmental Science and Technology,* 44(19): 7315-7320.

Nair, R., Varghese, S.H., Nair, B.G., Maekawa, T., Yoshida, Y. and Sakthi Kumar D (2010). Nanoparticulate material delivery to plants. *Plant Science,* 179: 154–163.

Rico. C.M., Majumdar, S., Duarte-Gardea, M., Peralta-Videa, J.R. and Gardea-Torresdey, J.L. (2011). Interactions of NPs with edible plants and their possible implications in food chain. *Journal of Agriculture and Food Chemistry*, 59(8): 3485-3498.

Servin, A., Elmer, W., Mukherjee, A., Torre-Roche, R., Hamdi, H., White, J.C., Bindraban, P. and Dimkpa, C. (2015). A review of the use of engineered nanomaterials to suppress plant disease and enhance crop yield. *Journal of Nanoparticle Research*, 17:92.

Wang, H., Kou, X., Pei, Z., Xiao, J.Q., Shan, X. and Xing B. (2010). Physiological effects of magnetite NPs on perennial ryegrass and pumpkin plants. *Nanotoxicology,* 5(1):30-42.

Zhu, H., Han, J., Xiao, J.Q., and Jin, Y. (2008). Uptake, translocation and accumulation of manufactured iron oxide nanoparticles by pumpkin plants, *Journal of Environmental Monitoring*, 10: 713–717.

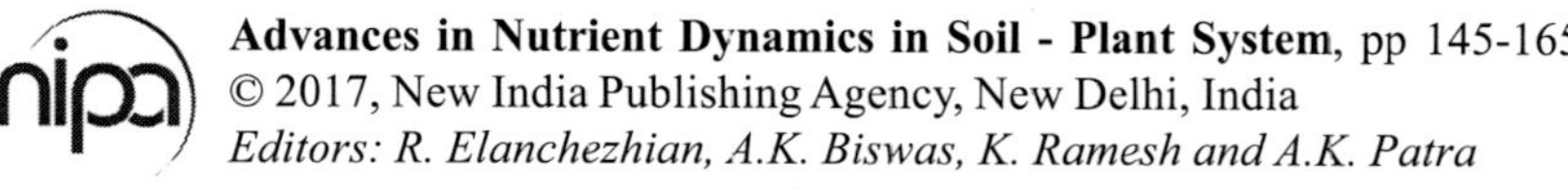
Advances in Nutrient Dynamics in Soil - Plant System, pp 145-165
© 2017, New India Publishing Agency, New Delhi, India
Editors: R. Elanchezhian, A.K. Biswas, K. Ramesh and A.K. Patra

14

Nano Rock Phosphate Absorption and Utilization in Plants

T. Adhikari

ICAR-Indian Institute of Soil Science, Nabi Bagh, Bhopal – 462038, India

In plants, mineral uptake is the process in which minerals enter the cellular material, typically following the same pathway as water. The most normal entrance portal for mineral uptake is through plant roots. Some mineral ions diffuse in-between the cells. In contrast to water, some minerals are actively taken up by plant cells. Mineral nutrient concentration in roots may be 10,000 times more than in surrounding soil. During transport throughout a plant, minerals can exit xylem and enter cells that require them. Mineral ions cross plasma membranes by a chemiosmotic mechanism. Plants absorb minerals in ionic form: nitrate (NO_3^-), phosphate (HPO_4^-) and potassium ions (K^+); all have difficulty crossing a charged plasma membrane. It has long been known plants expend energy to actively take up and concentrate mineral ions. Proton pump hydrolyzes ATP to transport H^+ ions out of cell; this sets up an electrochemical gradient that causes positive ions to flow into cells. Negative ions are carried across the plasma membrane in conjunction with H^+ ions as H^+ ions diffuse down their concentration gradient.

Ion uptake by Plants

The essential plant mineral nutrient elements that are absorbed by plant roots exist in the soil solution as ions. The ionic forms for the essential plant mineral nutrient elements are:

Major Element Ionic Form(s)

Nitrogen (N) NH_4^+ and NO_3^-

Phosphorus (P) $H_2PO_4^-$, HPO_4^{2-}

Potassium (K) K^+

Calcium (Ca) Ca^{2+}

Magnesium (Mg) Mg^{2+}

Sulfur (S) SO_4^{2-}

Micronutrient

Boron (B) BO_3^{3-}

Chlorine (Cl) Cl^-

Copper (Cu) Cu^{2+}

Iron (Fe) Fe^{2+}, Fe^{3+}

Manganese (Mn) Mn^{2+}

Molybdenum (Mo) MoO_4^{2-}

Zinc (Zn) Zn^{2+}

Ion absorption by plant roots is determined by both root physiology and the mobility of ions in the soil itself and within the soil solution. The nature of the movement of these ions in the soil and soil solution affects plant growth, crop yield, and product quality. The three processes involved in ion movement in the soil that determine their availability for plant root uptake are: mass flow, diffusion and root interception.

a. ***Mass flow*** is the movement of dissolved ions in water within the soil profile, downward movement due to rainfall and applied irrigation water, or upward movement by water evaporation from the soil surface, with this downward/upward movement occurring in the soil mass through its pores. The major ions moved primarily by mass flow are the nitrate (NO_3^-) and chloride (Cl) anions, and to a moderate degree the sulfate (SO_4^{2-}) anion, plus the cations, potassium (K^+) and magnesium (Mg^{2+}). Also, other ions may be carried in the moving soil water, depending on the physical and chemical characteristics of the soil and the concentration and characteristic of that element in the soil.

b. ***Diffusion*** is the movement of ions within water films that exist around soil particles, the driving force being the ion concentration gradient, always moving from an area of high concentration to an area of lower concentration. Most ions in solution are moved by

c. ***Root interception*** is exchange of nutrients as exchangeable ions on root surfaces contacts exchangeable ions on soil surfaces.

Quantity of ions intercepted depends on root mass or volume of roots

- Amount of exchangeable ions

- Maximum contact occurs when root volume = soil volume. Average root volumes rarely exceed 2% of soil volume.
- All the Ca could be obtained by root interception because the concentration in most soils is high
- Root interception of nutrients can be enhanced by mycorrhiza, due to larger nutrient absorbing surface provided by the fungi absorption of water and nutrients by roots

Absorption

In chemistry, **absorption** is a physical or chemical phenomenon or a process in which atoms, molecules, or ions enter some bulk phase – gas, liquid, or solid material. This is a different process from adsorption, since molecules undergoing absorption are taken up by the volume, not by the surface (as in the case for adsorption). A more general term is *sorption*, which covers absorption, adsorption, and ion exchange. Absorption is a condition in which something takes in another substance. In many processes important in technology, the chemical absorption is used in place of the physical process, e.g., absorption of carbon dioxide by sodium hydroxide – such acid-base processes do not follow the Nernst partition law.For some examples of this effect, see liquid-liquid extraction. It is possible to extract from one liquid phase to another a solute without a chemical reaction. Examples of such solutes are noble gases and osmium tetroxide. The process of absorption means that a substance captures and transforms energy. The absorbent distributes the material it captures throughout whole and adsorbent only distributes it through the surface. The reddish color of copper is an example of this process because it is caused due to its absorption of blue light.

If absorption is a physical process not accompanied by any other physical or chemical process, it usually follows the Nernst distribution law:

"the ratio of concentrations of some solute species in two bulk phases in contact is constant for a given solute and bulk phases":

$$\frac{[x]_1}{[x]_2} = \text{constant} = K_N(x,12)$$

The value of constant K_N depends on temperature and is called partition coefficient. This equation is valid if concentrations are not too large and if the species "x" does not change its form in any of the two phases "1" or "2". If such molecule undergoes association or dissociation then this equation still describes the equilibrium between "x" in both phases, but only for the same form – concentrations of all remaining forms must be calculated by taking into account all the other equilibria.[1]

In the case of gas absorption, one may calculate its concentration by using, e.g., the Ideal gas law, c = p/RT. In alternative fashion, one may use partial pressures instead of concentrations.

Types of absorption

Absorption is a process that may be chemical (reactive) or physical (non-reactive).

Physical absorption: Physical absorption or non-reactive absorption is made between two phases of matter: a liquid absorbs a gas, or a solid absorbs a liquid.When a liquid solvent absorbs a gas mixture or part of it, a mass of gas moves into the liquid. For example, water may absorb oxygen from the air. This mass transfer takes place at the interface between the liquid and the gas, at a rate depending on both the gas and the liquid. This type of absorption depends on the solubility of gases, the pressure and the temperature. The rate and amount of absorption also depend on the surface area of the interface and its duration in time. For example, when the water is finely divided and mixed with air, as may happen in a waterfall or a strong ocean surf, the water absorbs more oxygen.

When a solid absorbs a liquid mixture or part of it, a mass of liquid moves into the solid. For example, a clay pot used to store water may absorb some of the water. This mass transfer takes place at the interface between the solid and the liquid, at a rate depending on both the solid and the liquid. For example, pots made from certain clays are more absorbent than others.

Chemical absorption: Chemical absorption or reactive absorption is a chemical reaction between the absorbed and the absorbing substances. Sometimes it combines with physical absorption. This type of absorption depends upon the stoichiometry of the reaction and the concentration of its reactants.

Nano Rock Phosphate Synthesis and Utilization

Phosphorus (P) deficiency is a common phenomenon in soils of India. The situation is worse in the red soil regions where Fe, Al in the soil tend to fix more phosphorus and reduced the efficiency of common phosphate fertilizers. P is an essential plant nutrient and the deficiency of P severely restricted crop yields. Results from fertilizers trials had shown that yields can be double, in some cases tripled with phosphorous application. However, the high costs of high grade water soluble P-fertilizers coupled with the high fixing capacities of these soils for phosphorous prevents its application to crop production in the region. Nowadays direct application of Rock phosphate (RP) can be an effective alternative to the use of more expensive water-soluble P (WSP) fertilizers to provide P nutrition in adequate amounts for crop production under certain soil and crop productions. The agronomic effectiveness of PR depends on several factors involving RP reactivity, soil properties, crop species, and climate. In contrast to WSP fertilizers (TSP, SSP etc.) that dissolve rapidly and completely in soils, RP is insoluble in water, and the driving force for its dissolution in soils is mainly dependent on the proton supply and the gradient of P and Ca concentrations in solution. Crop species also vary greatly in RP use efficiency. For example crops can modify the pH at the root-soil interface or rhizosphere so that it differs from that of bulk soil. Crops such as canola, pigeon pea, and lupin are known to exude organic acids, for example, malic, oxalic, and citric acid, which reduce pH around the roots, thereby increasing the use of RP. The cation-anion imbalance caused by plant uptake of some crop species, including legume crops, can result in changes of the rhizosphere pH that can in turn affect RP dissolution and agronomic effectiveness. The density or structure of crop roots also varies, and thus, the total contact area with RP for P uptake varies among crops. Crops with longer growth duration, for example, perennials versus annuals, provide more time for RP dissolution, and hence, perennial crops in general have higher relative agronomic effectiveness (RAE) of RP with respect to WSP compared with annual crops. The effectiveness of rock phosphate was generally low compared with fertilizers such as triple super phosphate, single super phosphate etc. This is attributed to the relatively low solubility of the rocks as opposed to the readily water soluble phosphorus fertilizers. However, the effectiveness of rock phosphate fertilizers may be influenced by particle size and generally increases with decreasing particle size. In India the use of RP appears to be promising. The pressing need for economically acceptable fertilizer sources has prompted the search for an indigenous source. However, as P deficiency is so acute in the area and high solubility P fertilizers are beyond the reach of the farmers a question was posed, in the process of studying different P sources, as to whether or not this hard rock phosphate could at all contribute to the P nutrition of crops in the region. The fact that crop can absorb P from nano rock phosphate source leaves us with much room for research. Rock phosphate offers a good alternative or complement to triple super phosphate. In volcanic soils of Chile it is recommended to apply rock phosphate in combination with triple super phosphate because of the high phosphorus-retention capacity of these soils. However, in Western Australia it has been shown that rock phosphate response is low. Possible explanations are slow dissolution of rock phosphate due to the moderately acid pH (5.5 –6.5), low pH-buffering capacity, low calcium absorption in soil, which increases concentration of calcium in soil solution, and poor water holding capacity of these soils. The response of crops to rock phosphate depends not only on the solubility of rock phosphate but also on the type of soil. Gillepsie and Pope (1990) showed that in Argiaquoll, increase in rock phosphate solubility promoted plant uptake whereas in Hapludult, increased solubility resulted in less plant uptake, probably due to increased fixation in the acid, highly weathered soil. Rock phosphate solubilization is also mediated by microorganisms. Halder *et al.*, (1990) showed that nearly all strains of *Rhizobium* and *Brady Rhizobium* solubilized rock phosphate; maximum solubilization was achieved 3 days after inoculation. Lalljee and Facknath (1999) have shown the presence and evaluated the activities of the solubilising organisms in soils of Mauritius in a recent study. The manufacture of P fertilizers is currently based on the treatment of phosphate rock by H_2SO_4, H_3PO_4, or HNO_3 to make them more soluble. In the beginning of the twentieth century, some workers conducted long term field experiments to investigate the use of raw PR as such in sopdozolic zone of Russia, but because low quality PR was used, they produced disputable agronomic results. Subsequently, interest in the direct application of PR diminished considerably. During the last 20 years, a new stimulus has been given to the efforts on this subject by the increase in the demand for P fertilizers, especially for high P-fixing soils, coupled with rising energy and manufacturing costs. Nitrogen, Phosphorus and potash the three primary nutrients are required in large quantities by plants for sustaining life and their healthy growth. Phosphorus is very minutely present in the soil as much. It is applied to the soil in the form of phosphatic fertilizer, which is produced after the acid treatment of high-grade rock phosphate or its beneficiated concentrate. It is found that rock phosphate is the only major source of phosphorus for plants. The consumption of chemical fertilizers in our country has grown at a very rapid rate since independence. Currently, India rank third amongst the fertilizer consuming countries in the world. The continuously increasing domestic requirements of fertilizers are now being partially met by indigenous production and the balance is met through imports. The raw material for the production of phosphoric acid and phosphatic fertilizer is imported since indigenous rock phosphate is of low grade and mining activity is insufficient. Out of 145 million tones of indigenous reserves of rock phosphate including proven, probable and possible categories, the reserves of high grade (+30% P_2O_5) rock phosphate are approximately 13.5 MT only. This is mined by different government and private agencies for commercial purposes for any years. The beneficial low-grade reserves of rock phosphate, estimated to be 81.2 MT, have not been exploited so far. However, the production of fertilizer grade P_2O_5 can be increased through exploitation of large deposits of beneficial low-grade rock phosphate. The only possibility of increasing the contribution of P_2O_5 from indigenous rock phosphate reserve is to enrich/beneficiate the low-grade rock phosphate. The beneficial low-grade reserves may contribute significantly to the production of phosphatic fertilizers and hence result in saving of sizable amount of foreign exchange every year. Madhya Pradesh ranks second after Rajasthan in terms of rock phosphate reserves. The total reserve of phosphate rock in Madhya Pradesh is 36 MT. Indian Bureau of Mines, Nagpur has developed different beneficiation processes to treat different types of indigenous rock phosphate ore (grade varying between 16%to 25% P_2O_5) of jaguar successfully on laboratory scale. Except some recent efforts by RRL, Bhopal, no attempts had been made in the past to exploit the low-grade rock phosphate (<15% P_2O_5) of Madhya Pradesh for the production of useful phosphate concentrate for fertilizer industries. The exploitation of these low-grade deposits is of immense importance to increase the contribution of indigenous marketable grade P_2O_5. Therefore, development of economically viable beneficiation schemes for low grade ores in order to make them suitable for use in our fertilizer industry has assumed importance.

Location and Occurrences

The rock phosphate deposits in Madhya Pradesh are mainly concentrated in two areas. One is in Sagar and Chhatarpur districts, which is better known as Hirapur phosphate deposit and the other one is located in Jhabua district, called as Jhabua phosphate deposit. Another new deposit of phosphate has been located at Barwaha by G.S.I. The location and occurrences of the two main deposits are discussed below:

Hirapur Rock Phosphate Deposit

Location: The Hirapur rock phosphate deposit belt is situated around the village Hirapur In Banda Tehsil of Sagar district extending into the adjoining Chhatarpur district. Hirapur is situated 80 Km north of Sagar-Chhatarpur highway. Sagar and Damoh are the nearest equidistant railway station on Bina-Katni section of Center Railway. The phosphorite deposits of Hirapur are approachable by road from Sagar. Occurrences: All the phosphorite deposits are associated with the Bijawar Group of rocks and their occurrences have been divided into the following four sectors on the basis of geographic distribution and for convenience of exploration from west to east:

i. Basai sector ii. Mardeora sector

iii. Hirapur sector iv. Kachhar sector

These four sectors are again subdivided in blocks such as A, B, C, D, E with varying grade and reserves of rock phosphate. Basai sector, Mardeora sector and Hirapur sector have been subdivided into four, five and three blocks respectively.

The phosphorites are mainly of siliceous type i.e. silica is associated with rock phosphate as a major gangue mineral.

Jhabua Rock Phosphate Deposit

Location: The Jhabua rock phosphate deposit was discovered in 1973. The area is bounded by latitude 22’ 58’32" and longitude 74’25’15" to 74’26’2". The area is accessible through a 20 Km long road from Meghnagar, which is the nearest railway station on broad gauge line of western Railway between Ratlam and Baroda.

Occurrences: Jhabua rock phosphate deposit of sedimentary origin represents a southern extension of the early to middle Proterrozoic Aravali belt from Rajasthan into western Madhya Pradesh. The deposits extend for about 4.5 Km lengths from Amaliamal in the north to Khatamba in the south and are divided into the following five major blocks from north to south:

i. Amaliamal block
ii. Kelkua Nala block
iii. Kelkua block

iv. Khatamba north block
v. Khatamba south block

Two types of phosphorites present in Jhabua are:

i. The high-grade cherty type (around 20-30% P_2O_5).
ii. A low-grade carbonatic (dolomite/calcareous) type (around 10-20% P_2O_5).

Present Status of the Indigenous Rock Phosphate Reserves

The known and proven phosphate reserves of 145 Mt India is shown in Table 1. Classification of rock phosphate on the basis of grade and their use is provided in Table 2. In our country, the limited reserve of high grade (+30% P_2O_5) rock phosphate has another problem of high ore to overburden ratio, varying in the range of 1:6 to 1:8. Due to this reason the prospects of increasing the production of high-grade rock phosphate is very limited. On the other hand there are huge deposits of low grade (<20% P_2O_5) rock phosphate, which are unexploited. The production of lower grade of rock phosphate can be increased for their use either in the existing fertilizer plants that are presently using the imported rock phosphate or new fertilizer plants specially set up for utilizing the rock phosphates containing high impurities.

The rock phosphate assaying 20-25% P_2O_5 is currently being used for different purposes such as:

(i) Blending with high-grade rock phosphate ore.
(ii) Direct application to soil as fertilizer.
(iii) PAPR (Partially Acidulated Rock phosphate) production.
(iv) Direct application admixed with phosphate solubilizing bacteria
(v) Beneficiation to produce a useful concentrate (+30% P_2O_5)

The present economic limit of the grade of rock phosphate for direct application to soil is 18% P_2O_5. In respect of PAPR production, economic grade of rock phosphate is found to be 18% P_2O_5. For blending operations, lowest workable grade has been assessed as 25% P_2O_5. For beneficiation of low-grade rock phosphate, the economic limit with the present indigenous technology works out at 15 to 20% P_2O_5.

Jhabua and Hirapur: the two rock phosphate deposits of Madhya Pradesh have vast reserves of low grade (<15% P_2O_5) rock phosphate which have not received attention so far for its utilization. Keeping this aspect in view, the flow sheet for beneficiation of these low grade (around 13% P_2O_5) rock phosphate deposits have been successfully developed by RRL, Bhopal at laboratory as well as pilot plant scale which are discussed in next section.

Future Prospects of Madhya Pradesh Rock Phosphate Reserves

Beneficiation of low-grade ore is promising. Considerable progress has been made in this field and significant success has been achieved at pilot plant stage. Rajasthan State Mineral Development has set up already beneficiation plants: Corporation at Jhamarkotra and Hindustan Zinc limited at Maton. There appears to be a dire necessity to set up additional beneficiation plants in another area such as Jhabua and Hirapur for utilization of vast reserves of low-grade. rock phosphates of Madhya Pradesh.

Table 1: Reserves of Indian rock phosphate

State/District	Location	Grade % P_2O_5.	Reserves (in million tons)	Present mining agency
Rajasthan/	Jhamarkotra	30-35	79.0	RSMML,
Udaypur	Maton	25-30		RSMDC
	Kanpur	25-30		HZL
Madhya Pradesh/	Jhabua-Cherty	25-30	36.0	RSMDC
Jhabua	Jhabua-carbonate	25-30		MPSMC
Sagar and	Surajpur			MPSMC
Chhatarpur	Hirapur	25-30		MPSML
Uttar Pradesh	Mussoorie	18-20	30.7	MPSMC
Dehradun	(Maldeota and Durmala)			PPCL
Lalitpur	Lalitpur	25-30		UPSMDC

Table 2: Grade wise recoverable rock phosphate reserves and usage in India (as on 1.4.1995.

Grade %	Reserves(million tones)	Usage
1. > 30	13.5	Chemicals & Fertilizers Blendable
2. 25-30	18.8	Direct application
3. 15-20 & above	12.5	Beneficiation
4. 11-20	81.2	Benificiation
5. Unclassified & others	19.4	
Total	144.4	

What is Rock Phosphate?

Phosphate rock is formed in oceans in the form of calcium phosphate, called phosphorite. It is deposited in extensive layers that cover thousands of square miles. Originally, the element phosphorus is dissolved from rocks. Some of this phosphorus goes into the soil where plants absorb it; some is carried by streams to the oceans where it is precipitated. Rock Phosphate is a natural rock mined from phosphorus-rich deposits. The rock is washed free from clay impurities and heated to remove moisture. It is then pelletized for easy application. Phosphate rock is the raw material used in the manufacture of most commercial phosphate fertilizers. In its unprocessed state, phosphate rock is not suitable for direct application, since the phosphorus it contains is insoluble. This rock is derived from naturally-occurring ores. To transform the phosphorus into a plant-available form and to obtain a more concentrated product, phosphate rock is processed using sulphuric and/or phosphoric acid. The process begins by grinding phosphate rock to a fine material. Primary size reduction generally is accomplished by grinding mill media.

Specifications of rock phosphate as prescribed in the Fertilizer Control Order:

(i) Particle size:

The material shall completely pass through 6.3 mm IS sieve and not less than 20 per cent of the material shall pass through 150 micron IS sieve.

(ii) Total P_2O_5 content (20 to 40%) to be guaranteed by the dealer.

The ground rock phosphates are prepared for market from rock phosphate imported in India mainly from Egypt, Morocco, the USA, Jordan, France and UK. These imported rock phosphate vary in composition. As such, the ground prepared from them as fertilizer contains 20 to 40 % P_2O_5 depending on the quality of rock phosphate. When finely ground rock phosphate is applied to soils containing a high percentage of organic matter, carbonic acid and nitric acid present in the soil act on rock phosphate, which contains tricalcium phosphate, and convert unavailable phosphate to monocalcium phosphate or water soluble phosphate which is easily available to growing plants. The reactions can be expressed:

(i) $[Ca_3(PO_4)_2]_3 + 6H_2CO_3 = 3Ca(H_2PO_4)_2 + 7\ CaCO_3$

(ii) $[Ca_3(PO_4)_2]_3 + 14\ HNO_3 = 3Ca(H_2PO_4)_2 + 7\ Ca(NO_3)_2 + H_2CO_3$

Phosphate rock and elemental sulphur or alternatively phosphoric acid are the basic raw materials required to manufacture phosphatic fertilizer which by and large, have to be imported. Entire requirement of sulphur in India is met through imports,60-65% of which goes for fertilizer production. However, 85-88% of rock phosphate requirement of the country is currently being imported, the balance being met through indigenous sources. Consequently, any attempt to promote the use of PAPR especially in acid neutral soils, would be helpful to save foreign exchange and to economize on the cost of imported raw materials required for the manufacture of phosphatic fertilizers. Another reason in Indian perspective is that though we have sizeable rock phosphate reserves (259.47 million tonnes), but these are by and large, of low grade (Table 1) and are not easily amenable to allow their use in the phosphate industry. Nevertheless, a good proportion of known indigenous rock deposits having low P_2O_5 grade but amply reactive in nature could be advantageously utilized on being partially acidulated. Deposits of mussoorie rock, Purulia rock, Madhya Pradesh RP (Mehnagar deposits) and part of Jhamar

Kotra rock having absolute Citrate solubility 4.0 or more, and low organic matter and low carbonate content, may be considered reasonably suitable for use by partial acidulation. PAPR as fertilizer is cheaper than the completely acidulated processed phosphates. This is simply because in their case much less acid is required than what is for complete conversion of tri-calcium phosphate of rock to mono calcium phosphate in processed phosphates. Sporadic attempts to manufacture PAPR had been going on for quite some time is different countries. The first such successful attempt was perhaps made in Bulgaria. Garbouchev standardizes the production of a high grade PAPR having 40% P_2O_5 by employing new Dorr Oliver process. The product was manufactured on the pilot plant by keeping MCP/PR ratio of approximately 60/40 by using 26-30% acidulation North African Gafsa type of phosphate rock. The product required 68-70% less phosphoric acid than the amount normally required for the manufacture of triple super phosphate (TSP). Production of PAPR on pilot scale has also started in France, Germany, and Israel. In India to a pilot plant for the trial production of PAPR was started pyrites phosphates and chemicals Ltd (PPCL) in 1988. In India, recently Yadav *et al.,* of projects and development India Limited (PDIL) have developed a process based on the nitric acid. acidulation of non-standard indigenous rock phosphates ground only to16-20 mesh instead of the conventional 200-300 mesh size, followed by the formation of urea adducts. The fertilizer obtained by this process is claimed to have higher water soluble and citrate soluble phosphate content than the conventional nitrophosphate process, and economically competitive with urea and DAP.

Nano Rock Phosphate

The nutrient demand of Indian agriculture is going to increase tremendously in coming years. Among the major nutrient fertilizers, phosphatic fertilizers are expensive largely because of the import of good quality rock phosphate and sulphur. In India, it is estimated that about 260 m.t. of phosphate rock deposits are available, of which hardly 40 m.t. is of good quality and is being used for the production of fertilizer. Hence, we need to explore the possibility of utilizing this vast deposit of rock phosphate as a source of P in crop production using the modern tool of nano-science and nano-technology. In conventional approaches, the mineral sources of plant nutrients (such as P) is converted into water soluble form (more precisely, ionic form) through chemical processes, so as to make it more available to the plant. In the general perception of science, any material is said to be dissolved when it passes through a 0.45 mm (450 nm) filter. Hence, most of the nano-particles are often included in the dissolved fraction even though they are clearly distinct from molecules or ions. The size range of nano-particles of a number of minerals greatly affect their properties with respect to reactivity, transport and other geochemical characteristics. Hochella *et al.* (2008) introduced two terminologies, namely *"nano-minerals"* and *"mineral nano-particles"*. They defined "nano-minerals" as minerals that exist only in the nano-particle size range (eg. Ferrihydrates) whereas "mineral nano-particles" are minerals that exist both in nano-size and larger size range. Hence, most of mineral resources (of plant nutrients) can be converted to mineral nano-particles so as to enhance their chemical and biological reactivity in soil vis-à-vis bioavailability.

Benefit of Nano Rock Phosphate

Most of the indigenously available rock phosphates are having P content 7-15%, and if these rock-phosphates are converted to nano – size range through Top-down approach using high energy ball mills, it is expected to have following benefits to compensate the cost involved in conversion to nano size range.

1. Transport cost reduced by half as compare to SSP
2. No need of importing S for the manufacture of Phosphate fertilizer
3. Solution culture proved that nano rock phosphate particle can be directly utilized by plants
4. Also provides nano-particles of host of compounds containing calcium, silicon, magnesium, iron, manganese etc.
5. Less interaction with Fe, Al and Ca in soil
6. Very high solubilizing capability through soil microbes
7. Help usage of indigenous rock phosphate
8. Use of the nano rock phosphate help reducing pollution.

Size Reduction of Rock Phosphate by Rotor Mill

In the pursuit of developing a nano-rock phosphate and evaluating its effect on soil and crop, rock phosphate (RP) samples have been collected from Sagar (6 samples), Meghnagar (6 samples) and Udaipur (19 samples) for their characterization with regard to total P, water soluble P, citrate soluble P and heavy metal content apart from some basic physico-chemical parameters. Total P content in Sagar rock phosphates varied from 5.58 to 14.52 %, in Meghnagar rock phosphate total P content varied from 4.46 to 13.12 %, and in the Udaipur rock phosphate total P content varied from 10.38 to 15.26 %. The heavy metals namely, selenium (0- 114 ppm), Arsenic (Tr – 52 ppm) and Chromium (0-13 ppm) are also present in rock phosphate samples. Effectiveness of rock phosphate as a direct application fertilizer is determined by its chemical reactivity, which intern, depends upon the size of the particles and the degree of carbonate substitution for phosphate in apatite structure. Direct application of rock phosphate to the field to supplement P requirement of crops, is not a new thing but the use was limited due to uncertain agronomic response, inconvenience of handling and application of the fine dusty material to the soil. The uncertainty of agronomic response is mainly attributed to the sizes of the rock – phosphate particles. Generally, for field application, rock-phosphates are ground to pass through 100 mesh screen (160 mm size) and in most cases, the sizes are bigger than 160 mm. It is worthy to consider the following relationship between particle size and surface area what we have known for a long- time about how the properties of particles change as they decrease in size to the micron to sub-micron size range. For spherical particles, the ratio of surface area 'A' ($A=4pr^2$) to volume 'V' ($V= 4/3pr^3$) is inversely proportional to the particle radius 'r', i. e. A/V = 3/r. This relationship tells us that as a particle becomes smaller, its surface area becomes an increasingly larger component of its overall form. Thus, reduction of particle sizes of the naturally occurring rock-phosphate is an important means to increase their reactivity in soil vis-à-vis availability of P to the growing plants.

Size Reduction of Rock Phosphate by Laboratory Scale Ball Mill

In our initial attempt to further down size the rock phosphate particles, we were able to convert a small amount (20 g) of two rock phosphates (HGRP – 3 and Stone – 3 from Udaipur) into nano-size range using laboratory scale high-energy ball mill at ISI, Kolkata. Milling increased the solubility of rock phosphate by increasing the proportion of X-ray amorphous material and reducing the size of remaining apatite crystals. Rock phosphates were ball milled at ambient temperature and high energy intensities, which induces phase changes through solid-solid reactions. During this milling process repeated collisions between ball and powder continuously exposes new reactant surfaces. After ball milling these rock phosphate particles (HGRP3 and Stone 3) were analyzed by Photon Collision Spectroscopy (Dynamic light scattering techniques) to know the size distribution of the particles. The obtained results pointed out that produced rock phosphate powder is a highly dispersed, nano- scaled mixture of small particles that is crystallites with sizes in the range of 10-100 nm. Maximum portion of HGRP3 is in the size of 28 nm and Stone-3 is in 42 nm (Fig 1).

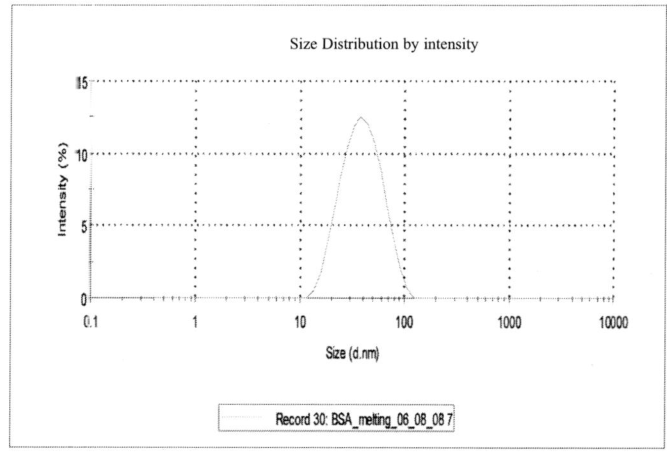

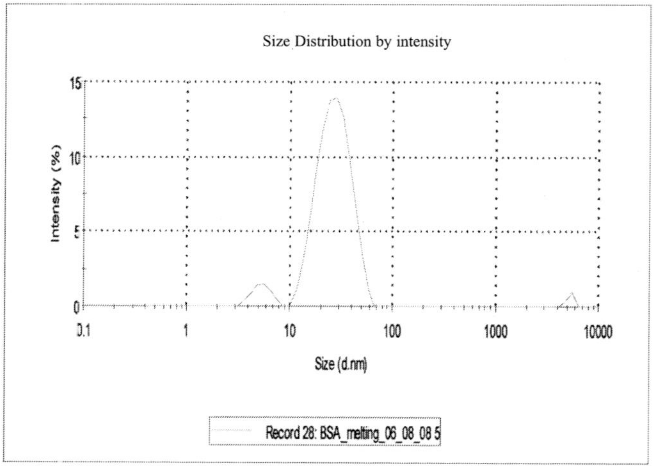

Fig. 1: Stone 3 (Av. Size 42 nm; Intensity 100%) and HGRP-3 (28.20 nm; Intensity 93.1%)

Size Reduction of Rock Phosphate by Pot Mill

Thereafter, a high energy ball mill (pot mill) with zirconium oxide balls (approximately 10 mm in diameter) and bowls (1000 mL) were used for grinding of rock phosphate particle, to get bulk amount (1 kg) of nanoparticles. The grinding process was performed in a continuous regime in air during 6hrs at the basic rotation speed of 120 rpm and rotation speed of balls of 300 rpm. We could get 1 Kg each of the four different rock phosphates (namely, HGRP – 3, BRP, Stone – 3 and SRP – II) after 6 hour of milling. The results of particle size analysis showed (Fig. 2) that after uniform milling time, the different rock phosphate attained different sizes, in case of Udaipur rock phosphate (HGRP-3) majority particles achieved sizes ~ 70.89 nm while in case of other rock phosphate from Udaipur (BRP), maximum particles achieved size ~106.6 nm. The Sagar rock phosphate also achieved particle size ~110.1 nm. It is need less to mention that the resultant particles are the mixture of particles of different sizes and composition as is clear from the photograph (Fig 3) of Scanning Electronic Microscope (SEM).

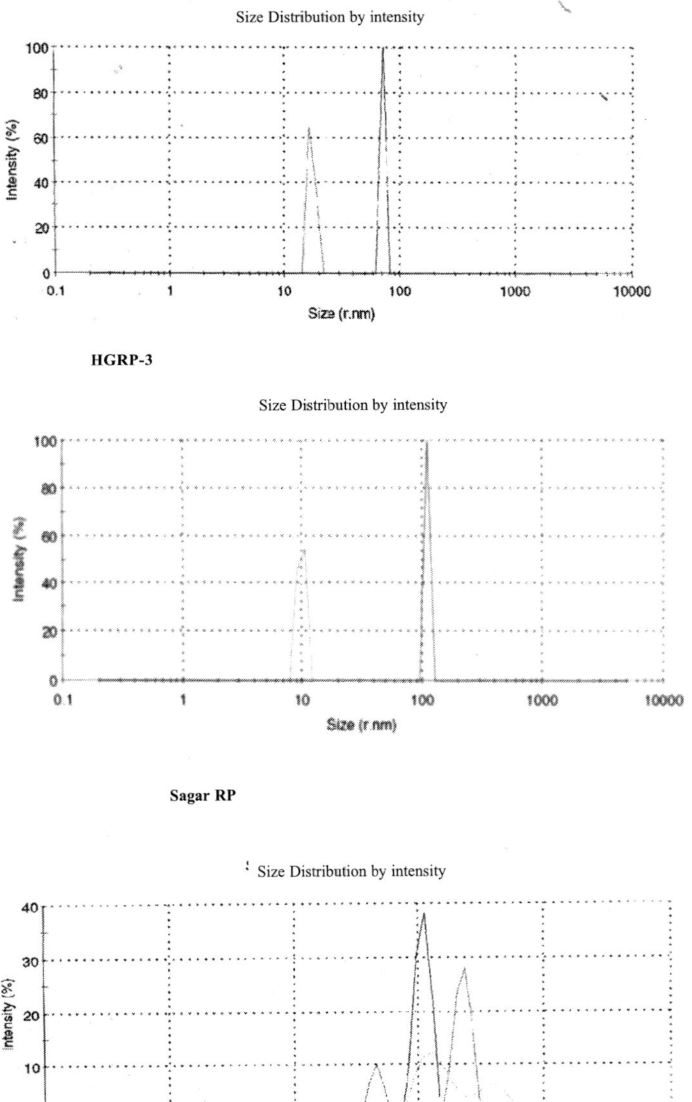

BRP

Fig 2. Particle size distribution of three rock phosphates after grinding in Pot Mill

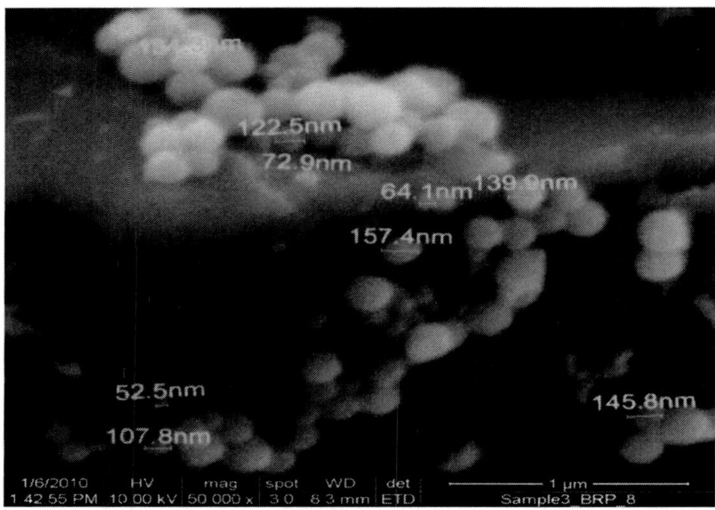

Fig 3. Scanning Electronic microscope photograph of BRP rock phosphate nano-particles (*See colour version on page 377*)

Analytical Problems of Nano-Particulate P

Since the particle sizes of these rock phosphates are small enough, the dilute solution of these rock phosphates appeared to be clear. We were interested to see the interaction or behavior of these particles with the chemical reagents used conventionally for the estimation of P. We prepared a stack solution of nano particles (containing 2 ppm P) of HGRP – 3 rock phosphate (~70.89 nm) and synthetic hydroxy apatite (200 nm). Required amount of these solution were dried in water and treated with concentrated HCL (1 ml) and made up to original volume by distilled water. These solution along with the original Stack solution (without treating with HCL) were used for measurement of P in Inductively Coupled Plasma Optical Emission Spectrophotometer (ICPOES). The results showed that ICP failed to measure the P in nano-particles of rock-phosphate and hydroxy apatite when they are suspended in aqueous system. However, 81 and 74 % of the P content of HGRP-3 rock phosphate and hydroxy apatite could be detected when these nano particles were treated with HCL acid. This suggests that ICP could detect P only in ionic form (due to acid solubilization) and not in particulate form (Table 3).

Table 3. Recovery of P from nano particles in ICP

Sl. No.	Sample Details (Nano Particle)	P (ppm)
1.	HGRP3 (2 ppm Solution) Pretreatment with HCL	1.622
2.	HGRP 3 (2 ppm Solution) without Pretreatment	0.01
3.	Hydroxy Apatite (2 ppm Solution) Pretreatment with HCL	1.467
4.	Hydroxy Apatite (2 ppm Solution) without Pretreatment	0.314

Similarly, P content of the aqueous solution of nano rock phosphate and hydroxy apatite nano-particles was measured by Vanadomolybdo phosphoric Yellow Color Method and Chlorostannous- Reduced Molybdophosphoric Blue Color Method and the recovery of P by these methods is given in above table 3. The results clearly indicate that these nano-particles behaved in a different way with respect to method of analysis and chemical nature of the particles. In general, the recovery of P was higher in Yellow color method, which may be attributed to the higher concentration of HNO_3 acid present in the color development reagent. In yellow color method, the recovery of P was higher in case of Stone –3, followed by HA, HGRP – 3 and TCP, while in case of blue color method the recovery of P was maximum with Stone – 3, followed by HGRP – 3, TCP and HA. Therefore, for recovery of P from any of nano-particles, we followed the process of evaporating the aqueous solution and acid dissolution using concentrated HCL. Thereafter P was estimated by standard colorimetric method. The concentration of nano-particles is also very important and it affects the behavior when applied to the soil. We studied the recovery of P applied to a soil (Aridsoil of Jodhpur) in the form of nano-particles of rock phosphate (HGRP-3, 70.89nm).

In another solution culture experiment with maize grown for 30 days, P was supplied through synthetic nano-particles of calcium phosphate (<100 nm) and Hydroxy apatite (<200 nm) as well as through two different rock phosphate nano-particles (HGRP3 and Stone 3) prepared through top-down approach using ball mill (Table 4). During the growth period, the nutrient solution (Hogland solution) containing P at the rate of 31 ppm in the form of different nano-particles was changed at 3 day interval. All the time, the nutrient solution was sonicated for 30 minutes before use in growth culture experiment. Like spirulina, the growth of maize was excellent. with Hydroxy apatite (<200 nm) followed by calcium phosphate (<100 nm), Stone 3 rock phosphate (~42 nm) and HGRP3 rock phosphate (~28nm), respectively, indicating gradual reduction of plant growth of maize due to the reduction in sizes of the nano-particles. Similar trend was also observed in case of P content in shoot and root of maize. Comparative analysis of growth behavior of maize under hydroxy apatite and HGRP-3 rock phosphate nano-particle clearly showed that there was 25.7% reduction in shoot biomass yield (Fig 4) while root biomass yield and P content in root decreased by 50% due to HGRP-3 rock phosphate nano-particle (~28nm).

Table 4. Effect of P Containing Nano-Particles on Growth of Maize

Sl. No.	Treatments	Root length (cm)	Root Volume (cc)	Shoot length (cm)	DMY – Shoot (g)	DMY – Root (g)	P Content - Shoot (%)	P Content - Root (%)	Shoot Uptake (mg)	Root Uptake (mg)
1.	Control	400	10	13	0.411	0.26	0.29	0.16	0.4	0.41
2.	HA (<200nm)	2479	60	61	12.46	4.54	0.55	0.31	68.5	14.0
3.	TCP (<100nm)	2132	50	57	10.85	3.71	0.52	0.25	56.4	9.2
4.	Stone3 (42nm)	2045	45	52	9.98	3.01	0.45	0.20	44.9	6.0
5.	HGRP3 (28nm)	1850	40	45	9.25	2.27	0.40	0.15	37.0	3.4
6.	HGRP3 (53mm)	830	25	26	3.94	0.72	0.20	0.19	7.8	1.36

Effect of Different size Nano-particles on Plant Growth Parameters of Maize

The roots of maize grown with hydroxy apatite was more healthy (with more thicker roots) than the roots observed with HGRP-3 nano-particles (Fig 5) moreover the roots of maize under HGRP-3 rock phosphate nano-particles showed some root injury/damage and possibly because of that the stored water, containing roots, after detachment from shoot became dark colored overnight. The results apparently suggest that the finer particles (<50 nm) might be causing some hindrance in root proliferation and it is very difficult to explain this behavior with the present state of knowledge in soil science.

Fig. 4. Response of maize to HGRP-3, Stone-3 and Synthetic Hydroxy Apatite NPs (200nm)
(*See colour version on page 378*)

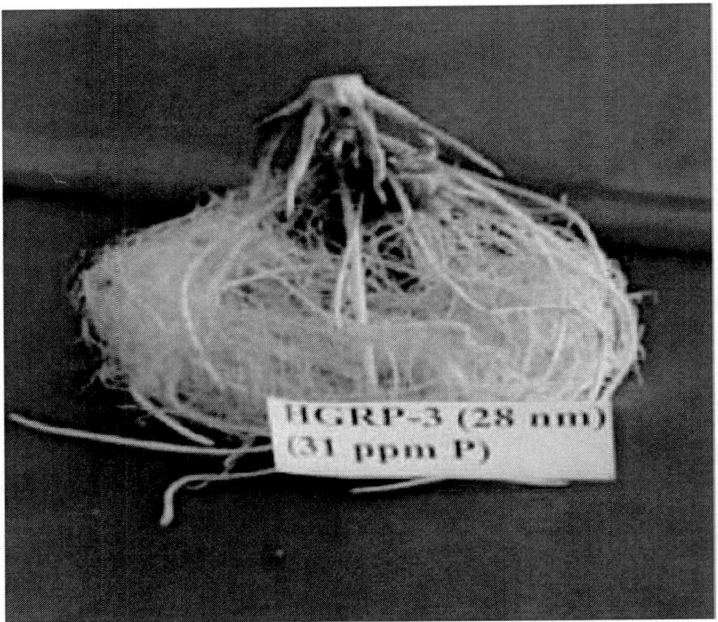

Fig. 5. Growth of Maize roots under nano-particles of HGRP-3 and Stone-3 Rock phosphates
(*See colour version on page 378*)

References

Banfield, J. F. and Zhang, H. (2001). Nanoparticles in the environment. In "Nanoparticles and the Environment". (J. F. Banfield and A. Navrotsky, Eds), pp. 1-58. Mineralogical Society of America, Washington DC Chapter 1.

Buffle, J. (2006). The key role of environmental colloids/ nanoparticles for the sustainability of life. *Environmental Chemistry*, 3: 155-158.

EPA (2007). Nanotechnology White paper. U. S. Environmental Protection Agency Report EPA 100/B-07/001, Washington DC 20460, USA.

Franklin N.M., Rogers N.J., Apte S.C., Batley G. E., Gadd G. E. and Casey P.S. (2007). Comparative Toxicity of Nanoparticulate ZnO, Bulk ZnO, and $ZnCl_2$ to a freshwater microalga *(Pseudokirchneriella subcapitata):* The importance of particle solubility. *Environmental Science and Technology,* 41(24):8484-8490.

Gao, F. Q., Hong, F. H., Liu, C., Zheng, L., Su, M. Y., Wu, X., Yang, F. and Wu, Yang, P., (2006). Mechanism of nano-anatase TiO_2 on promoting photosynthetic carbon reaction of spinach – including complex of Rubisco – Rubisco activase. Biol. *Trace Element Research,* 111: 239-253.

Handy R. D., Owen R. and Valsami-Jones E. (2008). The ecotoxicology of nano-particles and nano-materials: Current status, knowledge gaps, challenges and future needs. *Ecotoxicology,* 17:315-325.

Hochella, M. F., Jr., Lower, S. K. Maurice, P. A., Penn, R. L., Sahai, N., Sparks, D. L., and Twining, B. S. (2008). *Nanominerals, minerals nanoparticles and Earth chemistry. Science,* 21: 1631-1635.

Hong, F. H., Yang, F., Liu, C., Gao, Q., Wan, Z. G., Gu, F. G., Wu, C., Ma, Z. N., Zhou, J., and Yang, P. (2005). Influence of nano – TiO_2 on the chloroplast aging of spinach under light. Biol. *Trace Element Research,* 104: 249-260.

Itoh, K., Pongpeerapat, A., Tozuka, Y., Oguchi, T. and Yamamoto, K. (2003). Nanoparticle Formation of Poorly Water-Soluble Drugs from Ternary Ground Mixtures with PVP and SDS. *Chemical and Pharmaceutical Bulletin*, 51: 171-174.

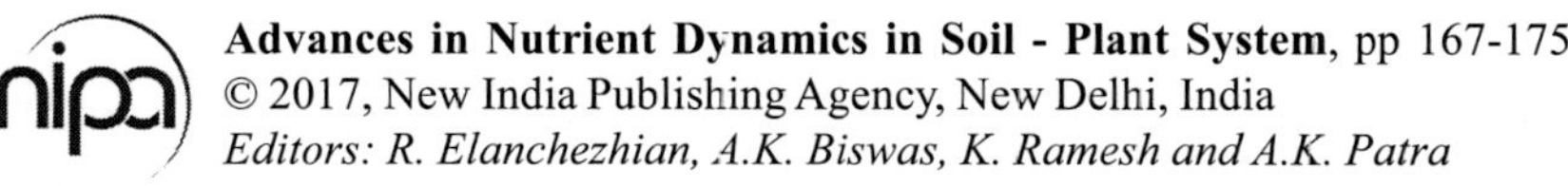
Advances in Nutrient Dynamics in Soil - Plant System, pp 167-175

Editors: R. Elanchezhian, A.K. Biswas, K. Ramesh and A.K. Patra

15

Advances in Genetic Basis of Nutrient Use Efficiency of Plants

C.N.Neeraja and S.R.Voleti

Indian Institute of Rice Research, Hyderabad 500030, India

Nutrient use efficiency of crops is being improved through temporal and spatial management of the form and amount of nutrient inputs as being demonstrated by various agronomic practices. In addition, understanding of the genetic basis of nutrient efficiency would lead to the development of nutrient use efficient varieties thus reducing the application of nutrients. Earlier studies on genetics of nutrient use efficiency were limited as increasing productivity with heavy nutrient inputs was the main focus. But in the backdrop of environmental degradation due to the excessive nutrients and its impact in climate change, nutrient use efficiency in crops is need of the hour for sustainable and eco-friendly agriculture.

Of the various essential nutrients required for crops, nitrogen (N) and phosphorus (P) are fundamental to crop development because they form the basic component of nucleic acids, proteins and many organic molecules. Till now varieties responding to nutrients with high uptake efficiency and utilization efficiency manifested in terms of yield were selected. Since uptake and utilization are interdependent, the efficiency for nutrients was not studied earlier.

Germplasm of major food crops has been screened under low inputs to identify the promising genotypes, however the mechanism of nutrient use efficiency was complex to be deciphered by explanation of single major genes and the effect of environment has confounded the studies. The genetic studies for nutrient use efficiency have been further complicated mostly by low heritability, high environmental variability, difficulty of field screening and absence of clear selection criteria. From the reported studies, it can be inferred that nutrient use efficiency should be targeted for optimum nutrient inputs rather than zero inputs as the nutrients are essential building blocks for realizing the output. With high throughput phenotyping and genotyping approaches along with the availability of genome sequences, the genetics of nutrient use efficiency is now being interpreted.

Being major nutrients, the advances in genetic basis of nitrogen and phosphorus use efficiency in rice and maize are being compiled in this chapter.

Nitrogen

Among the major fertilizer inputs, nitrogen (N) is the key nutrient element required in large quantities by rice. Nitrogen is a primary constituent of the nucleotides, amino acids, proteins, chlorophyll and several plant hormones and is a crucial macronutrient essential and rate-limiting for the growth and development of plants. It comprises 1.5-2% of plant dry matter and approximately 16% of total plant protein (Frink *et al.,* 1999). As nitrogen is crucial for cell/tissue expansion and multiplication, limitation of N would impose constraint in total biomass and therefore yield (Hirel 2007). Reduced growth, gradual chlorosis of older leaves followed by abscissions, altered root architecture, an increased root to shoot ratio and increased root surface have been reported for N deficient plants. Plant height is the most affected by N limitation in rice as observed by 20-30% height reduction under low N. The number of tillers and productive tillers were also relatively less in low N situation owing the less meristematic activity. The panicle weight also been reported to be reduced by approximately 40%. Under low N situation, there is a decrease in the panicle weight owing to the decrease in the number of filled grains per panicle. Several genotypes have been identified with promising performance for most of the parameters under low N and the performance of these genotypes needs to be validated under 50% of recommended dose of N (Rao *et al.,* 2014). The medium doses of nitrogen application have a positive effect on the activities of enzymes of physiological importance, thereby increasing the grain size and promoting grain filling by remobilizing assimilates towards panicles to increase grain yield by accelerating the endosperm cell number, grain length and grain width.

QTL analysis based on molecular linkage maps is now proven methodology for identification of genomic regions associated with complex traits. The initial studies for QTL identification for nitrogen metabolism and use efficiency were based on physiological parameters of nitrogen metabolism. Using backcross inbred lines (BILs) between Nipponbare and Kasalath, seven QTL for GS1 protein content and six QTL for NADH-GOGAT protein content were detected. Some of these QTLs were co-located in QTL regions for various biochemical and physiological traits affected by nitrogen recycling. A structural gene for GS1 was also mapped on chromosome 2 co-located in the QTL region for spikelet weight and a QTL region for NADH-GOGAT protein content was also coincided with the physical position of NADH-GOGAT gene on chromosome 1 (Obara *et al.,* 2001; Yamaya *et al.,* 2002). QTL controlling the ratio of Rubisco to total leaf N have been identified (Ishimaru *et al.,* 2001). Two QTL associated with uptake of nitrogen from ammonium source have been identified on chromosomes 2 and 5 and two QTL associated with uptake of N from nitrate source have been identified on chromosomes 5 and 6. In the same study, a QTL for N use efficiency has also been identified on chromosome 12 (Fang *et al.,* 2001). QTL for plant height under high and low N levels in nutrient solution and soil solution culture experiments (Fang and Wu 2001). A QTL on chromosome 2 associated with the protein content of cytosolic glutamine synthetase (GS1; EC 6.3.1.2) in senescing leaves, panicle number and panicle was characterized and substitution line with this QTL showed more active tillers during vegetative stage and more panicle number and total panicle weight (Obara *et al.,* 2004). Using 239 recombinant inbred lines (RILs) screened under low and normal N solutions, several QTLs for low N tolerance in seedling stage have been identified, however very few QTLs were found to be common for low and normal N conditions (Lian *et al.,* 2005). Under two nitrogen levels, QTL have been identified for plant height, panicle number per plant, chlorophyll content, shoot dry weight at late developmental stage (Tong *et al.,* 2006). Screening of DH population under three nitrogen regimes and mapping of QTLs led to the identification of seven QTL on chromosome 3 associated with nitrogen use, plant yield and associated traits (Senthilvel *et al.,* 2008). Using 166 RIL population, 22 single QTL and 58 pairs of epistatic QTL associated with physiological nitrogen use efficiency in rice have been identified (Cho *et al.,* 2007). With the same mapping population, 28 main effect QTL and 23 pairs of epistatic QTL were detected (Piao *et al.,* 2009). Several QTL for yield components were reported in chromosomal segment substitution lines of Nipponbare and 9311 grown under nitrogen and phosphorus deficiency conditions (Wang *et al.,* 2009). A set of RILs grown in four different seasons in two locations with three nitrogen fertilizations were analyzed for QTL for grain yield components and two main effect QTL were detected viz., grain yield per panicle on chromosome 4 and grain number per panicle on chromosome 12 under N zero level (Tong *et al.,* 2011). Four QTLs for trait differences of plant height and heading date between two N levels have been mapped on chromosomes 2 and 8 co-locating with reported QTLs for NUE (Feng *et al.,* 2011). In response to low nitrogen application for two years, 33 QTL have been identified in RIL population, out of which only ten QTL were consistent under low N (Wei *et al.,* 2012). QTL mapping for NUE and nitrogen deficiency tolerance traits in RIL population for two years resulted in four common QTL on chromosomes 1, 3, 4 and 7 (Wei *et al.,* 2012).

Though the candidate genes involved in nitrogen metabolism in rice are well characterized (Hirel and Gadal 1980; Wang *et al.,* 1993, Kronzucker *et al.,* 2000; Von Wiren *et al.,* 2000; Tabuchi *et al.,* 2007; Yuan *et al.,* 2007), the molecular response to N cannot be exactly attributed to candidate genes as shown by rapid induction/repression of genes and transcription factors (Lian *et al.,* 2006). The sources of nitrogen for rice in field, either ammonium or nitrate are absorbed through transporters. These transporters are divided into high-affinity transporter system (HATS) and low-affinity transport system (LATS). Under low nitrogen concentration (< 1 mM), HATS mediate most of the N uptake while under high concentration of N (>1 mM), LATS play role in N uptake (Forde Clarkson 1999; Glass *et al.,* 2001; Williams and Miller 2001). Both root architecture and the activities of ammonium and nitrate transporters regulated by N form and concentration, diurnal fluctuations, and temperature fluctuations after N acquisition by roots (Garnet *et al.,* 2009).

For the uptake of ammonia, initially three ammonium transporter (AMT) genes have been identified in rice (Sonoda *et al.,* 2003). Now ~ 12 putative rice AMT proteins have been identified and grouped into five sub-families (AMT1-AMT5) with one to three gene members (Suenaga *et al.,* 2003; Deng *et al.,* 2007; Li *et al.,* 2009). OsAMT1;1 is constitutively expressed in shoots and roots (Ding *et al.,* 2011), OsAMT1;2 show root specific expression and induced by ammonium and OsAMT1;3 is root specific and show ammonium derepressed expression (Sonoda *et al.,* 2003). Recently, a high affinity urea transporter (OsDUR3) has been identified in rice roots which is upregulated under N deficiency (Wang *et al.,* 2012). For nitrate transporters, low affinity nitrate transporter OsNRT1 has shown to contribute to N uptake in the root epidermis and root hairs (Lin *et al.,* 2000) and high affinity nitrate transporter OsNRT2 is nitrate inducible (Cai *et al.,* 2008). Recently it was found that rice OsNAR2.1 interacts with OsNRT2.1, OsNRT2.2 and OsNRT2.3a nitrate transporters to provide uptake over high and low concentration ranges of N (Yan *et al.,* 2011).

After uptake into the plant, the nitrate is reduced to nitrite catalyzed by nitrate reductase (NR) and then to ammonium by nitrite reductase (NiR). The primary assimilation takes both in shoots and in roots and ammonium is incorporated into organic molecules by GS and GOGAT. Out of two major forms of GS viz., cytosolic GS1 expressed in roots and shoots, and plastidic GS2 expressed in chloroplasts and plastids. GS1 is complex gene family comprising three genes in rice and GS2 detected in mesophyll cells (Sakurai *et al.,* 1996; Obara *et al.,* 2000), plays the major role in the photorespiratory nitrogen metabolism (Ireland and Lea 1999). Among three genes in rice, OsGS1;1 was expression in all tissues with higher expression in the leaf blade during vegetative stage of growth (Tabuchi *et al.,* 2005). OsGS1;2 transcripts have been found in all tissues with higher expression in root following the supply of ammonium at seedling stage and OsGS1;3 exclusively expressed in spikelet. Transcripts of OsGS1;1 accumulated in the dermatogens, epidermis and exodermis under ammonium limited conditions whereas transcripts of OsGS1;2 was abundantly expressed in the same cell layers under ammonium sufficient conditions. Cytosolic GS1;2 is was shown to be responsible for the primary assimilation of ammonium in rice roots (Funayama *et al.,* 2013).

There are two GOGAT molecular species in rice plants. One is the ferredoxin (Fd)-dependent GOGAT and the other, NADH-dependent. Fd-GOGAT is known to be involved in photorespiration. NADH-GOGAT occurs as a single gene in rice (Goto *et al.,* 1998). Cell type specific and age dependent expression of the NADH-GOGAT gene was confirmed by promoter analysis in transgenic rice (Kojima *et al.,* 2000). Over expression of NADH-GOGAT in sink organs of *indica* genotype and the subsequent increase in grain weight strongly supported the hypothesis that NADH-GOGAT in spikelets at the early stage of ripening is important to reutilize glutamine. Not only that, but also the grain filling process is co-regulated by the status of the re-utilization in which the precise mechanism is largely unknown (Yamaya *et al.,* 2002). Glutamate is a major free amino acid in the leaf blades (Kamachi *et al.,* 1991), whereas glutamine and asparagine, which are synthesized from glutamine (Lea *et al.,* 1990, Sechley *et al.,* 1992, Ireland and Lea 1999), are major forms of total amino acids in phloem sap of rice plants (Hayashi and Chino 1990).

The glutamate amino group can be transferred to aminoacids by a number of different aminotransferases (Lam *et al.,* 1996). Asparagine synthetase (AS) catalyzes the formation of asparagine and glutamate from glutamine and aspartate. Thus AS with GS plays an important role in primary N metabolism (Xu *et al.,* 2012). Mitochondrial NADH-glutamate dehydrogenase (GDH) can also incorporate glutamate under high levels of ammonium (Masclaux-Daubresse *et al.,* 2010). Map based cloning using rice mutant for reduced growth revealed loss of function in arginase gene (OsARG) and was shown to play crucial role in conditions of insufficient exogenous nitrogen (Ma *et al.,* 2012). In rice, OsIPT4, OsIPR5, OsIPT7 and OsIPT8 (adenoside phosphate-isopentenyltransferease) were upregulated in response to exogenously applied nitrate and ammonium with accumulation of cytokinins (Kamada-Nobusada *et al.,* 2013).

Several genes and various metabolic and regulatory pathways appear to be involved in the adaptive response of low N. Thus, genome-wide investigation of gene expression by microarray represented an effective approach for analysing gene regulatory networks in rice. Expression profiles of *indica* variety were studied after low N with normal N as control using a microarray of 11,494 rice ESTs representing 10,422 genes. No significant differences were found in the leaf tissue and 471 ESTs were detected in root tissues with 115 ESTs up-regulated and 358 ESTs down-regulated. The up regulated genes comprised early response genes involved with biotic and abiotic stress and some transcriptions factors and signal transduction. The down regulated genes included photosynthesis and energy metabolism, stress response, transcription factors and signal transduction. Under microarray analysis, no differential expression was found in the genes known to be involved in N uptake and assimilation (Lian *et al.,* 2006). Using AffymetrixGenechip rice arrays, the dynamics of rice transcriptome under N starvation situation, 3518 induced/suppressed genes belonging to cellular metabolic pathways including stress response, primary and secondary metabolism, molecular transport, regulatory process and organismal development representing 10.88% genome were identified. 462 or 13.1% transcripts for N starvation expressed similarly in root and shoot (Cai *et al.,* 2012). Differential expression indicates the potential target genes for nitrogen-use efficiency improvement of rice.

Proteomics is a high-throughput biotechnological approach being used to understand the biological function of proteins in response to different biotic or abiotic stresses (Agarwal *et al.,* 2002; Kim*et al.,*2004). Considering the complexity of NUE trait, it is important to identify the signal transduction pathways and the elements that function to regulate genes involved in N uptake and assimilation. Since signal transduction and gene regulation are based on proteins at large, protein-expression pattern has been studied to identify and understand the role of various proteins at a given point in time. Quantitative as well as qualitative differential expression of protein spots may help in identifying the essential molecules (enzymes) responsible for N uptake and assimilation. Comparative proteome analyses of proteins isolated from leaves under N starvation and N sufficient conditions through matrix assisted laser desorption/ionization time of flight (MALDI-TOF) mass spectrometry and electron spray ionization quadruopole TOF identified N starvation responsive proteins belonging to protein synthesis, metabolism and defence (Kim *et al.,* 2011). In another study at JamiaHamdard University, New Delhi, root proteome of nitrogen efficient and nitrogen inefficient rice cultivars was analysed using two dimensional gradient electrophoresis (2-DGE) based proteomics approach. 504 protein spots were identified with a positive correlation observed between physiological parameters and the concentration of a number of root proteins. Analyses showed that glutamine synthetase, cysteine proteinase inhibitor-I, porphobilinogendeaminase (fragment) and ferritin were involved in conferring N efficiency to N-efficient rice cultivars/genotypes (Hakeem *et al.,* 2013).

Maize

Extensive research has been done in maize for NUE. Variation in NUE among maize genotypes is well documented providing opportunities for genetic improvement. Breeding for low-N-tolerant maize has long been a target of CIMMYT and other major maize breeding companies. Using maize hybrids, the proof of concept of NUE was demonstrated with yield increase of 8-10% and reduction of N input by 16-21% (Chen *et al.,* 2013). With availability of datasets for QTL, genome wide association mapping, transcriptome, proteome, and metabolome, a comprehensive understanding of nitrogen regulation is being attempted in maize (Simons *et al.,* 2014). Models are being developed to provide a framework for understanding of the metabolic processes underlying the more efficient use of N-based fertilizers.

Phosphorus

Phosphorus (P) is a key element in the production offood crops and the demand for P fertilizer is increasing worldwide. With the apprehension about phosphate rock, the source of P fertilizer being a non-renewable resource and the cost of phosphorus fertilizers, development of crop varieties with high productivity under low P is one of long term strategies to address the problem of low P. Phosphorous is a major component of energy currency of cell and a macronutrient essential for plant growth and development.Phosphorus acquisition and requirement are highest during the earlygrowth stages.P deficiency delays plant development and reduces tillering causing significant yield losses. The deficiency also induces root growth encouraging a large root surface area while long root hairs and highly branched rootsystems, especially in the top soil are needed for uptake of phosphorus. Screening for P efficient varieties has its own problems is confounded as the phenotypic screening in problem soils, particularly acidic soils is often limited due to other stresses (eg. iron toxicity, aluminium toxicity) which retard root growth and restrict phenotyping.

Rice

Approximately 5.7 billion hectares of arable land lack sufficient P available for plant and almost 50% of rice soils are P-deficient worldwide (Batjes, 1997). In rice, a major QTL*Pup1*located on chromosome 12 explaining 78.8% of the total phenotypic variance for phosphorus uptake has been found to be associated with tolerance to P deficiency and efficient P uptake in low phosphorus soil (Wissuwa *et al.*, 1998; 2002). Using a marker-assisted backcrossing approach, *Pup1* has been successfully introgressed into two irrigated rice varieties, namely IR64 and IR74 and three Indonesian upland varieties, namely Dodokan, Situ Bagendit, and Batur (Chin *et al.*, 2011). Preliminary data from this study indicates the potential of*Pup1* to work across different genetic backgrounds and environmental conditions. Analysis of functional mechanism of Pup1 locus has shown the presence of a Pup 1 specific protein kinase gene named as phosphorus-starvation tolerance 1 (PSTOL1) acting as enhancer of early root growth and its over expression enhanced grain yield in phosphorus-deficient soil (Gamuyao *et al.,* 2012). In addition, a QTL on chromosome 6 was mapped in two independent studies (Ni *et al.,* 1998; Wissuwa *et al.,* 1998) and was shown to contain a cluster of P-responsive genes (Heuer *et al.,* 2009), including the transcription factor gene OsPTF1 which confers tolerance to P deficiency (Yi *et al.,* 2005). Some of the candidate genes involved in P transport have been characterized in rice viz., OsPT2 encodes a low-affinity transporter; OsPT6 a high-affinitytransporter (Ai *et al.,* 2009) and OsPht1; 8 also enhances Puptake (Jia *et al.,* 2011). Several genes responding to low P situation were identified through microarray gene expression studies in rice (Wasaki *et al.,* 2003; Pariasca-Tanaka *et al.,* 2009). The regulatory network of P homeostasis is also being studied to identify positiveregulators of P transporters and suppressors of P starvation genes.

Maize

For phosphorus, germplasm from various sources has been screened and promising genotypes have been identified (Zhang *et al.,* 2014). QTL for traits involved with phosphorus deficiency tolerance, transcriptome profiles and next generation sequencing data were generated. Attempts are in progress to develop a marker system for phosphorus deficiency tolerance in maize.

Conclusions and Future Prospects

Development of nutrient use efficient varieties is inevitable for sustainability of environmental friendly and economical agricultural practices. While several management practices are being studied for increasing efficiency of spatial and temporal inputs of nutrients, several attempts are being made to identify genotypes with differential nutrient use efficiency for Indian situation. An attempt should also be made to evaluate reported exotic germplasm for nutrient use efficiency and use the sources to develop nutrient use efficient varieties with multidisciplinary approach. As the major nutrients viz., nitrogen and phosphorus are the building blocks of biomass, an optimum quantity is required for realizing the yield. So, the strategy should be maximum uptake, maximum utilization and maximum remobilization of the optimum nutrient inputs to give maximum possible yield. From the observations of the reported studies across the world, the genotypes do exist in major food crops with differential ability for maximum uptake, utilization and remobilization. However, all the three traits are not usually observed in a single genotype. Therefore using multidisciplinary approach, pyramiding of possible mechanisms for nutrient use efficiency should be a possible strategy. With advent of genome sequencing and next generation sequencing, the identification of allelic variation for nutrient use efficiency appears to be a promising strategy. Earlier, several efforts were made to develop nutrient use efficient varieties using transgenic technology. Several candidate genes associated nitrogen and phosphorus metabolism were targeted for transgenic development and the proof of concept of enhancement of nutrient use efficiency and yield were shown. However, the transgenic experiments could not be taken to field level considering the constraints for the adoption of the technology. With the resources of information of candidate genes associated with nitrogen and phosphorus metabolism from genome sequencing studies, their expression pattern using transcritptomics and proteomics, the genomic regions and the alleles of candidate genes associated with nitrogen and phosphorus are being identified at several national and international research institutes and the information generated is being deployed to develop breeding lines with nutrient use efficiency.

References

Fang P and Wu P (2001). QTL × N-level interaction for plant height in rice (*Oryza Sativa* L.). *Plant and Soil,* 236(2): 237-242.

Fang P, Tao QN and Wu P (2001). QTLs underlying rice root to uptake NH_4^-N and NO_3^-N and rice N use efficiency at seedling stage. *Plant Nutrition and Fertilizer Science*, 2: 159-165.

Frink CR, Waggoner PE and Ausubel JH (1999). Nitrogen fertilizer: Retrospect and prospect. *PNAS* USA, 96: 1175-1180.

Hirel B, Le Gouis J, Ney B, Gallais A (2007). The challenge of improving nitrogen use efficiency in crop plants: towards a more central role for genetic variability and quantitative genetics within integrated approaches. *Journal of Experimental Botany*, 58: 2369–2387.

Ishimaru K, Kobayashi N, Ono K and Yano M (2001). Are contents of Rubisco, soluble protein and nitrogen in flag leaves of rice controlled by the same genetics? *Journal of Experimental Botany,* 52: 1827-1833.

Rao PR, Neeraja CN, Vishnukiran T, Srikanth B, Vijayalakshmi P, Subhakara Rao I, Swamy KN, Kondamudi R, Sailaja N, Subrahmanyam D, Surekha K, Prasadbabu MBB, Subba Rao LV and Voleti SR (2014). In Nitrogen Use Efficiency in Irrigated Rice for Climate Change– A Case Study. Directorate of Rice Research, Rajendranagar, Hyderabad.

Yamaya T, Obara M, Nakajima H, Sasaki S, Hayakawa T and Sato T (2002). Genetic manipulation and quantitative-trait loci mapping for nitrogen recycling in rice. *Journal of Experimental Botany,* 53: 917-925.

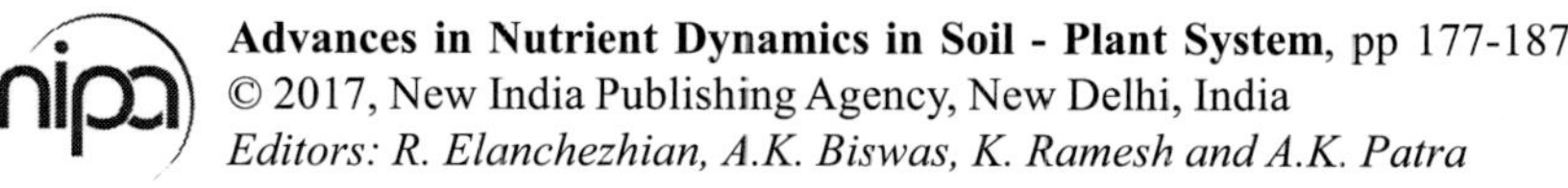

Advances in Nutrient Dynamics in Soil - Plant System, pp 177-187
© 2017, New India Publishing Agency, New Delhi, India
Editors: R. Elanchezhian, A.K. Biswas, K. Ramesh and A.K. Patra

16

Advances in Phosphorus Use Efficiency of Plants

Renu Pandey[1] and R. Elanchezhian[2]

[1]Division of Plant Physiology, IARI, New Delhi-110012
[2]ICAR-Indian Institute of Soil Science, Nabi Bagh, Bhopal – 462038, India

Low bioavailability of phosphorus (P) in soils is one of the major limiting factors influencing crop production throughout the world. Plants, however, are known to possess potential adaptive mechanisms at morphological, physiological, biochemical, and molecular levels to overcome P deficiency (Vance *et al.,* 2003). Such adaptive mechanisms mainly include an increase in total root length and root hair growth (Bates and Lynch, 2000), enhanced organic acid (Singh and Pandey, 2003), acid phosphatase (Pandey, 2006) and ribonuclease (RNase) secretion into the rhizosphere (Hocking, 2001), increase in expression of proteins such as phosphatase, inorganic phosphate (Pi) transporter, RNase and phosphoenolpyruvate carboxylase (PEPcase) in plant tissues.

Existence of genotypic variation in P uptake and utilization by cereals is known from early studies (Smith, 1934) and confirmed later (Nielsen and Barber, 1978; Gahoonia and Nielsen, 1996). The difference in plant genotypes in terms of nutrient acquisition efficiency lies in their mechanisms adopted under nutrient deficient condition. Nutrient-efficient genotypes may have an increased capacity (i) to exploit the soil (large root surface area), (ii) to convert non-available nutrient forms into plant-available forms, and/or (iii) to take up nutrients across the plasmamembrane. Developing plants with improved P efficiency through conventional plant breeding or genetic engineering is the need of the hour. The various physiological, biochemical and molecular plant processes that could be modified for efficient P uptake under P-stress are given below.

Modifications in Root Morphology

Root morphological characters such as architecture (spatial arrangement) and branching, size, density and length of root hairs have profound effect on plant acquisition of nutrients from soil. The size and/or rate of growth of root system into unexploited soil are particularly important for the uptake of those nutrient elements which are less mobile in soil such as P. Root growth provides greater root-soil contact for nutrient uptake. More roots, however, also mean higher demand for photosynthates. Experimental results have revealed large differences in root morphology and distribution between the species (O'Toole and Bland, 1987) and between the genotypes within the species (Atkinson 1991). Unlike mobile nutrients, like nitrate, which can be depleted at low rooting density (Pan *et al.,* 1985), the less mobile or immobile ions like P and micronutrients, uptake is closely related to root length (Romer *et al.,* 1988). The spatial arrangement of root system is another important factor in determining uptake of P. It has been reported that the P-efficient genotypes of common bean have a vigorous and highly branched root systems with large numbers of apices (Lynch and van Beem, 1993).

Root diameter is important because it defines the maximum volume of soil that can be exploited by roots with a given amount of photosynthate. A smaller root diameter suggests that a larger soil volume per unit root surface area can deliver nutrients. It has been reported that the root diameter varies between species and cultivars and changes with plant age (Atkinson, 1985). Though crop genotypes with thinner roots may be more effective in absorbing nutrients effectively from soil but the maintenance carbon cost of producing finer roots will be higher as these will have to be replaced more frequently (Persson, 1982).

Another important feature of root is fine root hairs which can substantially increase the root surface area available for ion uptake at relatively low carbon cost (Gahoonia and Nielsen, 1998). Regardless of plant species, P uptake was found to be a function of soil volume exploited by the root hairs (Misra *et al.,* 1988) and P uptake correlated to the number and length of root hairs of plant species (Fohse *et al.,* 1991). Genotypic variation in seven wheat cultivars was reported in terms of root surface area under low P soil which was directly correlated with P uptake (Pandey and Gahoonia, 2004). Though the root volume increased at P_0 (no applied P), significant decrease in root mass without a corresponding decrease in volume under P deficiency indicates a change in the density of the root *per se.*

A good example of root modification is the formation of 'cluster' roots (also termed as 'proteoid' roots) which are bottlebrush-like clusters of rootlets that arise from the pericycle of the lateral roots. Cluster roots are commonly found in Proteacea like *Hakeae* spp., *Leucadendron laureolum*, *Grevillea robusta* as well as in the legume *Lupinus albus*. The accompanying increase in root hair density of clustered rootlets results in an increase of surface area of greater than 100-fold as compared to normal roots. These roots secrete large proportion of organic acids under P stress which include malate and citrate as the major components. Among species examined for organic acid production in response to phosphorus stress, lupines exhibit the strongest trends (Dakora and Phillips 2002). Mature proteoid roots increase organic acid production and decrease organic acid metabolism compared to non-proteoid roots, resulting in much higher levels of organic acid exudation (1.16 compared to 0.09 μmol h^{-1} g^{-1}) (Kania *et al.,* 2003; Uhde-Stone *et al.,* 2003). As a result, phosphorus uptake can be as much as 50% greater in proteoid than non-proteoid lupine roots (Neumann and Romheld 1999).

Modifications in Root Physiology

Root-induced modification of rhizosphere is a key factor affecting plant uptake of less-mobile nutrient ions. Root exudation clearly represents a significant carbon cost to the plant and the magnitude of photosynthates secreted as root exudates varies with the type of soil, age, and physiological state of the plant, and nutrient availability. Root exudations has two possible effects: (1) reduction in rhizosphere pH that will increase the bioavailability of P and micronutrients in soils with high pH, such as calcareous soils and; (2) chelation of metal ions by organic anions thereby releasing fixed nutrients from solid phase into the soil solution.

Phosphorus uptake kinetics

Importance of kinetic parameters such as the affinity of the P uptake systems (K_m), maximum ion influx rate (I_{max}) or threshold concentration (C_{min}), the minimum P concentration at which plant roots can deplete phosphorus from the external solution, in regulating P-acquisition efficiency of cereal genotypes and the native intra- and inter-species variability for the kinetic attributes among cereals are a lesser probed aspects. Significance of C_{min} in ion uptake, especially on nutrient poor soils is known. Studies evaluating a relationship between kinetic attributes of P-uptake and the P-uptake efficiency of maize and barley (Nielsen and Barber 1978, Neilsen and Schjørring 1983) showed that a combination of low K_m, low C_{min} and high I_{max} could result in higher P-uptake and suggested that these parameters could be used to select phosphorus efficient genotypes. Under conditions in which the rate limiting step is located in the plant root, the mean net influx (I_n) of ions into plant root can be expressed by (Nielsen 1976):

$$I_n = I_{max} (C_0 - C_{min})/ K_m + C_0 - C_{min}$$

I_{max} is the mean maximum influx, mol/cm/s

C_0 is the concentration of ion at root surface, mol/cm^3

C_{min} is the concentration at which $I_n = 0$

K_m is the Michaelis-Menten factor, mol/cm^3

$I_n = ½$ if $C_0 = K_m + C_{min}$

Organic acids and P acquisition

Organic acid anions have a role in solubilisation of mineral nutrients in the rhizosphere. Typical carboxylates (organic acid anion) released from roots of axenic grown plant species include citrate, malate, malonate, acetate, fumarate, succinate, lactate and oxalate. Organic acids can also act as metal chelators in the rhizosphere, but are thought to have more important effects on phosphorus availability than on micronutrient availability (Dakora and Phillips, 2002). Several plants have been reported to increase root-secretion of organic acids in response to P deficiencies such as *Lupinus alba* (Jhonson *et al.,* 1994; Neumann and Romheld, 1999), *Brassica napus* (Hoffland *et al.,* 1992) and *Medicago sativa* (Lipton *et al.,* 1987). Differences in root exudation between crop species as well as cultivars were observed. Maize and green gram cultivars were studied for root exudation using ^{14}C (Singh and Pandey, 2003). The average amount of root exudation by green gram, based on ^{14}C activity measurements, was almost 3-fold higher as compared to maize. However, in this experiment maize exudates contained a higher level of free amino acids and reducing sugars.

Phosphatase activity

Acid phosphatase enzyme secreted mainly by the plant roots, causes hydrolysis of organic phosphates present in the soil as phosphate esters resulting in the release of inorganic P for plant uptake. Exudation of phosphatases increases when plants are grown under phosphorus stress (Radersma and Grierson, 2004). It was reported that when *Triticum aestivum* grown in an acidic P-deficient soil amended with Fe-P, the P-efficient genotypes had a greater acid phosphatase activity in the rhizosphere than the inefficient genotype, with phosphatase activity correlating positively with growth and phosphorus uptake (Marschner *et al.,* 2006). In an another study involving wheat, rye and triticale under phosphorus stress, showed that wheat plants exuded more acid phosphatase into the rhizosphere followed by rye and triticale. A significant positive correlation was also found between acid phosphatase activity (both extra-and intra-cellular) and root, shoot and total plant dry weight (Pandey, 2006).

Root induced-pH changes

It has been demonstrated that differences of as much as 2 units in soil pH around roots can take place. The reasons for such pH changes can be many, including imbalance of cation-anion uptake which is particularly affected by sources of nitrogen, enhanced efflux of protons as a result of P, Fe, and Zn deficiency, Al sensitivity of genotypes and secretion of organic acids. Whatever be the reason of pH changes, large variation exists among the plant species and along the root system which affects the bioavailability of soil P and micronutrients. Drop in rhizosphere pH below 5 leads to increase in solubility of Al which can be taken up at a level toxic by plants. Availability of phosphorus increases as it mainly enters plant roots in $H_2PO_4^-$ form which is again related to soil solution pH. Plant species or genotypes inducing rhizosphere acidification may absorb more P by these mechanisms. Hinsinger *et al.,* (2003) reported that rhizosphere acidification by proton extrusion causes dissolution of poorly available P forms, such as Ca-P, in calcareous soils. Rhizosphere acidification was shown to be more prominent in P-efficient *Phaseolus vulgaris* genotypes than in P-inefficient ones whereas no difference in rhizosphere acidification was found for a range of *Triticum aestivum* and *Hordeum vulgare* genotypes in P efficiency (Rengel, 2005).

Mycorrhizal associations

Arbuscular mycorrhizal fungi (AMF) and plant roots form associations in more than 80% of terrestrial plants. Mycorrhizal symbiosis usually represents a significant photosynthetic carbon cost. Mycorrhizal hyphae extending from the roots are able to absorb and translocate P and certain other nutrients from distant areas which are otherwise not accessible to plant roots. This enhanced P absorption by roots is achieved by increasing the effective root area or by improved P influx, that is, P uptake per unit root length. The mycorrhizal responsiveness is related to the inherent rate of growth, the plasticity of root/shoot ratios, root architecture and root hair development. In barley plants, Baon *et al.* (1992) found that the responsiveness of different cultivars to AMF was positively correlated with their response to P application, indicating the importance of AMF in plant P efficiency. Among crop species as well as genotypes, differences in the degree of arbuscular mycorrhizal colonization have been found to exist. It has been shown that the P utilization efficiency increases in some wheat genotypes when inoculated with AMF without any phosphatic fertilizer application (Pandey *et al.,* 2006). The potential value of breeding plants for greater susceptibility to colonization will depend on the cost-benefit of AMF for the specific crops, soil and environmental conditions.

Molecular Approaches to Improve P Acquisition Efficiency

Microarray studies have made it possible to explore the global gene expression profiles under P deprivation in Arabidopsis (Misson *et al.,* 2005; Chen *et al.,* 2011) and other crop species such as maize, potato, bean and rice. These studies revealed alterations in several biochemical and signalling pathways resulting in changes at the gene, protein and metabolite level. Recently, the RNA-seq technology has been employed to study the global gene expression in *Lupinus albus* wherein 2,128 differentially expressed sequences were identified in response to P deficiency with a 2-fold or greater change (O'Rourke *et al.,* 2013). Transcriptome analyses indicated rapid changes in the expression of several genes encoding Pi transporters, phosphatases and RNases and other genes involved in the re-programming of metabolic pathways of lipid recycling, nitrate assimilation and carbohydrate mobilization. However, altered gene expression regulating photosynthesis, carbohydrate and secondary metabolism were not reversed when P was resupplied. These studies suggest that specific mechanisms evolved in particular plant species to cope with P stress.

Root system architecture regulated at molecular level

Alteration in root development in response to P starvation is a complex trait. However, a few genes associated with this process have been identified in several plant species. In maize, six genes were reported to control root morphology *viz.* *Rtcs* (rootless concerning crown and seminal roots), *Bk2* (brittle stalk-2), *Bk2-L3* (brittle stalk2-like protein-3), *Rth1* (Root hairless1), *Rth3* (roothairless-3) and *Scr* (scarecrow), were studied under P stress. The expression of *Scr* controlling endoderm and cortex formation was not influenced by P starvation. Efficient lines exhibited higher expression of the genes *Rtcs*, *Bk*2 and *Rth*3 relative to P inefficient line (DeSousa *et al.,* 2012). The transcriptome analysis of lateral root primordium zone revealed that the auxin signalling is responsible for causing modification in root morphology under P stress.

A recent landmark development was the identification and functional analysis of *PSTOL1* (P starvation tolerance1) gene in rice (*Indica* cv. Kasalath) which regulates root development and growth under P stress (Gamuyao *et al.,* 2012). However, elucidation of the molecular mechanisms and downstream tar-gets of *PSTOL1* need to be further investigated. This gene can be used to improve P efficiency in rice crops by use of targeted inter-variety breeding. Progress made so far in deciphering the molecular and genetic control of root system development enriches our understanding of root morphogenesis in response to phosphate stress.

Induction of high affinity Pi transporters

The Pi transporters belong to Pht1 (high-affinity) and Pht2 (low-affinity) families expressed under low and high external P concentration, respectively (Lin *et al.,* 2009). The Pht1 genes are involved in P uptake against a sharp concentration gradient, as the root cells may contain 10,000-fold higher soluble P concentration than the soil solution. These transporters consists of 12 transmembrane domains and proton (H^+)/P symporter, a large hydrophilic loop between transmembrane 6 and 7 with both N and C termini located in the cytoplasm. The members of Pht1 family are more abundant in roots than shoots and enhanced transcripts are noted under P starvation. In Arabidopsis, nine members of Pht1 (AtPht1;1 to AtPht1;9) family have been identified and all are responsive to P nutrition. Significant roles of AtPht1;1 and AtPht1;4 are observed in plant Pi acquisition under both deficient and sufficient Pi conditions. Each member of Pht1 exhibits a tissue specific expression, such as in root epidermal, root hair or stellar cells while others were expressed in Golgi apparatus or endoplasmic reticulum. AtPht1;9 and AtPht1;8 were expressed in roots in response to P starvation.

The high affinity Pi transporter genes have also been identified in crop plants in response to P starvation. Five Pht1 genes in maize and 13 putative Pht1 genes (OsPT1 to OsPT13) in rice were identified which contributed to P uptake and allocation. Eight Pht1 genes have been described in barley out of which HvPHT1;1 and HvPHT1;2 were found to be expressed in root hairs, cortex and epidermal cells, and root vascular tissues. Liu *et al.* (2013) isolated and functionally characterised a high affinity P transporter in wheat, TaPht1;4, exclusively expressed in roots under P deficiency. A high affinity Pi transporter, GmPT5, identified in soybean expressed in the junction area between roots and young nodules and regulates the entry of Pi from roots to nodules (Qin *et al.,* 2012). Transport proteins are key targets for improving P uptake efficiency in various crops, however, genotypic variation in the expression pattern and regulation of P transporters is yet to be explored.

Induction of Purple acid phosphatases

Purple acid phosphatases (APases) or PAPs comprise the largest class of plant APases. Most PAPs are non-specific APases and catalyze Pi from phosphomonoesters over a wide pH range. The genes encoding PAPs are highly induced in response to P starvation and are secreted into the rhizosphere to utilize available Pi in the soil or transported to organelles to recycle Pi. The Arabidopsis genome encodes for 29 putative PAP isozymes which are transcriptionally expressed under various developmental and environmental factors (Tran *et al.,* 2010). The first purified and characterized PAP was AtPAP17 from Arabidopsis. Besides influencing P mobilization, it is also involved in the metabolism of reactive oxygen species during incidence of salinity, senescence and oxidative stresses. Another APase, AtPAP26, with a major phosphatase activity is localised in the vacuole. AtPAP26 recycles intracellular Pi and also remobilizes Pi from the organic pool during P starvation (Robinson *et al.,* 2012).

Several PAPs have been identified and characterized from crop species such as LaSAP2 in lupin (*Lupinus albus*), LePS2 in tomato (*Solanum lycopersicum*), GmPAP in soybean (*Glycine max*), VrPAP1 in mungbean (*Vigna radiata*), APase in *Brassica juncea*, PvPS2;1 in bean (*Phaseolus vulgaris*) and NtPAP12 in tobacco. PAPs play a vital role in recycling and scavenging of Pi, therefore are obvious targets for engineering P efficient crops. It is observed that the external application of phosphatic fertilizers influences the P_{org} and Pi content of agricultural soils, thereby altering the amount of P_{org} available for PAP hydrolysis (Richardson *et al.,* 2009). Therefore, it is possible to considerably improve the P uptake efficiency by overexpressing the secreted PAPs in crop plants.

Expression of anion efflux transporters

There are two separate transport processes causing efflux of organic anions into the rhizosphere *viz.* an *active H^+ efflux* which involves a plasma membrane

H^+-ATPase, and the *passive efflux* through channel-like transporters. An important strategy for efficient P acquisition is the organic anion efflux through channels induced during P starvation. However, the organic anion efflux transporters are greatly induced under aluminium (Al) toxicity than P starvation in several crops. Studies on root Al and P interactions provided evidence that P deficiency induces exudation of oxalate and malate, while Al activated roots exude citrate in soybean. Al^{3+}-activated anion channels (ALAACs) permeable to malate and/or citrate were predominantly expressed in the root tips of wheat and maize (Zhang *et al.,* 2001). The first gene, TaALMT1, identified in wheat conferring tolerance to Al-toxicity was overexpressed in highly Al-sensitive transgenic barley seedlings and tobacco. Overexpression of barley gene (HvAACT1) responsible for Al-induced citrate secretion in tobacco enhanced citrate secretion and Al resistance. This provided a direct evidence for the role of root organic acid exudation in plant Al tolerance. Further, many transgenic have been developed in various crops by identifying the homologs of TaALMT1 and overexpressing them. Other genes, encoding type I H^+-pyrophosphatase (AVP1, Arabidopsis Vacuolar Pyrophosphatase1) and a type II H^+- pyrophosphatase (AVP2) also have a role in organic acid exudation. Overexpression of AVP1 in Arabidopsis enhanced the citrate and malate secretion from roots, enabling plants to tolerate Al toxicity (Drozdowicz *et al.,* 2000, Yang *et al.,* 2007). Arabidopsis plants overexpressing malate transporter (GmALMT1) localized to root plasmamembrane exhibited malate efflux in response to P starvation in an extracellular pH-dependent and Al-independent manner (Liang *et al.,* 2013). Since Al and P stresses co-exist in acid soils, most of the genes conferring tolerance towards Al toxicity might also provide tolerance against P starvation.

Strategies to develop P efficient crop plants

There are three approaches to develop crops which can efficiently acquire P from soil, *viz.* conventional breeding, marker-assisted breeding and genetic engineering. Conventional methods such as backcross breeding and recurrent selection have resulted in soybean varieties with superior root traits and other agronomically important traits that helped them out-perform in acid soils with low P. In recent years, molecular marker assisted breeding has gained popularity after the identification of several quantitative trait loci (QTLs) for various traits in response to P stress. QTLs imparting tolerance to P stress have been identified in crops such as rice, common bean, soybean, *Brassica oleracea* and maize. Most of the traits used to map the QTLs associated with P efficiency were based on root characters. Although, QTLs are known but the challenge still lies in utilizing these QTLs in marker assisted selection. Because the genes underlying these QTLs has not been identified except the *Pup1* locus in rice. Therefore, a strategy combining fine mapping and transcriptome analysis may facilitate earlier gene identification for development of P stress tolerant cultivars.

References

Atkinson D. (1985). In: Ecological interactions in soil. A.H. Fitter, D. Atkinson, D.J. Read and M.B. Usher (eds) Blacwell, Oxford. Pp 43-66.

Atkinson, D. (1991). In: Plant Roots: The Hidden Half. Marcel Dekker, Y Waisel, A. Eshel and U. Kafkaki (eds.). Inc. New York, Basel, Hongkong. Pp 411-451.

Baon, J.B., Smith, S.E., Alston, A.M. and Wheeler, A.D. (1992). Phosphorus efficiency of three cereals as related to indigenous mycorrhizal infection. *Australian Journal of Agricultural Research,* 43: 479-491.

Bates, T.R. and Lynch, J.P. (2000). The efficiency of *Arabidopsis thaliana* root hairs in P acquisition. *American Journal of Botany,* 87: 964–970.

Chen, J., Li,u Y., Ni, J., Wang, Y., Bai, Y., Shi, J., Gan, J., Wu, Z., and Wu, P. (2011). OsPHF1 regulates the plasma membrane localization of low- and high-affinity inorganic phosphate transporters and determines inorganic phosphate uptake and translocation in rice. *Plant Physiology,* 157: 269-278.

Dakora, F.D., Phillips, D.A. (2002). Root exudates as mediators of mineral acquisition in low-nutrient environments. *Plant and Soil*, 245:35–47.

Delhaize, E., Ryan, P.R., Hebb, D.M., Yamamoto, Y., Sasaki, T. and Matsumoto H. (2004). Engineering high-level aluminum tolerance in barley with the *ALMT1* gene. *Proceedings of National Academy of Sciences USA,* 101: 15249–15254.

DeSousa, S.M., Clark, R.T., Mendes, F.F., Oliveira, A.C., deVasconcelos, M.J.V., *et al.,* (2012). A role for root morphology and related candidate genes in P acquisition efficiency in maize. *Functional Plant Biology*, 39: 925–935.

Drozdowicz, Y.M., Kissinger, J.C., Rea, P.A. (2000). AVP2, a sequence-divergent, K^+ insensitive H^+-translocating inorganic pyrophosphatase from Arabidopsis. *Plant Physiology*, 123, 353–362.

Gahoonia, T.S. and Nielsen, N.E. (1992). The effect of root induced pH changes on the depletion of inorganic and organic phosphorus in the rhizosphere. *Plant and Soil:* 143: 185-189.

Gahoonia, T.S. and Nielsen, N.E. (1998). Exploring rhizosphere in search for nutrient efficient crop genotypes for low input agriculture-phosphorus and microutrient. In: Crop Improvement for Stress Tolerance. RK Behl, DP Singh, GP Lodhi (eds) CCSHAU, Hisar & MMB, New Delhi, pp 142-156.

Gamuyao, R., Chin, J.H., Pariasca-Tanaka, J., Pesaresi, P., Catausan, S., Dalid, C., *et al.,* (2012). The protein kinase Pstol1 from traditional rice confers tolerance of phosphorus deficiency, *Nature*, 488: 533-539.

Hinsinger, P., Plassard, C., Tang, C., Jaillard, B., Tang, C.X. (2003). Origins of root-mediated pH changes in the rhizosphere and their responses to environmental constraints: a review. *Plant and Soil*, 248: 43–59.

Hoffland, E., Findenegg, G., Nelemans, J., van den Boogaard, R. (1992). Biosynthesis and root exudation of citric and malic acids in phosphate-starved rape plants. *New Phytology,* 122: 675–80.

Liang, C., Pineros, M.A., Tian, J., Yao, Z., Sun, L., *et al.,* (2013). Low pH, aluminium, and phosphorus coordinately regulate malate exudation through GmALMT1 to improve soybean adaptation to acid soils. *Plant Physiology*, 161: 1347–1361.

Liu, X., Zhao, X., Zhang, L., Lu, W., Li, X., Xiao, K. (2013). *TaPht1;4*, a high-affinity phosphate transporter gene in wheat (*Triticum aestivum*), plays an important role in plant phosphate acquisition under phosphorus deprivation. *Functional Plant Biology*, 40: 329–341.

Marschner, P., Solaiman, Z., Rengel, Z. (2006). Rhizosphere properties of Poaceae genotypes under P-limiting conditions. *Plant and Soil*, 283: 11–24

Misra, R.K., Alston, A.M., Dexter, A.R. (1988). Role of root hairs in phosphorus depletion from a macrostructured soil. *Plant and Soil*, 107: 11-18.

Misson, J., Raghothama, K.G., Jain, A., Jouhet, J., Block M.A., *et al.* (2005). A genome-wide transcriptional analysis using *Arabidopsis thaliana* Affymetrix gene chips determined plant responses to phosphate deprivation. *Proceedings of National Academy of Sciences USA,* 102: 11934–11939.

Neumann, G., Römheld, V. (1999). Root excretion of carboxylic acids and protons in phosphorus-deficient plants. *Plant and Soil,* 211: 121–130.

Nielsen, N.E., Barber, S.A. (1978). Difference among genotypes of corn in the kinetics of P uptake. *Agronomy Journal,* 70: 695-698.

Nielsen, N.E. and Schjorring, J.K. (1983). Efficiency and kinetics of phosphorus uptake from soil by various genotypes. *Plant and Soil,* 72: 225-230.

O'Rourke, J.A., Yang, S.S., Mille, S.S., *et al.,* (2013). An RNA-Seq transcriptome analysis of orthophosphate-deficient white lupin reveals novel insights into phosphorus acclimation in plants. *Plant Physiology,* 161: 705–724.

O'Toole JC, Bland W.L. (1987). Genotypic variations in crop plant root systems. *Advances in Agronomy*, 41: 91-145.

Pan, W.L., Jackson, W.A., and Moll, R.W. (1985). Nitrate uptake and partitioning by corn (*Zea mays*) root systems and associated morphological differences among genotypes and stages of root development. *Journal of Experimental Botany,* 36: 1341-1351.

Pandey, R., and ahoonia, T.S. (2004). Phosphorus (P) acquisition by wheat genotypes using genetic diversity to save non-renewable P sources. In: Proceedings, Jakobsen S.E., Jensen C.R., Porter J.R. (Eds.), VIII Congress of European Society of Agronomy, pp. 951-952.

Pandey, R., Singh, B. and Nair, T.V.R. (2005). Impact of mycorrhizal fungi on phosphorus efficiency of wheat, rye and triticale. *Journal of Plant Nutrition,* 28: 1867-1876.

Pandey R. (2006). Root-exuded acid phosphatase and ^{32}Pi-uptake kinetics of wheat, rye and triticale under phosphorus starvation. *Journal of Nuclear Agriculture and Biology*, 35 (3-4): 168-179.

Persson H. (1982). In: Root Ecology and its practical applications. International Symposium, Gumpenstein, Austria. Pp. 595-608.

Qin, L., Guo, Y., Chen, L., Liang, R., Gu, M., *et al.,*(2012). Functional characterization of 14 *Pht1* family genes in yeast and their expressions in response to nutrient starvation in soybean. *PLoS ONE,* 7, e47726. doi:10.1371/journal.pone.0047726.

Radersma, S., Grierson, P.F. (2004). Phosphorus mobilization in agroforestry: organic anions, phosphatase activity and phosphorus fractions in the rhizosphere. *Plant and Soil*, 259: 209–219.

Rengel, Z. (2005). Breeding crops for adaptation to environments with low nutrient availability. In: Ashraf, M., Harris, P.J.C., eds. Abiotic stresses: plant resistance through breeding and molecular approaches. New York, NY, USA: The Haworth Press, 239–276.

Robinson, W.D., Park, J., Tran, H.T., Vecchio, *et al.,* (2012). The secreted purple acid phosphatase isozymes AtPAP12 and AtPAP 26 play a pivotal role in extracellular phosphate-scavenging by *Arabidopsis thaliana*. *Journal of Experimental Botany*, 63: 6531–6542.

Romer, W., Augustin, J., Schilling, G. (1988). The relationship between phosphate absorption and root length in nine wheat cultivars. *Plant and Soil*, 111: 199-201.

Singh, B. and Pandey R. 2003. Differences in root exudation among phosphorus-starved genotypes of maize and green gram and its relationship with phosphorus uptake. *Journal of Plant Nutrition*, 26: 2391-2401.

Tran, L.S., Nishiyama, R., Yamaguchi-Shinozaki, K. and Shinozaki K. 2010. Potential utilization of NAC transcription factors to enhance abiotic stress tolerance in plants by biotechnological approach. *GM Crops*, 1: 32–39.

Uhde-Stone, C., Temple, S.J., Vance, C.P., Allan, D.L., Zinn KE, *et al.,* (2003). Acclimation of white lupin to phosphorus deficiency involves enhanced expression of genes related to organic acid metabolism. *Plant and Soil,* 248: 99–116.

Wang, B.L., Shen, J.B., Zhang, W.H., Zhang, F.S. and Neumann G. (2007). Citrate exudation from white lupin induced by phosphorus deficiency differs from that induced by aluminium. *New Phytolology,* 176: 581–589.

Yan, F., Zhu, Y.Y., Muller, C., Zorb, C. and Schubert S. (2002). Adaptation of H^+-pumping and plasma membrane H^+ ATPase activity in proteoid roots of white lupin under phosphorus deficiency. *Plant Physiology*, 129: 50–63.

Yang, J.L., You, J.F., Li, Y.Y., Wu, P., Zheng SJ. (2007). Magnesium enhances aluminum induced citrate secretion in rice bean roots (*Vigna umbellata*) by restoring plasma membrane H^+ATPase activity. *Plant Cell Physiology*, 48: 66–73.

Zhang, W.H., Ryan, P.R. and Tyerman, S.D. (2001).Malate-permeable channels and cation channels activated by aluminium in the apical wheat roots. *Plant Physiology*, 125: 1459–1472.

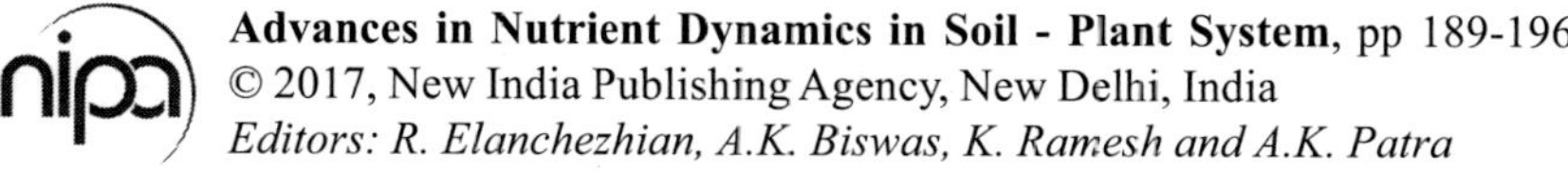
Advances in Nutrient Dynamics in Soil - Plant System, pp 189-196

Editors: R. Elanchezhian, A.K. Biswas, K. Ramesh and A.K. Patra

17

Development of Nutrient Use Efficient Genotypes

V. P. Bhadana, R.M. Sundaram, Brajendra
H.K. Mahadevaswamy, S. Kota and P. Senguttuvel

Indian Institute of Rice Research, Rajendranagar, Hyderabad – 500030, India

World population is expected to cross 8.5 billion by 2025 and it will be a gigantic task to produce enough to meet the future food demand by growing same set of crop varieties on anticipated reduced area. Most of the land that could be brought under cropping has been utilized with exception of some land in Sub-Saharan Africa and South America (Borlaug and Doswell, 1993). Therefore, it is warranted to produce more from less land, less water and less inputs which can be achieved by enhancing the efficiency of production system by devising more efficient crop management practices and developing input use efficient cultivars. Due to intensive cultivation for long time, most of the crop lands have become deficient in one or more elements essential to plants, which are being compensated in the form of costly chemical fertilizers to get desired level of yield. These deficiencies are increasing year after year and consequently use of fertilizers has also been increasing steadily. In last six decades, India has witnessed more than 5 fold increase in food grain production from 50.58 million tonnes in 1950-51 to 259.32 million tonnes in 2011-12. Credit of this increase goes to release and wide scale adoption of high yielding varieties and better management practices including fertilizer applications and most importantly policies of the government. On the other hand, there has been a dramatic increase worldwide in fertilizer applications in the last 50 years, and this has particularly been the case in India. Imbalanced and indiscriminate use of chemical fertilizers has not only greatly deteriorated the soil health but also resulted in widespread water eutrophication. In addition, high cost of fertilizers reduces the economic return and placed a heavy economic burden on the farmers. Ever increasing input cost is rendering the agriculture non-remunerative and that is why almost 76% of farmers want to do some other work other than farming, while 60% wanted their children to migrate and settle in a city according to a survey done by Centre For the Study of Developing Societies (CSDS). This situation would have been rather worst, had government not been bearing major chunk of spending on fertilizers in the form of subsidies. Demand for fertilizers is increasing while response of crop varieties is decreasing. So far, breeders have been breeding varieties that respond to heavy doses of fertilizers and nutrient use efficiency or tolerance to low nutrient status in soil has never attracted the attention of the breeders. As a result, the high yielding varieties could not become popular among the poor farmers who could not afford to apply the required quantity of chemical fertilizers. To sustain agriculture in future, to preserve environmental and soil health and to prevent drainage of natural nutrient resources, there is an immediate need for reduction in soil nutrient input (Vinod, 2014). Although skipping chemical fertilizers completely may not be possible in view of growing food demand, input reduction by growing more nutrient use efficient varieties is a feasible alternative. Therefore, developing crop varieties that use applied nutrient efficiently and perform well even under low input situation is essential for sustainability and profitability.

Mineral Nutrition in Crop Plants: Major Nutrients (NPK)

Crop plants require balanced supply of 17 essential elements for their normal growth of which carbon, hydrogen and oxygen are supplied naturally as they are present in the atmosphere and growing environment. Of the remaining 13, nitrogen (N), phosphorus (P) and potassium (K) are required in large quantities and also known as "fertilizer elements" and supplied through fertilizers. Other nutrients are required in minor quantities (secondary nutrient) or in micro quantities (micro elements) very often not applied from external source in majority of soils.

Nitrogen is an essential component of all proteins. Normally, nitrogen is taken up by plants in the forms of NO_3^- but in acidic environments where nitrification is less likely to occur, ammonium NH_4^+ is more likely to be the dominating source of nitrogen. Nitrogen deficiency results in slow and stunted growth and older leaves turn pale green to yellow due to inability to synthesize sufficient chlorophyll. Nitrogen deficient plants will also exhibit a purple appearance on the stems, petioles and underside of leaves from an accumulation of anthocyanin pigments. In general, nitrogen is the most limiting nutrient of high growth and supplied in large quantities in the form of fertilizers. Some plants require more nitrogen than others, such as corn (*Zea mays*). Because nitrogen is mobile, the older leaves exhibit chlorosis and necrosis earlier than the younger leaves. Soluble forms of nitrogen are transported as amines and amides.

Phosphorus is important in plant bioenergetics. As a component of ATP, phosphorus is needed for the conversion of light energy to chemical energy (ATP) during photosynthesis. Phosphorus can also be used to modify the activity of various enzymes by phosphorylation, and can be used for cell signalling. Since ATP is used for the biosynthesis of many plant biomolecules, phosphorus is important for plant growth and flower/seed formation. Phosphate esters make up DNA, RNA, and phospholipids. Most common in the form of polyprotic phosphoric acid (H_3PO_4) in soil, but it is taken up most readily in the form of H_2PO_4. Phosphorus is limited in most soils because it is released very slowly from insoluble phosphates. Phosphorus deficiency in plants is characterized by an intense green coloration in leaves. If the plant is experiencing high phosphorus deficiencies the leaves may become denatured and show signs of necrosis. Occasionally the leaves may appear purple from an accumulation of anthocyanin. Because phosphorus is a mobile nutrient, older leaves will show the first signs of deficiency.

Potassium regulates the opening and closing of the stomata by a potassium ion pump. Since stomata are important in water regulation, potassium reduces water loss from the leaves and enhances drought tolerance in plants. Potassium deficiency may cause necrosis or interveinal chlorosis. K^+ is highly mobile and can aid in balancing the anion charges within the plant. Potassium serves as an activator of enzymes used in photosynthesis and respiration· Potassium is used to build cellulose and aids in photosynthesis by the formation of a chlorophyll precursor. Potassium deficiency may render plants susceptible to biotic and abiotic stresses.

Need of Developing Nutrient Use Efficient Crop Genotypes

During last five years (2008-2012) global demand for fertilizers has increased at the rate of 1.7% per annum with demand for P fertilizers increased by ~2%, while N fertilizer demand is increasing by ~1.4% (Vinod and Heuer, 2012) which is more than average annual increase in food grain production (~1.0%). In India more than 60% of the total fertilizers is applied to rice and wheat therefore, substantial share of the increased fertilizer demand would also be utilized for rice and wheat production. This is of growing concern because above estimates indicate a declining trend in nutrient use efficiency as a consequence of the fertilizer consumption exceeding the grain production. Moreover, the amount of P available for use in agriculture is finite. Steen (1998) estimated that the depletion of current economically exploitable reserves would occur sometime in the next 60–130 years. It is now established reality rather than perception that yield of both rice and wheat has already reached the plateau. Further, jump in production will come by enhancing yield in resource poor ecologies by developing varieties tolerant to low nutrient conditions. Breeding strategies with objective: breeding for tolerance to low nutrient availability or for efficiency in their use has to be adapted in all the crops. The ability of the genotypes to develop, grow and reproduce under stress conditions, is understood by some authors as tolerance to low nutrients, by others as efficiency of nutrient use. Most of the released high yielding varieties are very good in response to applied fertilizers but lack tolerance to low input conditions. In order to maximize the production, breeders need to develop the varieties capable of using scarce resources to produce satisfactory yields under low input and high yield under ideal condition.

Infrastructure Required for Breeding P use Efficient Varieties

In order to select genotypes tolerant to low nutrient situation, germplasm has to be grown in the field deficient for a particular nutrient or combinations of nutrients. To begin with, one can do so in hydroponics by regulating the nutrient supply but it may not be suitable for screening large breeding populations. The only way to develop the experimental fields suitable for selecting low input tolerant genotypes is that the nutrient has to be exhausted by continuously growing crops without applying the fertilizers. Exhausting phosphorus from the field may take many years depending on the soil type and abundance of the initial P but it takes comparatively less time for nitrogen. It is appropriate to select the field wherein entry of water from neighbouring plots can be restricted. It is recommended that such field should be isolated from other plots by providing permanent barrier wall (Fig. 1).

Fig. 1: Screening of rice in low P plot (*See colour version on page 379*)

These low N/P plots will be used for selecting the genotypes tolerant to low N/P condition. However, to screen the genotypes for nutrient use efficiency and response to their application, the genotypes need to be grown under different gradients so as to compare their performance under different regimes of nutrient application. Therefore, low N/P plots should be divided into 3 to 4 sub-plots with different gradients and ideally need to be separated with some barrier which can block water movement from one gradient to another.

Physiology, Genes, Genetics and Genetic Variability of Nutrient use:

The genetic control of tolerance as well as use efficiency is quantitative and involves multiple loci distributed in different regions of the genome of crop species. During early vegetative stage, roots are expected to play a significant role in N absorption, with root density and distribution being the major determinants of better uptake efficiency. Many QTLs associated with root length, root density and distribution have been identified and few of these are known to be associated with the trait phenotype. In addition to the root related morphological traits, transporter proteins associated with transport of NO_3 and NH_4^+ are also known to influence uptake and transport of N from the soil and the roots to the shoots. There are two kinds of transport system for N in rice. They are (i) a high affinity transport system which works at low N concentrations (i.e. below 1 mM) and a (ii) low affinity transport system which works at higher N concentrations (i.e. above 1 mM). Low affinity transporters like *OsNRT1* and high affinity, NO_3 inducible transporters like NRT2 and NAR2 are known to play a key role in N uptake and transport with most of the transporter genes up-regulated under higher NO_3 and down-regulated under higher NH_4^+ concentrations. In addition to the above mentioned NO_3 transporters, high affinity NH_4^+ transporters like AMT1 and AMT2 and high affinity urea transporters like *OsDUR3* are also known in rice and some of these have been genetically engineered (*OsDUR3*) and shown to play a major role in increasing the N uptake and yield. Recently, an early nodulin gene called *OsENOD93-1* has been characterized with potential function in amino acid accumulation and transport. Once the applied N is taken up inside the roots of rice and transported, it is incorporated in the form of two amino acids, glutamine or glutamic acid, whose synthesis primarily takes place in the plastids via the glutamine synthetase (GS) and glutamate synthase (GOGAT) cycle. Through transgenic approach, over-expression of GS and GOGAT genes have shown to increase panicle weight and grain yield and mutants of *OsNADH-GOGAT2* have shown a significant decrease in yield, thus demonstrating the importance of both GS and GOGAT pathways in N use efficiency. In addition, stay-green trait is also known to be associated with enhanced N use efficiency and many QTLs associated with the trait have been identified in rice.

In majority of crops, P-fertilizers are generally applied only at the time of sowing/ transplanting, indicating that P acquisition and requirement are highest during the early growth stages. Under P-deficient conditions, plant development is delayed and P deficiency symptoms such as reduced tillering and darker leaves due to anthocyanin accumulation are easily recognized and cause significant yield losses. Another important factor is that P, in contrast to N and K, is not transported with the soil solution (mass flow) but mainly by diffusion, therefore it is established that large root surface area is important for P uptake since plants gain access to a larger soil area and thereby to P. In agreement with that, induction of root growth under P deficiency has been described in many species. For yield stability under P deficient conditions, long root hairs and highly branched root systems, especially in the top soil where P is mainly located, are considered beneficial. In agreement with this, the Phosphorus uptake 1 (Pup1) major QTL in rice for tolerance of P deficiency is an enhancer of root growth. The same phenomenon was also reported in other crops. Phosphorus is transported into the plant by P transporters located in the root plasma membrane. In the rice reference genome, 13 P transporter genes are present. Two of these transporters have been functionally characterized, revealing that OsPT2 encodes a low-affinity and OsPT6 a high-affinity transporter. High-affinity transporters are generally induced under low-P conditions and are therefore considered more important for P uptake under field conditions. However, almost all QTL mapping has focused on traits associated with efficient P acquisition rather than efficient internal use of. These traits include high total plant P content, large root systems, improved root architecture (increased lateral root production, improved topsoil foraging, greater root surface to volume ratio, greater root hair production, and greater root:shoot ratio), and the exudation of phosphatases and organic acids into the rhizosphere. Nevertheless, QTL that have the potential to influence internal PUE have been found in several crop species. However, internal PUE is generally lower in plants with high P-acquisition efficiency as a result of higher tissue P concentrations, making it difficult to separate QTL that affect agronomic PUE generally from QTL that may specifically influence internal PUE. Identification of QTLs for internal PUE requires studies where P acquisition is equal and metabolically non saturating among cultivars. As an alternative, explicit quantification of tissue P pools would allow a more specific evaluation of genotypes and identification of QTL that are related to the efficiency of P use in nucleic acids, phospholipids and P-esters. Calculation of PUE indices based on the main metabolically active P pools (i.e. growth or photosynthesis expressed per unit of nucleic acid, phospholipid, or P-ester), might yield better insights into the efficiency with which P is used at a cell physiological level. The P-pool-specific PUE indices are likely to be better suited to QTL mapping studies and should be targeted if the goal is to identify QTL and genes related specifically to internal PUE. There is much scope for future mapping of QTL that influence internal PUE. Advanced methods of genetic analysis have not yet been used extensively in the area of plant P relations. Approaches such as genome-wide expression (transcriptome) QTL analyses and genome-wide association studies supported by such emergent technologies as next generation sequencing hold much promise to identify loci related to and controlling plant P relations, including internal PUE. At Indian Institute of Rice Research, during last five years more than 300 rice genotypes of diverse origin were screened for their performance under different regimes of N/P and wide variation was observed for tolerance to low conditions as well as response to its application for both N and P.

Breeding for Enhanced Nutrient Use Efficiency

It has been observed and reported invariably that modern day varieties irrespective of the crops are not tolerant to low input situations but respond well to external application. However, tolerance to low nutrient status is reported in landraces but they hardly respond to external application. Combining both tolerance to low nutrients and response to its application is necessary to develop cultivars for sustaining future production. Most of our breeding programmes particularly are operating with the germ plasm having very narrow genetic base. Elite varieties which are closely related are being used as parents and use of landraces possessing characters such as low P and N tolerance in addition to tolerance to other biotic and abiotic stresses is diminishing. There is substantial genetic variation in traits associated with NUE within the crop plants that have been reported by various workers. Analysis of this variation has led to the identification of numerous genetic loci that influence NUE. The ability to identify these quantitative trait loci (QTL) suggests that improvements in nutrient use may be gained through conventional or marker-assisted breeding programmes, directed gene identification and genetic engineering, or a combination of these approaches. One promising step toward developing more P-efficient cultivars was the identification of a major QTL named *Pup1* in the indica rice variety Kasalath, which accounted for ~ 80% of the of the phenotypic variance for P-deficiency. The QTL has been transferred to the genetic background of Nipponbare, a variety which has low P uptake and the resulting near-isogenic line, (named NIL-C443), was genetically identical to Nipponbare by ~ 92%. Interestingly, when grown on P limiting soils, NIL-C443 had a threefold higher biomass than Nipponbare. This improvement in growth was due to the ability of the NIL to acquire more P from the soil. Higher root growth rates and more efficient P uptake per unit root size (uptake efficiency) were the main differences between NILC443 and Nipponbare. *Pup1* has been recently fine mapped on long arm of chromosome 12 of rice with the help of SSR markers. Four SSR markers RM28102, RM1261, RM277 and RM519 were observed to be very closely linked to *Pup1*. Now, it is established that *Pup1*-specific protein kinase gene6, which is named phosphorus-starvation tolerance 1 (*PSTOL1*) acts as an enhancer of early root growth, thereby enabling plants to acquire more phosphorus and other nutrients (Gamuyao *et al.,* 2012). This offers the opportunity for marker assisted deployment of the gene(s) into the popular varieties sensitive to low P conditions. Efforts should also be intensified for identification of low nutrient tolerant donors which are devoid of already reported genes/QTLs. Conservatively, breeding for low nutrient tolerance and its use efficiency is the same as it is for any other traits. As most of the studies revealed the involvement of QTLs in controlling these traits, therefore, breeding for nutrient use efficiency would be similar to breeding for tolerance to salinity and drought. Donors should be screened repeatedly under stringent conditions before involving them in crossing. Selection in early generations should be avoided and it is advisable to carry forwarded the maximum variability by advancing the generations following single seed descent method. Segregating material should be grown under moderate status of nutrients in the soil. Final selection has to be done based on performance under both nutrient deficient as well as ideal nutrient status and the lines which are tolerant to low nutrient situation and having good performance under non stress conditions will be selected .

References

Borlaug, N. E. and C .R. Dowswell (1993). Fertilizer: To nourish infertile soil that feeds a fertile population that crowds a fragile world. *Fertilizer News*, 38: 11–20.

Gamuyao, R., Chin, J.H., Parisca-Tanaka, J., Pesaresi, P., Dalid, C., Slamet-Loedin, I., Tecson-Mendoza, E.M., Wissuwa, M. and Heuer, S. (2012). The protein kinase OsPSTOL1 from traditional rice confers tolerance of phosphorus deficiency. *Nature,* 488: 535–539.

Steen, I. (1998). Phosphorus availability in the 21st century: Management of a non-renewable resource. *Potassium and Phosphorus,* 217: 25–31.

Vinod, K.K. (2014). The Need for Nutrient Efficient Crop Varieties. *Journal of Plant Biochemistry and Physiology,* 2: e123. doi:10.4172/2329-9029.1000e123

Vinod, K.K. and Heuer, S. (2012). Approaches towards nitrogen- and phosphorus-efficient rice. *AoB PLANTS,* 2012: pls028; doi:10.1093/aobpla/pls028.

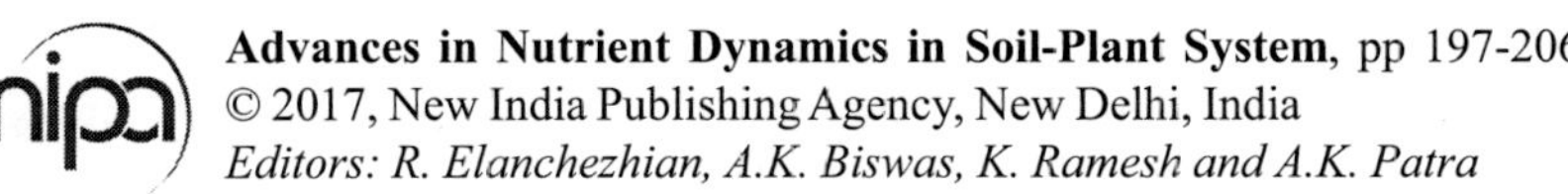
Advances in Nutrient Dynamics in Soil-Plant System, pp 197-206

Editors: R. Elanchezhian, A.K. Biswas, K. Ramesh and A.K. Patra

18

Biofortification of Crops for Improved Yield and Nutrient Use Efficiency

Ajay and R. Elanchezhian

ICAR-Indian Institute of Soil Science, Nabi Bagh, Bhopal – 462038, India

The diets of over two-thirds of the world's population lack one or more essential mineral elements. This can be remedied through dietary diversification, mineral supplementation, food fortification, or increasing the concentrations and/or bioavailability of mineral elements in produce (biofortification) involving aspects of soil science, plant physiology and genetics underpinning crop biofortification strategies, as well as agronomic and genetic approaches currently taken to biofortify food crops with the mineral elements most commonly lacking in human diets: iron (Fe), zinc (Zn), copper (Cu), calcium (Ca), magnesium (Mg), iodine (I) and selenium (Se). Two complementary approaches have been successfully adopted to increase the concentrations of bioavailable mineral elements in food crops. First, agronomic approaches optimizing the application of mineral fertilizers and/or improving the solubilization and mobilization of mineral elements in the soil have been implemented. Secondly, crops have been developed with: increased abilities to acquire mineral elements and accumulate them in edible tissues; increased concentrations of 'promoter' substances, such as ascorbate, â-carotene and cysteine-rich polypeptides which stimulate the absorption of essential mineral elements by the gut; and reduced concentrations of 'antinutrients', such as oxalate, polyphenolics or phytate, which interfere with their absorption.

The practice of plant husbandry (customized plant produce) began with early man. Even before we had the slightest concept of plant physiology, it was observed that certain conditions had an influence on plant health and vigor. Eventually we began to select and organize desirable plant types and contribute to their basic requirements. We suspected that there was something in the soil that plants consumed but it took thousands of years to stumble upon even the most basic understanding of the science of plant chemistry. Only in the last one hundred years has technology advanced to the point that we can accurately measure and manipulate plants internal functions. And even more recently isolate and identify plant substances genetically and at an atomic level. This knowledge has allowed us to breed and propagate specialized flower, food, and resource crops, more prolific and abundant than ever before. Our goal is to provide select and precise constituents to a plant growth environment, in a properly balanced manner, neither lacking or in excess.

What do we see? Do we see water and minerals moving across and into the roots, and traveling through stem and branches into living leaf cells? Or the water molecules evaporate out through the open leaf stomata? Do we see the CO_2 molecules diffusing into the chloroplasts, and then being fixed by photosynthesis into carbohydrates which then travel by vascular pathways to all points of demand? Do we see the ions from essential minerals being selectively absorbed and combined into coenzymes and organic compounds? Do we see the nourished cells split and reproduce to form the splendor of foliage and flower we see before us? Of course not, but it is happening, and we can gain at least a sense of awareness that it is. A plant, as any living organism, is a marvel of chemistry, yet it operates much like a machine, governed by the laws of physics.

Plant Composition

Plants consist of a complex arrangement of cell bodies working together, each in their own way to form a living organism. These cells are made up of, or contain many components such as proteins, polysaccharides, amino and organic acids, lignin's etc. These compounds are themselves comprised of principle elements which over 80% (as dry weight) consist of oxygen and carbon. Followed respectively in quantitative sequence is hydrogen, nitrogen, silicon, potassium, calcium, sulfur, phosphorus, magnesium, aluminum, iron, chlorine, sodium, manganese, boron, copper, zinc, molybdenum and other assorted scarce minerals. Different combinations of these constituents form molecules of many identities to construct new cells and tissues. This is why these essential elements must be readily available. These elements, along with water and sunlight, plants are able to synthesize all the compounds they require, as well as, vitamins and enzymes necessary to us as consumers. These essential elements can be provided in their elemental form, pure and immediately available, with the application of high quality inorganic fertilizers. These essential elements can also be found in organic molecules (such as found in soils) however, organic materials must be broken down into their pure elemental (inorganic) form before they can be utilized by the plant. Additional energy is required to disassociate these more complex organic commodities. Plants will absorb and accumulate numerous nonessential elements. Plants can uptake, breakdown, or retain many substances. If these substances are beneficial it could be desirable but if they are toxic it could be disastrous.

Many elements have been found in plant tissue which is not known to have any influence on plant metabolism. Lead, arsenic, mercury, gold and fluoride are among the more than 60 other known elements.

The Mystery of Roots

Roots not only provide a means of support, but they act as receptors providing pathways for select solutions and substances to be regulated into the plants circulatory systems. Root anatomy consists primarily of a xylem and phloem core of vascular tissue, surrounded by a cortex tissue and an outer layer of epidermal tissue. Microscopic projections called root hairs usually develop on the epidermal cell to further enhance the water absorption capability of the root surface.

Roots are specially adapted tissues which readily absorb aqueous substances and transport them into the plants main vascular system. Nutrient ions will diffuse into the root, between it's cells, through intercellular spaces called the apoplast, and interconnecting protoplasm called the symplast. These pathways allow water and solutes to pass across the cortex and through the endodermal layer and into the vascular bundles.

Nutrient assimilation

Within a plant's structure exist many types of cells. These cells vary in their ability to absorb solutes by the nature of their membranes. A solute could be anything dissolved in a solvent (water). Membranes are thin permeable tissues which surround the cell bodies. These cell membranes are designed to be specific to which elements are able to pass through them. The following forces control this flow (flux) process.

1. Osmosis is the tendency for a solvent (in our case water) to pass through a membrane from the side of less soluble salts to the side of higher concentration. It is attempting to dilute the solution to gain equilibrium on each side. This action is regulated by particle concentration, not by their properties. When this activity is measured it is termed the chemical potential.
2. The second type of membrane flow is called the electrical potential. This force is driven by the exchange of positive (cation) and negative (anions) ions creating a + or - potential within the cell. A positive affinity will generally attract negative ions to balance its polarity (and vise versa). This creates a flow of ions by electrical attraction.
3. Another method involves the use of a carrier molecule, often part of the membrane itself. If an ion is attracted to a site on a carrier molecule, it may then diffuse readily across the membrane to be released on the other side. This method controls ion selectivity by the ability of the carrier to combine with a specific element ion. This explains why only certain ions are able to pass through a given membrane tissue type.

Many factors and conditions affect these processes and their ability to absorb essential elements. Among these are, nutrient solution concentration, balance, pH, temperature, or the presence of incompatible chemicals which may bind and inhibit important minerals from being available to the plants.

Nutrient Sensing: Plant roots alter their root system architecture under nutrient deficiencies and respond to the local supply of certain nutrients with an enhanced formation of lateral roots. For instance, lateral root initiation and emergence is stimulated under local ammonium supply, whereas local supplies of nitrate or iron increase lateral root elongation. These observations support the view that nutrients interact with the root developmental program at specific points.

Stress adaptations of primary metabolism: In metabolic studies on the regulation of primary C and N metabolism under abiotic and biotic stress, to find target processes and underlying genes crucial for the maintenance of metabolic activity under stress conditions. In cereals such as barley this approach is used to better understand stress-sensitive processes involved in biomass accumulation or yield formation. In ornamental plants like Petunia we investigate the role of assimilates and nutrients in adventitious root formation to finally improve the rooting of cuttings.

Regulation of nitrogen transporters and retranslocation of nitrogen during senescence: Our previous studies have shown that high-affinity urea uptake relies on one transporter called DUR3, while the uptake of ammonium is mediated by six AMT-type ammonium transporters. These transporters are regulated at multiple levels, and specific regulatory features have been identified at the transcriptional and post-transcriptional levels as well as at the post-translational level, e.g. by the allosteric regulation of AMT subunits in homo- and heterotrimers. Uptake studies with $15NH_4^+$ in multiple AMT mutants and localization studies with GFP-labeled transport proteins have further shown that the contribution of individual transporters depends on their biochemical properties and their cell type-specific localization in the root.

Iron and heavy metal acquisition and transport in plants: In calcareous soils, iron is sparingly soluble, which often leads to reduced iron uptake, leaf chlorosis and impaired plant growth. To increase the solubility of iron in the rhizosphere, plants have developed different strategies to cope with low Fe availabilities. Dicotyledonous plant species like Arabidopsis decrease rhizosphere pH and reduce ferric Fe to ferrous Fe, which can then be taken up across the plasma membrane. By contrast, graminaceous plant species secrete phytosiderophores, which are hexa-dentate metal chelators, to form iron(III)-phytosiderophore complexes in the rhizosphere for subsequent uptake by the roots. Inside root cells, Fe undergoes probably multiple changes in its binding forms for further compartmentalization or long-distance transport. Our goal is to uncover plant-borne ligands for Fe and transport steps that are limiting for a more efficient allocation and use of Fe in plants. In our methodological approaches, we carry out screens of *Arabidopsis* mutants on calcareous substrates with low Fe availability or use yeast complementation to identify genes involved in iron efficiency in plants. As iron-binding substances also affect the susceptibility of the plants against plant pathogens, we further investigate the interaction between maize and the pathogenic fungus Colletotrichum.

Promotion of plant growth by rhizobacteria: Associative bacteria that colonize the rhizosphere can promote plant growth and thus potentially increase the grain yield in cereals. Certain rhizosphere bacteria are able to improve nutrient mobilization in the rhizosphere, fix atmospheric nitrogen or release hormone-like substances that stimulate root growth.

Organic fertilization

The scientific definition of organic is "any chemical compound containing carbon". A more common interpretation is any substance derived from living organisms, plant or animal. The concept of organic gardening usually implies that, the essential elements required for plant nutrition will be attained by dissociation from decomposing matter. This process occurs in nature when a plant or animal expires or sheds tissue which is then systematically acted upon by organisms and environmental conditions. The end result of all of this is to provide pure inorganic elements which are the building blocks of all life.

Inorganic fertilization

The term "inorganic" defines a substance as a non-living material neither of plant or animal origin - generally referring to a matter which is not having carbon. All organic structures are composed of inorganic compounds and will eventually degrade back to this original form. Pure inorganic elements and combinations thereof, are the foundation of all living things (and otherwise) on this planet. The mysterious interactions of these 103 elements somehow manage to create or at least sustain life and all things of substance. About fifteen (15) of these elements are known to be essential for normal plant growth. When these elements are in solution they become available (to some degree) for plants to assimilate, either in their pure form or as ions of simple compounds. When these elements are combined into compatible compounds, they are referred to as chemical fertilizers.

Chemical fertilizers have undeservingly been given a bad rap because they have been associated with large scale wasteful misuse. This has resulted in the contamination of soils and water supplies. This is not the fault of the chemical, rather the management thereof. Another unfair association is that of pesticides, fungicides, herbicides, inoculates, and preservatives etc., of which chemical fertilizers have no relationship. Chemicals compounds are not undesirable just because it has been refined or combined by man.

Biofortification: It is the process by which the nutritional quality of staple crops is enhanced, such as vitamin A, zinc, and iron. This is done through conventional plant breeding and/or modern technology. More research is needed, but it is hoped that people who consume biofortified crops will have an improved nutritional intake. These crops are "biofortified" by loading higher levels of minerals and vitamins in their seeds and roots during growth. Through biofortification, scientists can provide farmers with crop varieties that provide essential micronutrients and can naturally reduce anemia, cognitive impairment, and other malnutrition-related health problems that affect billions of people.

India's economic transformation and growth have received much attention in recent years. However, there is one area of human development where India has not fared particularly well: hunger and malnutrition. Child malnutrition rates in India are extraordinarily high – among the highest in the world, with nearly one-half of all children under 3 years old being either underweight or stunted. Why is combating hunger and malnutrition so important? Freedom from hunger and malnutrition is a basic human right, and until India can provide these freedoms, its claims to successful human development are questionable.

The Economic Costs of Hunger and Malnutrition

In addition to the human cost, there is a huge economic cost to hunger and malnutrition, which suggest that malnutrition may be costing the Indian economy the equivalent of 4%-5% of its GDP. Perhaps surprisingly, the problem of under-nutrition in India now coexists with the problem of over-nutrition and associated non-communicable diseases for a different segment of the population. Recent medical evidence suggests that the two might be related – low birth-weight children and children who are malnourished are more likely to develop chronic illnesses, such as diabetes, as adults. India has the largest number of adults with type 2 diabetes in the world and this number is growing rapidly – having doubled over the past 10 years. Indeed, India has a higher rate of diabetes than many Western countries with much higher levels of economic prosperity.

Adding more support to the view that child malnutrition is weakly correlated with income is the finding that among children of mothers with 10 or more years of schooling as well as among children of mothers from the top income quintile, around one-quarter are underweight. Even in a relatively prosperous and dynamic state like Gujarat, child malnutrition rates have been stagnant over the past decade.

The UN's Food and Agricultural Organisation (FAO) estimates the number of 'hungry' people in India at 230 million, which is remarkable given robust agricultural productivity growth during the last three decades. For instance, yields of food grains have doubled since the early 1970s and of 'coarse' cereals (such as maize, sorghum and pearl millet), which are traditionally the main foods of the poor in India, have more than doubled. Yet astonishingly, over the same period, mean calorie intake in the country has actually fallen – by about 10% in the rural areas and 4% in the urban areas.

Income and Food Intake

What does this all mean? Simply that we do not yet have a good understanding of how the poor in India make their food consumption and nutritional choices. A very large portion of the demand for food is thus based on the non-nutritive attributes of food, such as taste, aroma, variety, and status. This means that increases in household income do not always translate into improvements in calorie consumption.

The poor in India do not put as much of their money into obtaining more calories (or at least as many calories as the UN FAO might think are appropriate, see Banerjee and Duflo 2011). Indeed, quite surprisingly, as child malnutrition rates have stagnated and calorie consumption has actually fallen, mobile phone use – even among the Indian rural poor – has increased dramatically. This raises many questions, including the obvious one – why do the poor, when given an opportunity, choose to spend their additional income on luxury durables, such as mobile phones, than on the nutrition of their children? Is it because they are uninformed about the long-term economic benefits of child nutrition? Or is it because 'expert' assessments about the prevalence and economic cost of under-nutrition in India are essentially incorrect?

Reason for hope: Is Biofortification Cost-Effective? Unlike the continual financial outlays required for supplementation and fortification programs, a one-time investment in breeding-based solutions can yield biofortified crops for farmers to grow around the world for years to come. It is this multiplier aspect of biofortification across time and distance that makes it so cost-effective in reducing malnutrition.

Is Biofortification Sustainable? While government attention to malnutrition may fade, and international funding for micronutrient interventions may be substantially reduced, nutritionally improved biofortified varieties can continue to be grown and consumed year after year to reduce malnutrition in entire populations. Recurrent costs required for monitoring and maintaining these traits in crops will be far lower than the initial costs of developing biofortified crops.

Does Biofortification Require Genetic Engineering? No. In fact most of the work being done by traditional plant breeding techniques to increase the nutritional quality of staple foods. Seed and germplasm banks throughout the world are evaluated for varieties of staple food crops that have naturally occurring higher levels of micronutrients in their seeds. These varieties are then crossed with modern high yielding varieties in an attempt to breed new naturally biofortified varieties that perform well in the field, produce high yields, and that are also more nutritious. However, those essential nutrients that cannot be bred into key food staples through conventional plant breeding methods may require transgenic breeding approaches to provide enough nutrients to significantly improve human nutrition.

Do Biofortified Foods Require a Change in Consumer Behavior? Mineral micronutrients make up a tiny fraction of the physical mass of a seed, for example, only 5 to 10 parts per million in milled rice. Whether such small amounts will alter the appearance, taste, texture, or cooking quality of foods is being investigated. In the case of iron and zinc, increased levels of nutrients may not be noticeable, requiring no special intervention or marketing campaigns. This would be analogous to the practice of adding iodine to salt or fluoride to drinking water.

In contrast, higher levels of beta-carotene (converted in the body to Vitamin A) will often turn the color of grain, flours and roots/tubers from preferred white or light yellow colors to dark yellow and orange colors. Nutrition education programs will be needed to encourage malnourished populations to switch to more nutritious varieties and recognize the color changes as a food quality trait

How Will Biofortification Improve Agronomic Properties of Crops? Adequate nutrition is as important to plant health as it is in human health. Micronutrient deficiency in plants greatly increases their susceptibility to diseases, especially fungal root diseases of the major food crops. Efficiency in the uptake of mineral micronutrients from the soil is associated with disease resistance in plants, which leads to decreased use of pesticides and fungicides. Breeding for micronutrient efficiency can confer resistance to root diseases that had previously been unattainable.

Micronutrient-efficient varieties grow deeper roots in mineral-deficient soils and are better at tapping subsoil water and minerals. When topsoil dries, roots in the dry soil zone (which are the easiest to fertilize) are largely deactivated and the plant must rely on deeper roots for further nutrition. Roots of plant genotypes that are efficient in mobilizing surrounding external minerals not only are more disease resistant, but are better able to penetrate deficient sub soils and so make use of the moisture and minerals contained in sub soils. This reduces the need for fertilizers and irrigation. Plants with deeper root systems are also more drought resistant.

Micronutrient-dense seeds are associated with greater seedling vigor, which, in turn, is associated with higher plant yield A significant percentage of the soils in which staple foods are grown are "deficient" in these trace minerals, which has kept crop yields low. In general, these soils, in fact, contain adequate amounts of trace minerals, enough for hundreds or thousands of crops. However, because of chemical binding to other compounds, these trace minerals are "unavailable" to staple crop varieties presently used.

An important question in development of micronutrient efficient genotypes is the possibility of combining high yield with better micronutrient nutritional quality.

Previous studies showed that it is possible to combine micronutrient-rich traits with high yield. Both seedling vigor and nutritional quality can be improved through genetically modifying seeds with micronutrient enrichment traits. The highest micronutrient densities, which are approximately twice as high as those popular modern cultivars and indicating the existing genetic potential, can be successfully combined with high yield.

The combining of benefits for human nutrition and agricultural productivity, resulting from breeding staple food crops which are more eficient in the uptake of trace minerals from the soil and which load more trace minerals into their seeds, results in extremely high benefit-cost ratios for investments in agricultural research in this area. This approach would be more valuable and cost-effective by estimating the costs paid to heal micronutrient malnutrition in developing countries. Some adverse effects of Zn and Fe deficiency cannot be remediated by supplying adequate levels of Zn or Fe later in childhood, so prevention needs to be the focus of this seed improvement program.

High trace mineral density in seeds produces more viable and vigorous seedlings in the next generation, and efficiency in the uptake of trace minerals improves disease resistance, agronomic characteristics which improve plant nutrition and productivity in micronutrient deficient soils (Welch, 1999; Yilmaz *et al.,* 1998). Adoption and spread of nutritionally improved varieties by farmers can rely on profit incentives, either because of agronomic advantages on trace mineral deficient soils or incorporation of nutritional improvements in the most profitable varieties being released (Harris *et al.,* 2008).

It has been shown that wheat plants grown from seed with high Zn content can achieve higher grain yields than those grown from the low-Zn seed when Zn was not applied to the soil (Yilmaz *et al.,* 1998). Therefore, sowing seeds with higher Zn contents can be considered a practical solution to alleviate plant Zn deûciency especially under rainfed conditions, in spite of it being insufficient to completely overcome the problem (Yilmaz *et al.,* 1998).

Mineral-packed seeds sell themselves to farmers because these trace minerals are essential in helping plants resist disease. More seedlings survive and initial growth is more rapid. Ultimately, yields are higher, particularly on trace mineral "deficient" soils in arid regions. Because roots extend more deeply into the soil and so can tap more subsoil moisture and nutrients, the mineral-efficient varieties are more drought-resistant and so require less irrigation. And because of their more efficient uptake of existing trace minerals, these varieties require lesser chemical inputs. Thus, the new seeds can be expected to be environmentally beneficial as well. It is conceivable that seed priming by spraying seed fields with Fe and Zn fertilizers during grain filling will provide enough additional yield benefit to justify additional seed price. But this approach is insufficient to improve the density of bio-available micronutrients in grain of crops grown with "primed" seed. Several different commercial practices may be beneficial to improve plant production despite low levels of soil micronutrients in many nations.

Can Bio-fortified Crops Reduce Soil Fertility? A soil is said to be deficient in a given nutrient when the addition of fertilizer produces better growth—even though the amount of nutrient in the fertilizer added may be small compared with the total amount of the nutrient in the soil. This seeming paradox can occur when only a small part of the nutrient in the soil is available to plants, owing, for example, to the chemical properties of the soil.

Alternatively, the view can be taken that there is a genetic deficiency in the plant rather than a deficiency in the soil. Rather than adapting the soil to the plant, breeding can adapt the plant to the soil.

References

Chalk, P., Zapata, F. and Keerthisinghe, G. (2002). "Towards integrated soil, water and nutrient management in cropping systems: The role of nuclear techniques". Proceedings of the 17 World Soil Science Congress, Symposium 59, Paper no.2164, Bangkok, 14-20 August 2002.

Delhaize, E., Ryan P.R. and Randall P.J. (1993). "Aluminum tolerance in wheat (*Triticum aestivum* L.) 2. Aluminum stimulated excretion of malic acid from root apices". *Plant Physiology*, 103:695-702.

Dobermann, A. and Cassman. (2002). "Plant nutrient management for enhanced productivity in intensive grain production systems of the United States and Asia". *Plant and Soil,* 247: 153-175.

FAO (1998). Guide to efficient plant nutrition management. Rome, FAO, 18 p.

FAO/IAEA. (2001). Report of the 3rd Research Co-ordination Meeting of the FAO/ IAEACoordinated Research Project on Management of Nutrients and Water in Rainfed Arid and Semi-arid Areas for Increasing Crop Production, Vienna, 24-28 September, Vienna, 2001.

Heano, J. and Baanante, C. (2001). "Nutrient depletion in the agricultural soils of Africa.2020" Vision Brief 62, International Food Policy Research Institute, Washington, D.C.

Hocking, P.J., *et al.* (2000). "The role of organic acids exuded from roots in phosphorus nutrition and aluminium tolerance inacidic soils". IN: IAEA-TECDOC-1159:Management and Conservation of Tropical Acid Soils for Sustainable Crop Production, pp. 61-73.

Montengero, A. and Zapata, F. (2002). Rape genotypic differences in P uptake and utilization from phosphate rocks in an Andisol of Chile. *Nutrient Cycling in Agroecosystems*, 63: 27-33.

Pinstrup-Anderson, P. (1999). "Towards ecologically sustainable world food production". UNEP Industry and Environment, 22. No. 2-3. 10-18.

Poth, M. La Favre, J.S. and Focht. D.D. (1986)."Quantification by direct 15N dilution offixed N2 incorporation into soil by cajanuscajan (Pigeon pea)". *Soil Biology and Biochem*istry, 18: 125-127.

Vanlauwe, B., *et al.* (2002). Recent developments in soil fertility management of maize-based systems: The role of legumes in N and P nutrition of maize in the moist savanna zone of West Africa. http://www.cgiar.org/spipm /news/ ccropmtg /results/ ccb1 /CCBEN1.htm

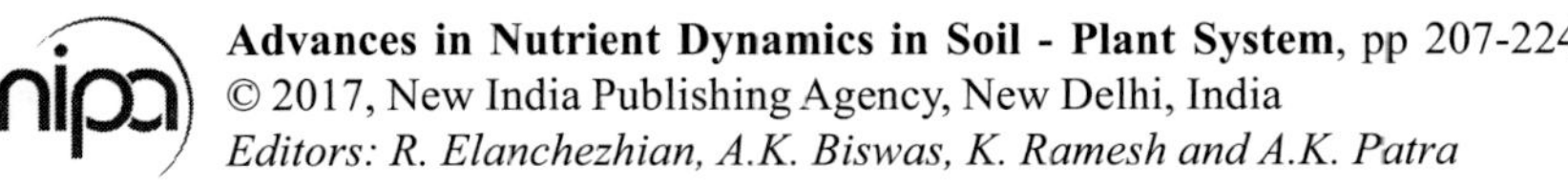
Advances in Nutrient Dynamics in Soil - Plant System, pp 207-224

Editors: R. Elanchezhian, A.K. Biswas, K. Ramesh and A.K. Patra

19

Rhizodeposition by Plants: A Boon to Soil Health

M.L. Dotaniya and B.P. Meena

ICAR-Indian Institute of Soil Science, Nabi Bagh, Bhopal – 462038, India

Under natural conditions, the largest source of labile carbon (C) inputs to the soil is root exudates (Bertin *et al.,* 2003; Hutsch *et al.,* 2002; Kuzyakov 2002). Roots exude a variety of low-molecular weight organic compounds, including sugars, amino acids, organic acids (Bertin *et al.,* 2003), but also contain hormones, vitamins, amino compounds, phenolics and sugar phosphate esters (Uren, 2001). These compounds are rapidly metabolized; for example, Hutsch *et al.,* (2002) found that 64–86% of maize root exudates were respired by soil microorganisms. The effect of root exudates depends on the distance that they can diffuse away from rhizoplane (Gupta and Mukerji, 2002). Previous studies have demonstrated that labile C inputs to the soil, such as root exudates, can stimulate the mineralization of existing soil C pools (Dalenberg and Jager 1989; Hogberg and Ekblad 1996; Mary *et al.,* 1993; Subke *et al.,* 2004), resulting in a "priming" effect. An example of soil priming comes from Subke *et al.* (2004), who combined 13C-labeled litter addition and girdling of Norway spruce trees, and found that the difference in CO_2 efûux between their litter-addition and litter-removal plots could not be accounted for entirely by root respiration, and suggested that rhizodeposition had a priming effect, increasing the decomposition component of soil respiration. His positive priming effect was responsible for 15- 20% of the total soil CO_2 ûux in June and July (Subke *et al.,* 2004). Kuzyakov (2002) hypothesized that elevated rates of decomposition are the ultimate cause of priming. According to this hypothesis, priming involves an enhancement of microbial C status and demand for other nutrients, which in turn stimulates decomposition. In support of this hypothesis, Subke *et al.* (2004) concluded that girdling slowed litter decomposition, indicating that rhizodeposition enhanced litter breakdown. Multiple studies, however, have concluded that primed C is derived from microbial C pools, not soil organic matter (SOM) breakdown, suggesting higher rates of microbial turnover and release of endocellular constituents as the principal sources (Dalenberg and Jager 1981, 1989; De Nobili *et al.,* 2001). Broadly, three distinct components recognized in the rhizosphere; the rhizosphere *per se* (soil), the rhizoplane, and the root itself. The rhizosphere is thus the zone of soil influenced by roots through the release of substrates that affect microbial activity. The rhizoplane is the root surface, including the strongly adhering root particles (Schweinsberg-Mickan *et al.,* 2012). The root itself is a part of the system, because certain endophytic microorganisms are able to colonize inner root tissues (Bowen and Rovira, 1999). The rhizosphere effect can thus be viewed as the creation of a dynamic environment where microbes can develop and interact. The number and diversity of microorganisms are related to the quantity and quality of the exudates but also to the outcome of the microbial interactions that occur in the rhizosphere (Somers *et al.,* 2004). In a review of the mechanisms of priming, Kuzyakov *et al.* (2000) defines rhizosphere priming as a short-term change in the intensity of soil organic matter (SOM) decomposition. The group of authors uses a broad deûnition of SOM that includes microbial biomass, plant and animal derived organic compounds, and freshly deposited litter (Kuzyakov *et al.,* 2000; Fontaine *et al.,* 2004). In this context, root exudation has been quantified by measuring the production of labelled CO_2 in the rhizosphere of "C-labelled plants and it has been estimated that 12-40% of the total amount of carbohydrates produced by photosynthesis is released into the soil by surrounding roots (Brimecombe *et al.,* 2007). Unlike plant litter, which is largely comprised of polymers too large for microbial uptake, much of the C in exudates can be taken up without extracellular enzymatic decomposition.

While, rhizodeposition has been found to stimulate additional decomposition of existing soil C pools, it is still unclear which pools and processes are affected. The number of microbial species present in soil may vary from thousands to millions. Many studies indeed suggest that the Proteobacteria and the Actinobacteria form the most common of the dominant populations (>1%, usually much more) found in the rhizosphere of many different plant species (Singh *et al.,* 2007). These groups contain many "cultured" members. They are the most studied of the rhizobacteria, and as such, contain the majority of the organisms investigated, both as beneficial microbial inoculants and as pathogens. The release of these low molecular weight compounds is a passive process along the steep concentration gradient which usually exists between the cytoplasm of intact root cells (millimolar range) and the external (soil) solution (micromolar range). Direct or passive diffusion through the lipid bilayer of the plasma membrane is determined by membrane permeability, which depends on the physiological state of the root cell and on the polarity of the compounds, facilitating the permeation of lipophilic exudates (Rudrappan *et al.,* 2007). The rhizosphere inhabiting microorganisms compete each other for water, nutrients and space and sometimes improve their competitiveness by developing an intimate association with plants. This process can be regarded as an ongoing process of micro-evolution in low-nutrient environments, which are quite common in natural ecosystems (Schloter *et al.,* 2000). Soil is a complex system (Fig.1); its formation by the human reactions on mineral matter is an important phenomena. The organic exudates secreted from plant roots as well as from microorganisms play a vital role in soil formation, fertility and soil health.

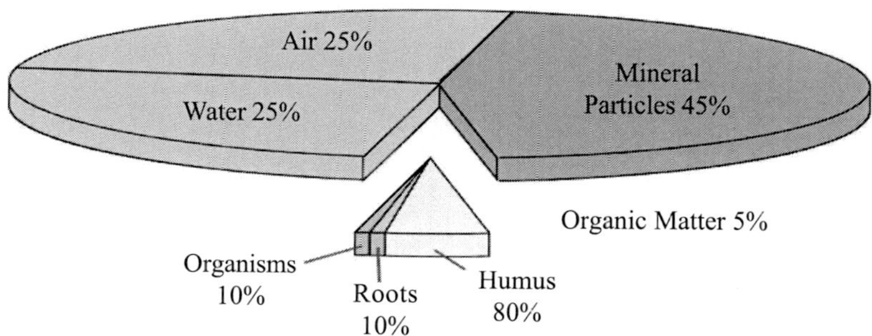

Fig. 1: Composition of soil *Source*: Pidwirny (2006) (*See colour version on page 379*)

The efficiency of the exudation process may thus be enhanced by stress factors affecting membrane integrity such as nutrient deficiency, temperature extremes, or exudation stress (Ratnayale *et al.,* 1978). Thus, it is uncertain whether primed C is derived from enhanced decomposition, greater C release from microbial pools, or both. Tree girdling has been shown to be an effective method of isolating the effects of rhizodeposition on soil processes without disturbing the soil environment (Bhupinderpal Singh *et al.,* 2003; Hogberg *et al.,* 2001; Scott-Denton *et al.,* 2006; Subke *et al.,* 2004). In another girdling study in a lodgepole pine forest in the Front Range of the Rocky Mountains in Colorado, USA, Scott-Denton *et al.,* (2006) found that tree roots prime the soil with sugar-rich exudates beneath the spring snowpack. Girdling offers two signiûcant advantages to soil trenching, the alternative and more commonly used method of inhibiting rhizodeposition. First, girdling does not result in the instantaneous death of roots, which causes a large root litter input to the soil; and second, girdling allows roots to maintain their ability to transport water, so that soil moisture is less affected by girdling than trenching (Subke *et al.,* 2004). While girdling will ultimately result in root and tree death, roots can remain alive with virtually no growth for a considerable period of time (Subke *et al.,* 2004). Based on observations in the subalpine forest ecosystem, it may take as long as 2–3 years for trees to die following girdling (Scott-Denton *et al.,* 2006). Plants modify the chemical, physical and biological properties of the soil environment surrounding their roots. Organic compounds released from living roots (rhizodeposits) are easily available sources of energy for microorganisms strongly affecting SOM dynamics. Although, rhizodeposition is a key driver of microbially mediated processes in the soils, it still remains the most uncertain component of the terrestrial C cycle. Root materials remain mixed within the soil as they decompose, providing gum-like materials that cement soil particles into aggregates (Melillo and Gosz, 1983; Tresder *et al.,* 2005).

Rhizodeposition

During their life, plant roots release organic compounds into their surrounding environment. The area of rhizosphere is mostly blessing with rhizo-deposition (Fig. 2). This is a wide and wise definition, already coined more than hundred years ago by Hiltner (1904). Some of the microbes that inhabit this area are bacteria that are able to colonize very efficiently the roots or the rhizosphere soil of crop plants. These bacteria are referred to as plant growth promoting rhizobacteria (PGPR). This process, named rhizodeposition, is of ecological importance because 1) it is a loss of reduced C for the plant, 2) it is an input flux for the organic C pool of the soil, and 3) it fuels the soil microflora, which is involved in the great majority of the biological activity of soils, such as the nutrient and pollutant cycling or the dynamics of soil-borne pathogens, for example. Marschner (1995) reported that major components released from plant roots *i.e.* low-molecular weight organic compounds (free exudates), high-molecular weight gelatinous materials (mucilage), and sloughed-off cells and tissues and their lysates).

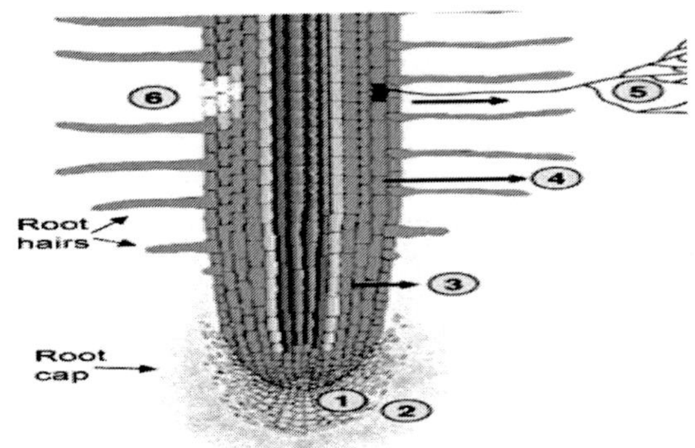

Fig. 2: Schematic of a root showing the 6 major regions of rhizodeposits. 1 loss of cap and border cells, 2 loss of insoluble mucilage, 3 loss of soluble root exudates, 4 loss of volatile organic carbon, 5 loss of C to symbionts, 6 loss of C due to death and lysis of root epidermal and cortical cells (*Source*: Jones *et al.*, 2009).

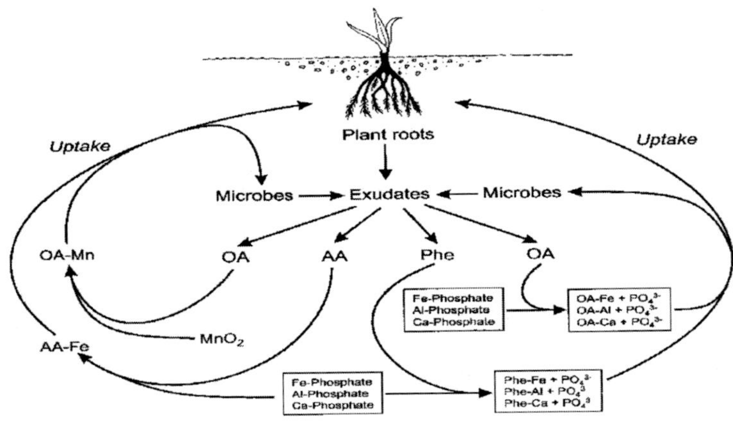

Fig. 3: Effects of root exudates components on nutrient availability and uptake by plants and rhizosphere microbes. OA = organic acids; AA = amino acids including phytosiderophores, Phe = phenolic compounds (Dakora and Phillips, 2002).

The present review described the composition and the mechanisms by which major rhizodeposits are released into the soil: the production of root cap cells, the secretion of mucilage, and the passive and controlled diffusion of root exudates. In a second part, its composition and its amount of secretion affected factors. On average, 17% of the net C fixed by photosynthesis is lost by roots and recovered as rhizosphere respiration (12%) and soil residues (5%), which corresponds to 50% of the C exported by shoots to belowground (Nihorimbere *et al.,* 2011). Direct plant growth promotion may result either from improved nutrient acquisition and/or from hormonal stimulation. Diverse mechanisms are involved in the suppression of plant pathogens, which is often indirectly connected with plant growth (Fig. 3). The amount of C and N supplied by roots significantly improved or maintaining SOM (Sainju *et al.,* 2005b). These soil C sources are important for microbial food during nutrient transformation in soil. The organic inputs from plant roots to the adjoin soil is the principal support of the biological activities and abundance and diversity of organisms in the rhizosphere (Cheng *et al.,* 1994; Kirchner *et al.,* 1993). As much as 7–43% of the total aboveground and belowground plant biomass can be contributed by roots (Kuo *et al.,* 1997a,b). Balesdent and Balabane (1996) measured that maize roots significantly contributed 1.6 times more C to soil organic C than stover. It enhanced the more stable aggregates than does shoot-derived C (Gale *et al.,* 2000a,b). A range researcher reported that the roots can supply from 400 to 1460 kg C ha^{-1} during a growing season (Kuo *et al.,* 1997a; Qian and Doran, 1996). But later on Liang *et al.,* (2002) reported that roots contributed 12% of soil organic C, 31% of water soluble C, and 52% of microbial biomass C within a growing season, which is a key reason for enhancing soil C during crop growth. The amount of root exudation also varies crop growth stages and soil fertility status (Dotaniya *et al.,* 2013a), increasing the phosphorus level decreasing the oxalic acid concentration (Table 1). By the root exudation, roots may play a prevailing role in soil C and N cycles (Gale *et al.,* 2000a; Puget and Drinkwater, 2001; Wedin and Tilman, 1990). The estimation of amount of root exudates organic carbon material is a deadly job. It requires a lot of time and accuracy during measurements. The roots may have a relatively larger influence on soil organic C and N levels than the aboveground plant biomass (Boone, 1994; Haider, 1993; Milchumas *et al.,* 1985; Norby and Cortrufo, 1998; Sanchez *et al.,* 2002).

Table 1: Effect of P concentration in solution and time interval on oxalic acid (µg/mL) exudation during sand culture

P concentration (µg/mL)	Time interval (days)				Mean
	5	10	20	30	
P_1(0)	1.84	1.67	2.57	1.42	1.88
P_2(0.2)	2.07	1.39	1.65	1.12	1.56
P_3(0.3)	1.18	1.83	1.33	0.97	1.33
P_4(0.4)	1.23	1.73	0.10	0.10	1.24
P_5(0.5)	1.12	1.36	1.09	1.21	1.20
Mean	1.49	1.59	1.52	1.15	

LSD (*P*<.05)P = 0.60 Time interval = 0.38 P × Time interval = NS

Composition of Rhizodeposition

The soil fertility determines the crop productivity and soil health. A healthy soil produces a healthy crop and healthy crops promote good soil health. The healthy crop secreted a range of organic compounds from the root as exudates. Root exudates include both secretions (including mucilage) which are actively released from the root, which are passively released due to osmotic differences between soil solutions and the cell, or lysates from autolysis of epidermal and cortical cells. The root secreted organic compounds can be further divided into high and low molecular weight (HMW and LMW, respectively) (Table 2). By weight, the HMW compounds which are those complex molecules that are not easily used by microorganisms (e.g. mucilage, cellulose) makes up the majority of C released from the root; however, the LMW compounds are more diverse and thus have a wider array of known or potential functions (Table 3). The rhizodeposite a significant amount of plant nutrient, which enhanced the microbial population and diversity. The more microbial diversity and population, enhanced the nutrient transformation, and nutrient availability to plants.

The proportion of carbon released from the roots has been estimated as much as 50% of the young plants (Whipps, 1990) but less in plants grown to maturity in the field (Jensen, 1993). Labeling of plants with ^{14}C has been widely used to quantify rhizodeposition as it allows for a distinction between root-derived organic C and native soil organic C (Whipps, 2001). The nature of exudates may also vary according to the growth stage of the plant. Helal and Sauerbeck (1987) estimated that the amount of C released from roots as rhizodeposit could be more than 580 kg C ha^{-1}. For instance, there are more carboxylates and root mucilage at the six leaf stage than earlier. On the other hand, nitrogen is also of considerable importance to nutrient cycling, usually as NH_4^+ & NO_3^- (Wacquant *et al.,* 1989), amino acids (Boulter *et al.,* 1966), cell lysates, sloughed roots, and other root-derived debris. It is estimated at the maturity that the rhizodeposition of N amounted to 20% of the total plant nitrogen (Jensen, 1997).

Table 2: Organic compounds and enzymes identified in root exudates of different plant species (Dakora and Phillips, 2002)

Amino acids α-alanine	Organic acid	Sugars	Vitamins	Purines/nucleosides	Enzymes	Inorganic ions and gaseous molecules
β-alanine	citric	glucose	biotin	adenine	acid/alkaline	HCO_3^-
oxalic	fructose	thiamin	guanine		phosphate	OH
asparagine	malic	galactose	niacin	cytidine	invertase	$^-H^+$
aspartate	fumaric	maltose	pantothenate	uridine	amylase	CO_2
cystine	Succininc	ribose	rhiboflavin		protease	H_2
cystein	acetic	xylose				
glutamate	butyric	rhamnose				
glycine	valeric	arabinose				
isoleucine	glycolic	raffinose				
leucine	piscidic	desoxyribose				
lysine	formic	oligosaccharides				
methionine	aconitic					
serinc	lactic					
threonine	pyruvic					
proline	glutaric					
valine	malonic					
tryptophan	aldonic					
ornithine	erythronic					
histidine	Tetronic					
arginine						
homoserine						
phenylalanine						
γ-aminobutyric acid						
α-aminoadipic acid						

Compiled from West (1939), Fries and Forsman (1951), Rovira and Harris (1961), Vancura (1964), Vancura and Hovadik (1965), Boutler *et al.,* (1966), Rovira (1969), Gardner *et al.,* (1983), Lipton *et al.,* (1987), Fox and Comerford (1990), Ae *et al.,* (1990), Ohwaki and Hirata (1992), Hoffland *et al.,* (1992) and Gagnon and Ibrahim (1998).

Table 3: Functional role of root exudates components in the rhizosphere [Modified from Hawes *et al.,* (1998; Dakora and Phillips, 1996)].

Component	Rhizosphere function
Phenolics	Nutrient source
	Chemoattractant signals to microbes
	Microbial growth promoters
	nod gene inducers in rhizobia
	nod gene inhibitors in rhizobia
	Resistance inducers against phytoalexins
	Chelators of poorly soluble mineral nutrients
	Detoxifiers of Al
	Phytoalexins against soil pathogens
Organic acids	Nutrient source
	Chemoattractant signals to microbes
	Chelators of poorly soluble mineral nutrients
	Acidifiers of soil
	Detoxifiers of Al
	nod gene inducers
Amino acids and phytosiderophores	Nutrient source
	Chelators of poorly soluble mineral nutrients
	Chemoattractant signals to microbes
Vitamins	Promoters of plant and microbial growth, Nutrient source
Purines	Nutrient source
Enzymes	Catalysts for P release from organic molecules
	Biocatalysts for organic matter transformation in soil
Root border cells	Produce signals that control mitosis
	Produce signals controlling gene expression
	Stimulate microbial growth
	Release chemoattractants
	Synthesize defense molecules for the rhizosphere
	Act as decoys that keep root cap infection-free
	Release mucilage and proteins

Effect on Soil Health

The plant secreted organic compounds affected the physical, chemical and biological properties (Fig. 4). The rhizodeposition mainly contributing (1) it is a loss of reduced C for the plant, (2) it is an input flux for the organic C pool of the soil and (3) it fuels the soil microflora, which is involved in the great majority of the biological activity of soils such as the nutrient and pollutant cycling or the dynamics of soil borne pathogens (Nguyen, 2009). The root exudates accelerate the soil reactions and nutrient transformation and output effect on various soil functions.

Recent interest in evaluating the soil quality has been stimulated by increasing awareness that soil is a critically important component of the earth, biosphere as it functions not only in the production of food and fiber, but also in the maintenance of environmental quality as related to agroecosystems management and in formulating and evaluating sustainable agricultural and land use policies

(Granatstein and Bezdicek, 1992). Plant-microbe interactions may thus be considered beneficial, neutral, or harmful to the plant, depending on the specific microorganisms and plants involved and on the prevailing environmental conditions (Bais *et al.,* 2006). Vigorous root systems are needed for the development of healthy plants and consequently, higher yields, due to secretion or more organic acid compounds from the roots. In the intensive crop cultivation areas roots that are left in the soil after crop harvest, which, improved SOM and contribute to plant nutrient cycles and microbial activity (Sainju *et al.,* 2005a). All these activities improve soil physical properties *i.e.* structure, aggregates size and durability, soil water holding capacity, water infiltration , reduce soil bulk density and soil erosion, ultimately leading to greater crop productivity (Jackson *et al.,* 1997). These positive effects enhanced the soil productivity and higher crop yield. The carbon cycling belowground is increasingly being familiar as one of the most significant components of the ecosystem C fluxes and pools, it is a basic component for soil fertility (Cheng and Kuzyakov, 2005; Zak and Pregitzer, 1998).

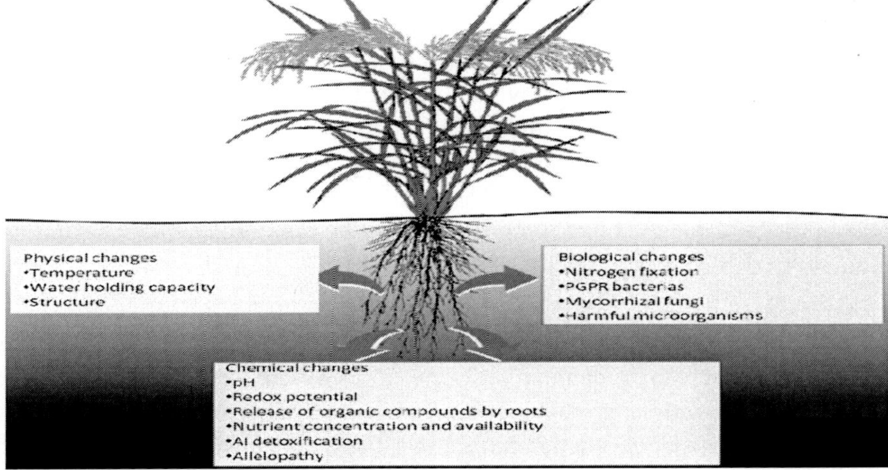

Fig. 4: Major physical, chemical and biological changes in the rhizosphere (Fageria and Stone, 2006).

Soil quality is the capacity of a soil to function within ecosystem boundaries to sustain biological productivity, maintain environmental quality and promote plant and animal health and thus has a profound effect on the health and productivity of a given ecosystem and the environment related to it (Doran *et al.,* 1994).The SOC, soil total nitrogen (STN) and soil microbial biomass C (SMBC) and N (SMBN) are some of the soil properties that are used as basic indicators in assessing soil quality (Doran and Parkin, 1994). Huge amount of rhozodeposited C material enhanced the microbial population and diversity and act as phytostimulation. It enhances plant growth in a direct way. In the processes of plant growth, phytohormones [*e.g.,* production of indole-3-acetic acid (IAA), auxins, cytokinins, and gibberellins] play an important role. These hormones can be synthesized by the plant themselves, but also by their associated microorganisms such as *Azospirillum* spp., besides having nitrogen-fixing ability (Steenhoudt *et al.,* 2000). The applications of silicon fertilizers enhanced the crop growth and indirectly root exudates in soil (Meena *et al.,* 2013).

The soil microbial biomass (SMB) is a small but key component of the active SOM pool and serves as a source and sink of soil nutrients (Smith and Paul, 1990). It governs the plant nutrient chemistry and nutrient transformation in soil. It has been used to understand soil nutrient dynamics and as an ecological marker (Paul and Voroney, 1980; Parton *et al.,* 1989). Soil biota (bacteria, fungi, micro-fauna and the plant root) are themselves embedded in food webs and thus interactions with consumers or predators in the microbial as well as macro- and mesofaunal world are important to understand rhizosphere processes. A high number of soil microbes attained properties enabling them to interact more efficiently with roots and withstand the quite challenging conditions of rhizosphere life. Bacterial communities in rhizosphere soils are thus not static, but will fluctuate over time in different root zones. The clearing of the forest undergrowth by burning and poor stocking of plantations may impair the quality of the soil. The SOM and STN are the major determinants and indicators of soil quality and fertility and are closely related to soil productivity and sustainability in an agricultural ecosystem (Reeves, 1997; Al-Kaisi *et al.,* 2005). The reduction of the SOC and STN will lead to a decrease in soil fertility, soil nutrient supply, porosity and an increase in soil erosion (Gray and Morant, 2003). The microorganisms were more active in the plantation soils due to their having more C and N immobilized in their microbial biomass (Adeboye *et al.,* 2011).

Factor Affecting Rhizodeposition

It is assumed that both the qualitative and quantitative compositions of root exudates are affected by various environmental factors, including pH, soil type, oxygen status, light intensity, soil temperature, nutrient availability and the presence of microorganisms. These factors may have a greater impact on root exudation than differences due to the plant species. The presence of organic matter enhanced the microbial diversity and population in the soil. Application of crop residue in soil enhanced the plant nutrient in soil (Dotaniya, 2013; Dotaniya and Datta, 2014) and also organic exudates (Dotaniya *et al.,* 2013d). In another experiment, application of bagasse and pressmud enhanced the plant nutrient uptake and root exudates in soil (Dotaniya, 2014). But in phytosiderophore condition, it is totally differed, its secretion was more in low fertility soil compared to high fertility status (Dotaniya *et al.,* 2013b)

Elevated atmospheric CO_2 directly affects ecosystems by increasing C assimilation in plants and microbial activities in soils (Mondal *et al.,* 2013). Field studies have shown that this greater C uptake results in an increase in plant production of on average 20% (Ainsworth and Long, 2005; de Graaff *et al.,* 2006a). Biomass production, rhizodeposition and cycling of root-borne N in maize genotypes were not affected by elevated CO_2. Elevated CO_2 stimulated above and below-ground biomass production of the wheat genotypes on average by 38%, and increased rhizodeposition and immobilization of root-derived N on average by 30%. Concurrently, elevated CO_2 reduced mineral ^{15}N and re-uptake of the root-derived N by 50% in wheat (de Graaff *et al.,* 2006b). The crop growth temperatures also decide the level of rhizodeposition in soil (Kushwah *et al.,* 2014; Kundu *et al.,* 2013); the vegetative or profuse growth of crop plant secreted the maximum level compared to other crop growth stages (Amrawat *et al.,* 2013; Pradhan and Dotaniya, 2013). In another experiment was conducted by Dotaniya *et al.,* (2014c) and observed that various summer season crops rhizosphere produced a different level of acid phosphatase activity in soil. The highest acid phosphatase activity was observed at different stages for different crops, but alkaline phosphatase and dehydrogenase activities were a maximum at 75 days after sowing in all the crops.

Plant rhizosphere is the soil nearest to the plant root system where roots release large quantities of metabolites from living root hairs or fibrous root systems. These metabolites act as chemical signals for motile bacteria to move to the root surface, but also represent the main nutrient sources available to support growth and persistence in the rhizosphere. A better understanding of the basic principles of the rhizosphere ecology, including the function and diversity of inhabiting microorganisms is on the way, but further knowledge is necessary to optimize soil microbial technology to the benefit of plant-growth and health of the natural environment. The plant roots exudates affected by exudation area and growth period of crops, which is the zone of maximum nutrient transformation and released organic substrates from plants (Dotaniya and Kushwah, 2013). The rhizoplane showing the highest rhizodeposition in soil and, its amount decreased with increasing distance from the roots (Dotaniya and Meena, 2013).

The root exudation is also largely dependent on the nutritional status of the plant regarding oligoelements. Low concentrations of some nutrients such as K^+, Na^+ and Mg^{++} readily stimulate the activity of major enzymes of the glycolytic pathway, namely phosphofructokinase and pyruvate kinase, which together regulate glycolysis in plant cells (Plaxton, 1996). Individual micronutrients are similarly important components of major enzymes, which regulate all biological processes in plants. It is clear from these considerations that low nutrient availability can constraint plant growth in many environments of the world, especially the tropics, where soils are extremely deficient in these oligoelement nutrients (Pinton *et al.*, 2007). Some species typically exude organic acid anions in response to P and Fe deficiency or phytosiderophores due to Fe and Zn deficiency (Haynes, 1990). Plant species, plant developmental stage and soil type have thus been indicated as major factors determining the composition of rhizosphere microbial communities. The rhizodeposition, such as root exudates, mucilages, and sloughed cells, may be a significant source of energy for microbes (Balesdent and Balabane, 1996; Buyanovsky *et al.,* 1986; Sainju *et al.,* 2005a). This amount of rhizodeposition increases microbial activity and influences C and N mineralization in the soil (Bakken, 1990; Texier and Biles, 1990).

Arable crop cultivation in both rainy and dry seasons with fertilizer application was a good measure in improving the quality of the soil, especially in terms of

SOC and STN (Adeboye *et al.,* 2011). The surface soil, 0-5 cm, was the main site for microbial meditated processes of nutrient cycling (Narolia *et al.,* 2013; Pingoliya *et al.,* 2013) and decomposition in all the agroecosystems. In 2013, Shukla *et al.,* reported that application of bio-organics enhanced the fertility of soil and crop yield, it might be due to higher amount of organic secretion from chickpea (*Cicer arietinum* L.) roots. In another, pot experiment Increasing the concentration of phosphorus fertilizers significantly decreased the root exudates *i.e.* oxalic acid. (Dotaniya *et al.,* 2013a). The application of silicon containing fertilizers also enhanced the crop growth, resistance to insect pest, diseases and indirectly roots exudates in soil (Meena *et al.,* 2013).

The plant nutrient uptake affected by the availability of toxic metal, species and concentration of a particular metal (Gaur and Adholeya, 2004). Continuous use of heavy metal contaminated water for irrigation, build up a significant amount of toxic metal (Dotaniya *et al.,* 2014c), which reduced the soil biological activities and plant root exudation (Wyszkowska, 2002). Detainee *et al.,* (2014a) observed that higher concentration of chromium reduced the root elongation adverse, which may reduce the area of crop exudates. The amount of oxygen is required for aerobic microbial population, which indirectly affected the rhizodeposition amount from root exudates. Imposition of stress factors on the microbial biomass will increase its maintenance, energy requirement which may reduce the yield efficiency of the biomass and increase the death rate of the biomass (Franzluebbers *et al.,* 1995).

Future Research Needs

- Rhizodeposition is beneficial for crop plant and soil, but its amount varies crop to crop and species, should needs more attention.
- With the help of biotechnological tools, genetic code responsible for its secretion can be modified and also transfer to other crops?
- More specific research on secretion period of root exudates and its synchronization with application of fertilizer.
- Quality and quantity of rhizodeposition in respect to particular crops.
- Effect of global climate change on rhizodeposition for a particular crop species and region.

Conclusions

The rhizosphere is the zone of soil surrounding a plant root where the biological and chemical parameters of the soil are influenced by the root exudates. Ranges of the process are working in the rhizosphere of crops, and affected the microbial diversity and nutrient transformation in soil. Through diverse interactive mechanisms exudates play a basic role in mineral acquisition of plants. The genetic code of a particular species controls the signal process during root exudations. If all the process identified and manipulation with the help of agronomic and biotechnological methods, it can be a good option for enhancing plant nutrient uptake and assimilation. The improvement in plant nutrient, especially in low fertility soil by root exudates can be a viable option for higher yield and crop sustainability.

References

Adeboye, M.K.A., Bala, A., Osunde, A.O., Uzoma, A.O., Odofin, A.J. and Lawal BA (2011). Assessment of soil quality using soil organic carbon and total nitrogen and microbial properties in tropical agroecosystems. *Agricultural Science*, 2(1): 34-40.

Ae, N., Arihara, J., Okada, K., Yoshihara, T. and Johansen, C. (1990). Phosphorus uptake by pigeon pea and its role in cropping systems of the Indian subcontinent. *Science,* 248: 477-480.

Ainsworth, E.A. and Long, S.P. (2005). What have we learned from 15 years of free-air CO_2 enrichment (FACE)? A meta-analytic review of the responses of photosynthesis, canopy properties and plant production to rising CO_2. *New Phytology*, 165: 351-372.

Al-Kaisi, M.M., Yin, X.H. and Licht, M.A. (2005). Soil carbon and nitrogen changes as influenced by tillage and cropping systems in some Iowa soils. *Agriculture, Ecosystem and Environment* 105: 635-647.

Amrawat, T., Solanki, N.S., Sharma, S.K., Jajoria, D.K. and Dotaniya M.L. (2013). Phenology growth and yield of wheat in relation to agrometeorological indices under different sowing dates. *African Journal of Agricultural Research*, 8(49): 6366 -6374.

Bais, H.P., Weir, T.L., Perry, L.G., Gilroy, S. and Vivanco JM (2006). The role of root exudates in rhizosphere interactions with plants and other organisms. *Ann Review of Plant Biology,* 57: 233-266.

Bakken, L.R. (1990). Microbial growth and immobilization/mineralization of N in the rhizosphere. *Symbiosis*, 9: 37-41.

Balesdent, J. and Balabane, M. (1996). Major contribution of roots to soil carbon storage inferred from maize cultivated soils. *Soil Biology and Biochemistry,* 28: 1261-1263.

Bertin, C., Yang, X. and Weston, L.A. (2003). The role of root exudates and allelochemicals in the rhizosphere. *Plant and Soil*, 256: 67-83.

Bhupinderpal-Singh, Nordgren, A., Lofvenius, M.O., Hogberg, M.N., Mellander, P.E. and Hogberg, P. (2003). Tree root and soil heterotrophic respiration as revealed by girdling of boreal Scots pine forest: extending observations beyond the first year. *Plant Cell and Environment*, 26:1287-1296.

Boone, R.D. (1994). Light-fraction soil organic matter: Origin and contribution to net nitrogen mineralization. *Soil Biology and Biochemistry*, 26: 1459-1468.

Boutler, D., Jeremy ,J. J. and Wilding, M. (1966). Amino acids liberated into the culture medium by pea seedling roots. *Plant and Soil*, 24: 121-127.

Bowen, G., Rovira, A. (1999). The rhizosphere and its management to improve plant growth. *Advances in Agronomy*, 66: 1-102.

Brimecombe, M.J., De, Leij FAAM, and Lynch, J.M. (2007). Rhizodeposition and microbial populations. In: Pinton R, Veranini Z, Nannipieri P (ed)The rhizosphere biochemistry and organic substances at the soil plant interface. Taylor & Francis Group, New York, USA.

Buyanovsky, G.A., Wagner, G.H. and Gantzer, C.J. (1986). Soil respiration in a wheat ecosystem. *Soil Science Society of America Journal,* 50: 338–344.

Cheng, W., Coleman, D.C., Carroll, C.R., and Hoffman CA (1994). Investigating short-term carbon flows in the rhizospheres of different plant species, using isotopic trapping. *Agronomy Journal*, 86: 782-788.

Cheng,W. and Kuzyakov, Y. (2005). Root effects on soil organic matter decomposition. In: Zobel RW, Wright SF (ed) Roots and soil management: interactions between roots and the soil. ASA, CSSA, and SSSA, Madison, WI, pp 119-143.

Dakora, F. D. and Phillips, D. A. (1996). Diverse functions of isoflavonoids in legumes transcend anti-microbial definitions of phytoalexins. *Physiolical and Molecular Plant Pathology,* 49: 1-20.

Dakora FD, Phillips DA (2002) Root exudates as mediators of mineral acquisition in low-nutrient environments. *Plant and Soil,* 245:35-47.

Dalenberg, J.W. and Jager, G. (1981). Priming effect of small glucose additions to C-14-labeled soil. *Soil Biology and Biochemistry*, 13:219-223.

Dalenberg, .J.W and Jager, G. (1989). Priming effect of some organic additions to C-14-labeled soil. *Soil Biology and Biochemistry,* 21:443-448.

de Graaff, M.A., Six, J. and Kessel, C.V. (2006b). Elevated CO_2 increases nitrogen rhizodeposition and microbial immobilization of root-derived nitrogen. *New Phytology*, 173: 778-786.

de Graaff, M.A., van Groeningen, K.J., Six, J., Hungate, B.A. and van Kessel, C. (2006a). Interactions between plant growth and nutrient dynamics under elevated CO_2: a meta analysis. *Global Change Biology*, 12: 1-15.

De Nobili, M., Contin, M., Mondini, C. and Brookes, P.C. (2001). Soil microbial biomass is triggered into activity by trace amounts of substrate. *Soil Biology and Biochemistry,* 33:1163.

Doran, J.W. and Parkin, T.B. (1994). Defining and assessing soil quality. Special Publication of the *Soil Science Society of America,* 35: 3-21.

Doran, J.W., Sarrantonio, M. and Janke, R. (1994). Strategies to promote soil quality and health. *Proc OECD Co-operative Res project on Biol Res Manage* 1: 230-237.

Dotaniya, M.L. (2013). Impact of various crop residue management practices o.n nutrient uptake by rice-wheat cropping system. *Current Advances in Agricultural Science,* 5(2):269-271.

Dotaniya, M.L., Das, H., Meena, V.D. (2014a). Assessment of chromium efficacy on germination, root elongation, and coleoptile growth of wheat (*Triticum aestivum* L.) at different growth periods. *Environmental Monitoring and Assessment*, 186:2957-2963.

Dotaniya, M.L. and Datta S.C. (2014). Impact of bagasse and press mud on availability and fixation capacity of phosphorus in an Inceptisol of north India. *Sugar Technology*, 16(1): 109-112.

Dotaniya, M.L., Datta, S.C., Biswas, D.R., Meena, H.M., and Kumar, K. (2013a). Production of oxalic acid as influenced by the application of organic residue and its effect on phosphorus uptake by wheat (*Triticum aestivum* L.) in an Inceptisol of north India. *National Academy of Science Letters*, DOI: 10.1007/s40009-014-0237-y

Dotaniya, M.L., Datta, SC., Biswas, DR., Meena, B.P. (2013c). Effect of solution phosphorus concentration on the exudation of oxalate ions by wheat (*Triticum aestivum* L.). *Proceedings of National Academy of Science Letters, India Sec B. Biol. Sci.* 83(3):305-309

Dotaniya, M.L. and Kushwah, S.K. (2013). Nutrients uptake ability of various rainy season crops grown in a Vertisol of central India. *African Journal of Agricultural Research,* 8(44): 5592-5598.

Dotaniya, M.L., Kushwah, S.K., Rajendiran, S., Coumar, M.V., Kundu, S., Rao, A. Subba (2014b). Rhizosphere effect of Kharif Crops on Phosphatases and Dehydrogenase activities in a Typic Haplustert. *National Academy of Science Letters*. DOI: 10.1007/s40009-013-0205-4.

Dotaniya, M.L., Meena, V.D. (2013). Rhizosphere effect on Nutrient Availability in soil and Its Uptake by plants -A review. *Proceedings of National Academy of Science Letters*, India Sec B Biol Sci. DOI 10.1007/s40011-013-0297-0.

Dotaniya, M.L., Prasad, D., Meena, H.M., Jajoria, D.K., Narolia, G.P., Pingoliya, KK., Meena, O.P., Kumar, K., Meena, B.P., Ram, A., Das, H., Chari, MS. and Pal, S. (2013b). Influence of phytosiderophore on iron and zinc uptake and rhizospheric microbial activity. *African Journal of Microbiological Research*, 7(51): 5781-5788.

Dotaniya, M.L., Saha, J.K., Meena, V.D., Rajendiran, S., Coumar, M.V., Kundu, S. and Rao, A. Subba (2014c). Impact of tannery effluent irrigation on heavy metal build up in soil and ground water in Kanpur. *Agrotechnol*ogy 2(4):77

Dotaniya, M.L., Sharma, M.M., Kumar, K. and Singh, P.P. (2013d). Impact of crop residue management on nutrient balance in rice-wheat cropping system in an Aquic hapludoll. *The Journal of Rural Agricultural Research*,13(1):122-123.

Fageria, N.K. and Stone, L.F. (2006). Physical, chemical, and biological changes in the rhizosphere and nutrient availability. *Journal of Plant Nutrition,* 29: 1327-1356.

Fontaine, S., Bardoux, G., Benest, D., Verdier, B., Mariotti, A. and Abbadie, L. (2004). Mechanisms of the priming effect in a savannah soil amended with cellulose. *Soil Sci Soc Am J* 68: 125-131.

Fox, T.R., Comerford, N.B. (1990). Low-molecular weight organic acids in selected forest soils of the southeastern USA. *Soil Science Society of America Journal,* 54: 1139-1144.

Franzluebbers, A.J., Hons, F.M. and Zuberer, D.A. (1995). Soil organic carbon, microbial biomass, and mineralizable carbon and nitrogen in sorghum. *Soil Science Society of America Journal,* 56: 460-466.

Fries, N. and Forsman, B. (1951). Quantitative determination of certain nucleic acid derivatives in pea root exudates. Physiol Plant 4: 210-234.

Gagnon, H. and Ibrahim, R.K. (1998). Aldonic acids: A novel family of nod gene inducers of *Mesorhizobium loti, Rhizobium lupini* and *Sinorhizobium meliloti. Molecular Plant Microbe Interaction,* 11: 988- 998.

Gale, W.J., Cambardell, C.A. and Bailey, T.B. (2000a). Root-derived carbon and the formation and stabilization of aggregates. *Soil Science Society of America Journal,* 64: 201-207.

Gale, WJ., Cambardell, CA., Bailey, T.B. (2000b). Surface residue and root-derived carbon in stable and unstable aggregates. *Soil Science Society of America Journal,* 64: 196-201.

Gardner, WK., Barber, DA. and Parbery, DG. (1983). The acquisition of phosphorus by *Lupinus albus* L. III. The probable mechanism by which phosphorus movement in the soil/root interface is enhanced. *Plant and Soil,* 70: 107-124.

Gaur, A. and Adholeya, A. (2004). Prospects of arbuscular mycorrhizal fungi in phytoremediation of heavy metal contaminated soils. *Current Science,* 86(4): 528-534

Granatstein, D. and Bezdicek, DF. (1992). The need for a soil quality index: Local and regional perspectives. *American Journal of Alternative Agriculture,* 17: 12-16.

Gray, L.C. and Morant, P. (2003). Reconciling indigenous knowledge with scientific assessment of soil fertility changes in southwestern Burkina Faso. *Geoderma,* 111: 425-437. doi:10.1016/S0016-7061(02)00275-6.

Gupta, R. and Mukerji, K.G. (2002). Root exudate-biology. In: Mukerji KG, Manoharachary C, Chamola BP (ed) Techniques in mycorrhizal studies. Kluwer Academic Publishers, Dordrecht, The Netherlands, pp 103-131.

Haider, K. (1993). Impact of growing crop plants on N-utilization from fertilizer, crop residues, and on organic N and C-mineralization. *Soil Science,* 1:183-191.

Hawes, MC., Brigham, LA., Wen, F., Woo, HH. and Zhu, Y. (1998). Function of root border cells in plant health: Pioneers in the rhizosphere. *Annu Review of Phytopathology,* 36: 311-327.

Haynes, RJ. (1990). Active ion uptake and maintenance of cation anion balance: A critical examination of their role in regulating rhizosphere pH. *Plant and Soil,* 126: 247-264.

Helal, HM. and Sauerbeck, D. (1987). Direct and indirect influences of plant root on organic matter and phosphorus turnover in soil. In: Couley JH (ed) Soil organic matter dynamics and soil productivity. INTECOl Bulletin 15, International Association for Ecology, Athens, GA, pp 49-58.

Hiltner, L. (1904). Uber neuere Erfahrungen und Probleme auf dem Gebiete der Bodenbakteriologie unter besonderer Beffícksichtigung der Grundiingung und Brache. *Arb Dtsch Landwirtsch Ges.* 98: 59-78.

Hoffland, E., Van Den, Boogaard, R., Nelemans, J. and Findenegg, G. (1992). Biosynthesis and root exudation of citric and malic acids in phosphate-starved rape plants. *New Phytology,* 122: 675-680.

Hogberg, P. and Ekblad, A. (1996). Substrate-induced respiration measured in situ in a C_3-plant ecosystem using additions of C_4-sucrose. *Soil Biology and Biochemistry*, 28:1131.

Hogberg, P., Nordgren, A., Buchmann, N., Taylor, AFS., Ekblad, A., Ho¨gberg MN., Nyberg, G., Ottosson-Lofvenius, M. and Read, DJ. (2001). Large-scale forest girdling shows that current photosynthesis drives soil respiration. *Nature,* 411:789-792.

Hütsch, BW., Augustin, J. and Merbach, W. (2002). Plant rhizodeposition – an important source for carbon turnover in soils. *Journal of Plant Nutrition and Soil Science*, 165: 397-407.

Jackson, RB., Mooney, HA. and Schulze, ED. (1996). A global budget for fine root biomass, surface area, and nutrient contents. *Proceedings of National Academy of Sciences USA* 94: 7362-7366.

Jensen, B. (1993). Rhizodeposition by $^{14}CO_2$-pulse-labelled spring barley grown in small field plots on sandy loam. *Soil Biology and Biochemistry*, 25: 1553-1559.

Jones, DL., Nguyen, C. and Finlay, RD. (2009). Carbon flow in the rhizosphere: carbon trading at the soil-root interface. *Plant and Soil* 321:5-33.

Kirchner, MJ., Wollum, AG. and King, LD. (1993). Soil microbial populations and activities in reduced chemical input agro ecosystems. *Soil Science Society of America Journal,* 57: 1289-1295.

Kundu, S., Dotaniya, M.L. and Lenka, S. (2013). Carbon sequestration in Indian agriculture. In: Lenka, S., Lenka, N.K, Kundu, S., Rao AS (ed) Climate change and natural resources management. New India Publishing Agency, India, pp 269-289.

Kuo, S., Sainju, UM. and Jellum, EJ. (1997a). Winter cover crop effects on soil organic carbon and carbohydrate. *Soil Science Society of America Journal,* 61: 145-152.

Kuo, S., Sainju, UM. and Jellum, EJ. (1997b). Winter cover cropping influence on nitrogen in soil. *Soil Science Society of America Journal,* 61: 1392-1399.

Kushwah, SK., Dotaniya, M.L., Upadhyay, A.K., Rajendiran, S., Coumar, M.V., Kundu, S. and Rao, A. Subba, (2014). Assessing carbon and nitrogen partition in kharif crops for their carbon sequestration potential. *Proceedings of National Academy of Science Letters*, DOI: 10.1007/s40009-014-0230-y

Kuzyakov, Y. (2002). Review factors affecting rhizosphere priming effects. *Journal of Plant Nutrition and Soil Science*, 165: 382-396.

Kuzyakov, Y., Friedel, J.K. and Stahr, K. (2000). Review of mechanisms and quantification of priming effects. Soil Biology and Biochemistry, 32: 1485-1498.

Liang, BC., Wang, XL. and Ma, B.L. (2002). Maize root-induced change in soil organic carbon pools. *Soil Science Society of America Journal,* 66: 845-847.

Lipton, D.S., Blanchar, R.W. and Blevins, DG. (1987). Citrate, malate, and succinate concentration in exudates from P-sufficient and P-stressed *Medicago sativa* L. seedlings. *Plant Physiology*, 85: 315- 317.

Mandal, A., Radha, TK. and Neenu, S. (2013). Impact of climate change on rhizosphere microbial activity and nutrient cycling. In: Lenka S, Lenka NK, Kundu S, Rao AS (ed) Climate change and natural resources management. New India Publishing Agency, India, pp 93-115.

Marschner, H. (1995). Mineral nutrition of higher plants. Academic Press, New York

Mary, B., Fresneau, C., Morel, J.L. and Mariotti, A. (1993). C and N cycling during decomposition of root mucilage, roots and glucose in soil. *Soil Biology and Biochemistry*, 25:1005.

Meena, V.D., Dotaniya, M.L., Rajendiran, S., Coumar, M.V., Kundu, S. and Rao, A. Subba (2013). A case for silicon fertilization to improve crop yields in tropical soils. *Proceedings of National Academy of Science*, India Sec B Biological Sciences. DOI 10.1007/s40011-013-0270-y.

Melillo, JM. and Gosz, JR. (1983). Interactions of biogeochemical cycles in forest ecosystems. In: Bolin B, Cook RB (ed) The major biogeochemical cycles and their interactions. John Wiley & Sons, Chichester, England, pp 177-220.

Milchumas, DG., Lauenroth, WK., Singh, JS. and Cole, CV. (1985). Root turnover and production by ^{14}C dilution: Implications of carbon portioning in plants. *Plant and Soil,* 88: 353-365.

Narolia, G.P., Jajoria, D.K and Dotaniya, M.L. (2013). Role of phosphorus, PSB and zinc in isabgol (Plantago ovata F.). Lambert Academic Publishing, Saarbrucken

Nguyen, C. (2009). Rhizodeposition of organic c by plant: mechanisms and controls. *Sustainable Agriculture*, DOI10.1007/978-90-481-2666-8_9.

Nihorimbere, V., Ongena, M., Smargiassi, M. and Thonart, P. (2011). Beneficial effect of the rhizosphere microbial community for plant growth and health Biotechnol. *Biotechnology, Agronomy, Society and Environment,* 15(2): 327-337.

Norby, RJ. and Cortrufo, MF. (1998). A question of litter quality. *Nature,* 206: 17-18.

Ohwaki, Y. and Hirata, H. (1992). Differences in carboxylic acid exudation among P-starved leguminous crops in relation to carboxylic acid contents in plant tissues and phospholipid level in roots. *Soil Science and Plant Nutrition,* 38: 235-243.

Parton, W.J., Sandford, RL., Sanchez, PA. and Stewart, KWB. (1989). Modelling soil organic matter dynamics in tropical soils. Uni Hawaii 240: 153-171.

Paul, EA. and Voroney, RP. (1980). Nutrient and energy flows through soil microbial biomass. *Contemperory Microbial Ecology,* 1: 215-237.

Pidwirny, M. (2006). Introduction to Soils. In: Fundamentals of physical geography. http://www.physicalgeography.net/fundamentals/10t.html.

Pingoliya, K.K., Dotaniya, M.L. and Mathur, A.K. (2013). Role of phosphorus and iron in chickpea (*Cicer arietinum* L.). Lambert Academic Publishing, Saarbrucken.

Pinton, R., Veranini, Z. and Nannipieri, P. (2007). The rhizosphere. In: Biochemistry and organic substances at the soil-plant interface. Taylor & Francis Group, New York, USA.

Plaxton, WC. (1996). The organization and regulation of plant glycolysis. *Annual Review of Plant Physiology and Plant Molecular Biology,* 47: 185-214.

Pradhan, Y., Dotaniya, M.L. (2013). Water management in european dill (*Anethum graveolens* L.). Lambert Academic Publishing, Saarbrucken.

Puget, P. and Drinkwater, LE. (2001). Short-term dynamics of root- and shoot-derived carbon from a leguminous green manure. *Soil Science Society of America Journal,* 65: 771-779.

Qian, JH. and Doran, JW. (1996). Available carbon released from crop roots during growth as determined by carbon-13 natural abundance. *Soil Science Society of America Journal,* 60: 828-831.

Ratnayale, M., Leonard, RT. and Menge, A. (1978). Root exudation in relation to supply of phosphorus and its possible relevance to mycorrhizal infection. *New Phytology*, 81: 543-552.

Reeves, DW. (1997). The role of soil organic manure in maintaining soil quality in continuous cropping systems. *Soil Tillage and Research,* 43: 131-167.

Rovira, A.D. (1969). Plant root exudates. *Botanical Reviews*, 35: 35-57.

Rovira, A.D. and Harris, J. R. (1961). Plant root excretions in relation to the rhizosphere effect. V. The exudation of B-group vitamins. *Plant and Soil*, 14, 199-214.

Rudrappan, T., Quinn, W.J., Stanley-Wall, N.R. and Bais HP (2007). A degradation product of the salicylic acid pathway triggers oxidative stress resulting in down-regulation of *Bacillus subtilis* biofilm formation on *Arabidopsis thaliana* roots. *Planta,* 226: 283-297.

Sainju, U.M., Singh, B.P. and Whitehead, W.F. (2005a). Tillage, cover crops, and nitrogen fertilization effects on cotton and sorghum root biomass, carbon, and nitrogen. *Agronomy Journal,* 97: 1279-1290.

Sainju, U.M., Whitehead, W.F. and Singh, B.P. (2005b). Carbon accumulation in cotton, sorghum, and underlying soil as influenced by tillage, cover crops, and nitrogen fertilization. *Plant and Soil*, 273: 219-234.

Sanchez, J.E., Paul, E.A., Wilson, T.C., Smeenk, J. and Harwood, R.R. (2002). Corn root effects on the nitrogen supplying capacity of a conditioned soil. *Agronomy Journal*, 94: 391-396.

Schloter, M., Lebuhn, M, Heulin, T. and Hartmann, A. (2000). Ecology and evolution of bacterial microdiversity. *FEMS Microbiological Reviews,* 24: 647-660.

Schweinsberg-Mickan, MS., Jörgensen, R.G. and Müller, T. (2012). Rhizodeposition: Its contribution to microbial growth and carbon and nitrogen turnover within the rhizosphere. *Journal of Plant Nutrition and Soil Science,* 175 (5): 750-760.

Scott-Denton, L.E., Rosenstiel, T.N. and Monson, R.K. (2006). Differential controls by climate and substrate over the heterotrophic and rhizospheric components of soil respiration. *Global Change Biology,* 12:205-216.

Shukla, M., Patel, R.H., Verma, R., Deewan, P. and Dotaniya, M.L. (2013). Effect of bio-organics and chemical fertilizers on growth and yield of chickpea (*Cicer arietinum* L.) under middle Gujarat conditions. *Vegetos,* 26(1):183-187.

Singh, S., Ladha, J.K., Gupta, R.K., Bhushan, L., Rao, AN., Sivaprasad, B. and Singh, P.P. (2007). Evaluation of mulching, intercropping with Sesbania and herbicide use for weed management in dry-seeded rice (*Oryza sativa* L.). *Crop Protection,* 26: 518-524.

Smith, J.L. and Paul, E.A. (1990). The significance of soil microbial biomass estimation. *Soil Biology and Biochemistry,* 6: 357-396.

Somers, E., Vanderleyden ,J. and Srinivasan, M. (2004). Rhizosphere bacterial signalling: a love parade beneath our feet. *Critical Reviews in Microbiology,* 30: 205-235.

Steenhoudt, O. and Vanderleyden, J. (2000). *Azospirillum*, a free-living nitrogen-fixing bacterium closely associated with grasses: genetic, biochemical and ecological aspects. *FEMS Microbiol Reviews,* 24: 487-506.

Subke, J.A., Hahn, V., Battipaglia, G., Linder, S., Buchmann, N. and Cotrufo, M.F. (2004). Feedback interactions between needle litter decomposition and rhizosphere activity. *Oecologia,* 139:551-559.

Texier, M. and Biles, G. (1990). The role of rhizosphere on C and N cycles in a plant–soil system. *Symbiosis,* 9: 117-123.

Tresder, K. K., Morris, S. J., and Allen, M. F. (2005). The contribution of root exudates, symbionts, and detritus to carbon sequestration in the soil. In: Zobel RW, Wright SF (ed) Roots and soil management: interactions between roots and the soil. ASA, CSSA, and SSSA, Madison, WI, pp 145-162.

Uren, N.C., (2001). Types, amounts and possible functions of compounds released into the rhizosphere by soil-grown plants. In: Pinton R, Varanini Z, Nannipieri P (ed) The rhizosphere: Biochemistry and organic substances at the soil-plant interface. Marcel Dekker, New York, pp 19-40.

Vancura, V. (1964). Root exudates of plants. I. Analysis of root exudates of barley and wheat in their initial phases of growth. *Plant and Soil,* 21: 231-248.

Vancura, V. and Hovadik, A. (1965). Root exudates of plants. II. Composition of root exudates of some vegetables. *Plant and Soil*, 22: 21-32.

Wacquant, J.P., Ouknider, M. and Jacquard, P. (1989). Evidence for a periodic excretion of nitrogen by roots of grass-legume associations. *Plant and Soil*, 116: 57-68.

Wedin, D.A. and Tilman, D. (1990). Species effects on nitrogen cycling: A test with perennial grasses. *Oecologia,* 84: 433-441.

West, P.M. (1939). Excretion of thiamin and biotin by the roots of higher plants. *Nature,* 144: 1050-1051.

Whipps, J.M. (2001). Microbial interactions and biocontrol in the rhizosphere. *Journal of Experimental Botany*, 52: 487-511.

Wyszkowska, J. (2002). Soil Contamination by chromium and its enzymatic activity and yielding. *Polish Journal of Environmental Studies,* 11(1): 79-84.

Zak, D.R. and Pregitzer, K.S. (1998). Integration of ecophysiological and biogeochemical approaches to ecosystem dynamics. In: Pace ML, Groffman PM (ed) Success, limitations, and frontiers in ecosystem science, Springer, New York, pp 372-403.

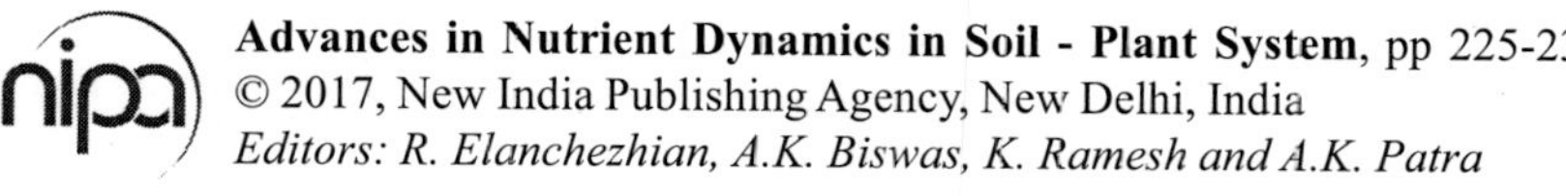
Advances in Nutrient Dynamics in Soil - Plant System, pp 225-238

Editors: R. Elanchezhian, A.K. Biswas, K. Ramesh and A.K. Patra

20

Climate Change and its Impact on Soil-Plant Systems

Rajendiran S., Vasanda Coumar, M. and S. Kundu

ICAR-Indian Institute of Soil Science, Nabi Bagh, Bhopal – 462038, India

Introduction

Climate change is a result from emission of greenhouse gases (e.g. CO_2, CH_4, N_2O, etc.) that will cause atmospheric warming (IPCC, 2007). Climate change may be due to natural internal processes or external forcing such as modulations of the solar cycles, volcanic eruptions, and persistent anthropogenic changes in the composition of the atmosphere or in land use. Human induced changes in atmosphere composition are a major concern. Burning of fossil fuels, like oil, coal, and natural gas is adding CO_2 to the atmosphere. The Fourth Assessment Report of the Intergovernmental Panel on Climate Change (IPCC) concludes, "That most of the observed increase in the globally averaged temperature since the mid-20th century is very likely due to the observed increase in anthropogenic greenhouse gas concentrations." The concentration of CO_2, CH_4 and N_2O have increased markedly by 40%, 145% and 15%, respectively as a result of human activity since the industrial revolutions (1750) (IPCC, 2013). The CO_2, CH_4 and N_2O concentration in the atmosphere were 280 ppm, 715 ppb and 270 ppb in 1750 AD. In 2013, these values have become 396 ppm, 1893 ppb and 326 ppb, respectively (IPCC, 2013). The atmospheric concentration of CO_2, CH_4 and N_2O accounting for about 63%, 15% and 5% of global warming respectively. Over the last 100 years (1906-2005) global mean surface temperature have increased in the order by 0.74 °C ± 0.18 °C. The rate of increase in temperature is highest in recent decades. Fourteen of the last seventeen years (1995-2011) rank the warmest years since 1850. IPCC has projected that by the end of the 21st century the temperature increase is likely to be in the range of 2 to 4.5 °C. This has been accompanied by warming of the ocean, a rise in sea level, a strong decline in Arctic sea ice, and many other associated climate effects.

Climate change involves complex interactions and changing likelihoods of diverse impacts. Climate change induces the change in earth's atmospheric weather events like precipitation, air circulation, cyclonic storms and wind speed and direction. Global warming is expected to intensify regional contrasts in precipitation that already exist: dry areas are expected to get even drier and wet areas even wetter. Warmer temperatures tend to increase evaporation from oceans, lakes, plants, and soil which exacerbates the impacts of drought in subtropics. In contrast, sub-polar and polar regions are expected to see increased precipitation, especially during winter. It also imparts other problems like reduction in food production, change in biodiversity and human health risks. Therefore addressing and predicting direct and indirect impact of climate change on vegetation, soil process, nutrient dynamics and soil health is very important for adaptation and mitigation and for making reliable strategies and policies to overcome the overall impact of climate change on soil, plant and human systems. In the following pages, impact of climate on soil-plant system is briefly discussed.

Impact of Climate Change on Plants

Effects of elevated CO_2 on plant growth and yield

The effects of elevated CO_2 on plant growth and yield will depend on photosynthetic pathway, species, growth stage and management regime, such as water and nitrogen (N) applications (Ainsworth and Long, 2005). On an average under unstressed conditions, crop yields increase in the range of 10-20% for C_3 crops and 0-10% for C_4 crops under elevated CO_2 of 550 ppm compared to current atmospheric CO_2 concentrations (Ainsworth and Rogers, 2007). However, global climate change, in the form of rising temperature and altered soil moisture, is projected to decrease the yield of food crops over the next 50 years (Thomson *et al.*, 2005). Importantly, plant physiologists and modelers alike recognize that the effects of elevated CO_2 measured in experimental settings and implemented in models may overestimate actual field- and farm-level responses, due to many limiting factors such as pests, weeds, competition for resources, soil, water and air quality, *etc.*, which are neither well understood at large scales, nor well implemented in leading models (Ainsworth and Long, 2005). Assessment studies should therefore include these factors where possible, while analytical capabilities need to be enhanced. Based on many studies covering a wide range of regions and crops, negative impacts of climate change on crop yields have been more common than positive impacts. The smaller number of studies showing positive impacts relate mainly to high-latitude regions, though it is not yet clear whether the balance of impacts has been negative or positive in these regions. Climate change has negatively affected wheat and maize yields of many regions and in the global aggregate. Effects on rice and soybean yield have been smaller in major production regions and globally, with a median change of zero across all available data, which are fewer for soybean compared to the other crops.

Change in biodiversity and phenology

Climate change is also likely to impact on agricultural genetic resources. For example, impacts on the distributions of crop wild relatives (themselves bearers of genes potentially critical for adaptation) are predicted to be significant, with 18% of species in important crop genepools likely to become extinct by 2050. Additionally, farmers may drop traditional landraces no longer adapted to the local climate, resulting in genetic erosion and a lowering of the diversity in crop genepools. Species respond in very different ways to climate change. Variation in the distribution, phenology and abundance of species will lead to inevitable changes in the relative abundance of species and their interactions. These changes will flow on to affect the structure and function of ecosystems (Walther *et al.* 2002). Through global warming, an anticipated increase in temperature can potentially have various effects, *e.g.* spikelet sterility in rice, reversal of vernalisation in wheat, reduced formation of tubers in potatoes, loss of pollen viability in maize. Yields can be severely affected if temperatures exceed critical limits for periods as short as 1 h during anthesis (flowering). Flowering is a very important event in crop development, as it is a phase which is particularly vulnerable to environmental stresses.

It is thought that extreme temperatures are more important than average temperatures in determining plant responses. Crop yields are affected by net primary productivity and also by the phenology of crop development. Increased temperature can speed phenological development, reducing the grain-filling period for crops and lowering yield. Crop yields were greater under elevated CO_2, but warmer temperatures reduced the duration of crop growth and, hence, the yield of determinate crops such as winter wheat and onion; but the yield of carrot, for example, an indeterminate crop, increased progressively with temperature (Wheeler *et al.,* 1996). In terms of temperature, a 12-day period of high temperature stress close to anthesis reduced spring wheat root biomass from 141 to 63 g m^{-2} (Ferris *et al.,* 1998) by the end of the elevated mean temperature period, whereas mean temperatures over the treatment period had no effect on either above-ground biomass or grain yield at maturity. Interestingly, it was increasing maximum temperatures over the mid-anthesis period which was related to a decline in the number of grains per ear at maturity. Grain yield and harvest index also declined sharply with maximum temperature. This suggests that high temperature extremes may reduce yields considerably.

Physiological changes

Increase in atmospheric CO_2 concentration affect photosynthesis, resulting in increases in plant water use efficiency, enhanced photosynthetic capacity and increased growth (Steffen and Canadell, 2005). Increased CO_2 has been implicated in 'vegetation thickening' which affects plant community structure and function (Gifford and Howden, 2001). Depending on environment, there are differential responses to elevated atmospheric CO_2 between major 'functional types' of plant, such as C_3 and C_4 plants, or more or less woody species; which has the potential among other things to alter competition between these groups (Jeffrey and Harold 1999). Increased CO_2 can also lead to increased Carbon: Nitrogen ratios in the leaves of plants or in other aspects of leaf chemistry.

Plant structures: leaves and roots

CO_2 enrichment of the air in which crops grow usually stimulates their growth and yield. Plant structure and physiology are usually markedly altered; this includes increased leaf expansion and cell wall extensibility and often cell turgor pressure, leading to increased leaf and root growth. If increased turgor pressure is alone insufficient to account for increases in leaf growth under elevated CO_2, then cell wall relaxation (extensibility), cell division or both may also be affected.

Simplistically, scientists have suggested that increased leaf size, if associated with larger cells, suggests that cell expansion has been stimulated, whilst increased leaf size, if associated with more cells, suggests that cell division has been stimulated. However, various studies have reported differences in the various cellular mechanisms driving leaf expansion, which can vary between species and seasons; thus, one or other mechanisms may play a larger role. Similar studies have focused on roots and such similar mechanisms for increased root length and/or biomass under conditions of elevated CO_2.

Carbon dioxide fertilization effect

The atmospheric concentration of CO_2 has two major short-term but direct effects on plants. It increases photosynthesis in C_3 plants and decreases stomatal conductance and transpiration in C_3 and C_4 plants (Long *et al.,* 2006). The increase in net photosynthetic C accumulation due to increase in atmospheric CO_2 concentration is called "CO_2 fertilization effect". The term literally indicates that with supply of CO_2, the production increases similar to the production increase observed when nutrients like N and P are added. This effect is pronounced in C_3 plants species. In C_3 plants, the first target of CO_2 is ribulose-1,5-bisphosphate corboxylase oxygenase (RuBisCO). The enzyme is not CO_2 saturated under the current level of atmospheric CO_2 concentration. Since O_2 and CO_2 compete each other for the same reaction sites on this enzyme, the oxygenase reactions of RuBisCO is suppressed under CO_2 enrichment (Long *et al.,* 2006). The estimates from free-air enrichment (FACE) and other studies show 14% (Lee *et al.,* 2011) to 26-31% (Ainsworth and Rogers, 2007) increase in light-saturated photosynthesis. The availability of additional photosynthates enables most plants to grow faster under elevated CO_2, with average increase in dry matter production by 17% for the above ground and more than 30% for the below ground biomass (Ainsworth and Long, 2005). The increased growth results in higher yield of crops, with wheat, rice and soybean accounts for yield increase of 12-14% in the elevated CO_2 conditions (Ainsworth and Long, 2005).

Increased water use efficiency in plants

Generally C_4 plants response to the elevated CO_2 level in the atmosphere with increased water use efficiency. A reduced stomatal conductance decreases the transpiration loss of water through stomatal opening. Stomatal conductance is reduced by 22% in field crops and graminacoeus species (Ainsworth and Rogers 2007, Lee *et al.,* 2011) and by 33-50% across broad range of C_3 and C_4 plant species (Kimball *et al.,* 2002). A possible mechanism may be that the ratio of inter cellular and to ambient CO_2 remains approximately constant, hence under elevated CO_2 conditions the stomata remains closed as there is no any additional requirement of CO_2, thus reduces the transpiration loss.

Reduction in leaf N and protein concentration

The primary mechanism of reduction of tissue N concentration seems to be from a reduction in RuBisCO content since much of the soluble leaf protein is associated with RuBisCO. The reduction in leaf N concentration can be as much as 10-20% in grasses and 13-31% in field crops. The reduced leaf N concentration also causes lowering of N in grains by 8-31% and protein content in wheat, rice, barley and potato by 5-14%. Under elevated CO_2 conditions leaves tend to accumulate more sugar and starch resulting in 30-40% higher non-structural carbohydrates on unit leaf area. Overall the leaf C:N ratio increase under elevated CO_2 conditions in both C_3 and C_4 plants. An exception is observed in case of nodulating legumes where almost no significant effect of elevated CO_2 on C:N ratio observed. For example in some comparative studies C:N ratio increased to 27-40% in grain crops and no changes were observed in legumes. Hence plants demand more N under elevated CO_2 conditions for higher productivity and increasing the protein content (Taub *et al.,* 2008). Under high N fertilized (more than 30 kg N ha^{-1}) treatments show 12% above ground and 25% below ground biomass yield than low N fertilized (0-30 kg N ha^{-1}) treatments under elevated CO_2 conditions across different plant species (de Graffe *et al.,* 2006). Another important change in plant tissue is an increase in the lignin content. On an average, there may be upto 6.5% increase in lignin content in leaf tissues in elevated CO_2 conditions.

Impact of Climate Change on Soil

Soil is a complex system and sustains life on the earth. Soils can also serve as a source of greenhouse gases. In fact, soils are a major source of anthropogenic non-CO_2 emissions from agriculture. Nitrous oxide emissions from soils constituted 38% and CH_4 emissions from rice production 11% of the total non-CO_2 greenhouse gas emissions from agriculture in 2005 (Smith *et al.,* 2007). Additionally, increasing soil temperatures have been shown to lead to increased CH_4 production in rice paddies and wetlands (Schutz *et al.,* 1990, Stêpniewski *et al.,* 2011). However, soils are not currently considered to be a major net source of CO_2. While agricultural soils produce large quantities of CO_2 each year, they also take up large quantities, such that agricultural soils are estimated to contribute less than 1% of net global anthropogenic CO_2 emissions (Smith *et al.,* 2007). Soils are intricately linked to the atmospheric-climate system through the carbon, nitrogen, and hydrologic cycles. Altered climate will, therefore, have an effect on soil processes and properties, and at the same time, the soils themselves will have an effect on climate. Study of the effects of climate change on soil processes and properties is still nascent, but has revealed that climate change will impact soil organic matter dynamics, including soil organisms and the multiple soil properties that are tied to organic matter, soil water, and soil erosion. The exact direction and magnitude of those impacts will be dependent on the amount of change in atmospheric gases, temperature, and precipitation amounts and patterns.

Soil processes in general has been broadly classified into structural (physical) and hydrological processes, and biogeochemical processes. These are often working simultaneously in the soil and involved in influencing the various soil quality parameters in any soil ecosystem. These key soil processes are likely to be influenced by global changes. All these are temperature and moisture dependent and microbial mediated processes. Thus elevated atmospheric CO_2 induced temperature and precipitation changes result in changes of these key soil processes which will subject the soils to physical and chemical degradation, soil erosion, salinization, decreased water availability and changes in C and N dynamics, decreased nutrient storage in soil and depletion of soil biodiversity. These unfavourable changes pose big threat to soil productivity, soil and water quality and sustainability of the production system (Emmett *et al.,* 2009, Benbi and Kaur, 2009). The potential impacts of elevated atmospheric CO_2 on soil processes and resultant unfavourable changes in soil quality parameters are briefed in the following discussions.

Soil Genesis

Climate, vegetation, parent material, topography and time are the factors which affect the soil formation. Among these factors climate and vegetation are considered as active factors. Changing global atmospheric CO_2 may affect the climate particularly temperature and rainfall, and vegetation in an ecosystem. These changes accelerate the soil forming process by modifying the processes of chemical, biochemical or mineralogical changes in soil parent material over a period of time at prevailing topographic conditions.

The most rapid processes of chemical or mineralogical change under changing climatic conditions would be loss of salts and nutrient cations by leaching, and salinization. Changes in the surface properties of the clay fraction, while generally slower than salt movement, can take place much faster than changes in bulk composition or crystal structure. Not only can the speed of soil formation be accelerated by human action, but also its very nature or direction. In most places, the natural soil-forming processes are not fundamentally changed, but there are certain threshold situations, generally with fragile soils, where even a small change in external conditions may cause a major, and adverse, change from one dominant soil-forming process to another. The four examples summarized below (Sombroek, 1990) illustrate a change from hydrolysis to cheluviation (Ferralsols to Podzols); clay illuviation forming dense subsoil in originally homogeneous, porous Ferralsols; irreversible hardening of the subsoil; and salinization. These changes in the long run may alter soil formation processes in different degree as per the local environmental conditions prevailing in particular region results in changes in world soil distribution scenario.

Soil hydrological cycle, erosion and runoff

Climate change can affect soil hydrological cycle through multiple pathways because many climatic variables such as precipitation, temperature, and CO_2 concentration as well as their interactions are often complex, dynamic, and nonlinear. The actual impacts of individual variables and/or their interactions, which may differ seasonally and geographically, can be adequately assessed. The rise in temperature, increase the rate of evapotranspiration causes soil drying and lowering of ground water table. Rapid melting of snow and large glaciers of Himalayas may increase the stream flow of some north Indian rivers, increase the availability of water for crop growth for short period but in the long run the scenario may change. The decrease in atmospheric precipitation in some parts will result in a decrease in water infiltration and water storage in the soil. The extreme variability in precipitation pattern, spatially and temporally may lead to higher surface run-off and erosion which radically modify the field water balance and its component.

Climatic factors have become more significant in recent times due to rapid climate changes induced by intensive anthropogenic activities affecting our ecosystem in multiple ways. It is projected that by the end of the 21st century rainfall over India will increase by 15-40%, and mean annual temperature by 3-6°C (NATCOM 2004). Changes in precipitation particularly increased rainfall intensities increase chance of run-off and soil erosion on crop and pasture lands. The recent unprecedented rainfall leading to floods in 13 districts of North Karnataka and estimate made after the calamity revealed that nearly 287 million tonnes (mt) of top soil, 8 lakh tonnes of soil nutrients and 39 lakh tonnes of soil organic matter were washed away and nearly 57.5 mt of soil microbial biomass were lost from the region during this short period (Natarajan *et al.,* 2010). Increase in frequency of extreme weather events will increase the run-off and erosion losses of the fertile surface soil in a considerable part of humid tropics. Each 1% increase in annual rainfall increases erosion by approximately 1.7%. On the other hand drought induced losses of biomass and vegetative land cover causes the removal of top soil and loss of fertility due to nutrient export. Changes in wind regimes where increased winds increase chance of soil erosion, particularly on fine textured soils. Erosion potential for crop lands is more than in pasture and rangelands depending on climate change scenarios.

Soil fertility and plant nutrient availability

Significant change in soil processes such as net nitrogen mineralization and soil respiration as a result of elevated soil and water temperatures would have profound implication for ecosystem functioning by affecting changes in nutrient availability and carbon storage. Across the range of ecosystem types, warming was found to increase soil respiration by mean of 20%, where as net N mineralization increased by an average of 46% (Rustad *et al.,* 2001). Accelerated decomposition and mineralization would bring more N in the soil and would have positive impact on N supply to crops and vegetation. However the rate of denitrification is very sensitive to soil temperature and its rate increases rapidly from 2 °C to 25 °C, and the rate of denitrification increases gradually after 60 °C. At higher temperature, the process of production of less soluble reaction products of P with soil components and P retention in soil increases. The increased microbial and root activity in the soil would entail higher CO_2 partial pressure in soil air and CO_2 activity in soil, hence increased rate of plant nutrient release from weathering of soil minerals particularly P, K, Mg and micronutrients. Similarly, the mycorrhizal activity would lead to better phosphate uptake. Increasing soil temperature also affects availability of K to plants. Consequently rapid increase in soil organic matter dynamics and soil microorganisms may cause the competition for plant nutrients. The changes in global variables that affect the various soil process lead to mineral stress are presented in Table 1.

Table 1: Potential interactions of global change variables on soil processes with mineral stress (adopted from Samuvel and Jonathan, 2010)

Process	Global change variables	Interaction with mineral stress
Erosion	Heavy precipitation and drought	Loss of soil nutrients and SOC
Soil redox status	Flooding	Mn, Fe, Al and B
Soil leaching	Heavy precipitation	NO_3, SO_4, Ca, Mg
Salinization	Soil moisture, temperature, CO_2	All nutrients
Mass flow	Drought, temperature, RH, CO_2	NO_3, SO_4, Ca, Mg, and Si
Soil organic carbon status	Precipitation, soil temperature	N, K, Ca, Mg
Soil microbes (N cycling)	Drought, soil temperature	N
Plant Biological N fixation	Drought, soil temperature	N
Mycorrhizas	CO_2	P, Zn (VAM) and N
Phenology	Soil temperature	P, N, K
Root growth &architecture	Drought, soil temperature, CO_2	All nutrients, especially P and K

Carbon cycle

Soil organic matter is elixir of life particularly to soil life. It is the food for many microbes which are living in the soil. It acts as reservoir of soil nutrients which becomes available to plant by mineralization processes. It acts as both source and sink of atmospheric CO_2. Soil contains large amount of carbon compared to atmosphere and mostly in the organic form. Human activities particularly clearing of forests, intense farming practices mediated release of CO_2 tilting the input-output balance considered as source. Application of large amount of organic residues to the soil through enhanced biomass production from forests, incorporation of crop residues, exogenous application of organic materials and decreased losses through adoption of improved management practices for reducing the soil respiration increase the net sink of C in soils. Changes in climate are likely to influence the rate of accumulation and decomposition of SOM, both directly through changes in temperature and moisture, and indirectly through changes in plant growth and rhizodepositions (Das and Hati, 2010).

Increased temperature and frequent rainfall stimulates microbial activity. Hence there will be an increase in mineralization (or decomposition of organic residues) in the soil releases the CO_2 to atmosphere. Increased atmospheric CO_2 increases plant water use efficiency (WUE) results in increased biomass production per mm of available water (Kimball *et al.,* 2002). But the decomposition rates are greater than primary production under increased water deficits. So it causes a reduction of biomass accumulation and depletion of soil C and decrease the C: N ratio of soil (Krischbaum 2000, Lal, 2004). Drier conditions favour the organic carbon reduction and thus reduction in annual and perennial vegetation. During summer periods increased risk of fires destroys vegetation cover leading to rapid defoliation and conversion of carbon source to atmospheric carbon (Hennessy *et al.,* 2005). Over all climate change has the negative impact on soil organic carbon and its storage in tropical climate. Globally, the net effect of climate change will be a decrease in stocks of organic carbon in soils, thus releasing additional CO_2 into the atmosphere and acting as positive feedback, further accelerating climate change.

In humid tropics and temperate region, climate change induced increase in rainfall will have positive influence on C storage in the soil. While some parts of the sub-tropics where scanty rainfall and higher temperature induced drought and desertification will reduce the vegetation cover and carbon sequestration potential of the soil. Climate change mediated soil organic matter dynamics is a complex process influenced by many factors and processes will differently interplay in long term basis is unpredictable.

Nitrogen cycle

In many natural and semi-natural ecosystems, the atmospheric N deposition has increased over the years. As compared to estimated inputs of 1-3 kg N ha^{-1} yr^{-1} in the early 1900s (Goulding 1998), the atmospheric N deposition rates of 20-60 kg N ha^{-1} yr^{-1} in non-forest ecosystems in Western Europe, and up to100 kg N ha^{-1} yr^{-1} in forest stands in Europe or the USA have been observed (Bobbink *et al.*, 2002). While there has been an increase in deposition rate across all the biomes at both temperate and tropical latitudes, the increase is the greatest in the northern hemisphere temperate ecosystems (Table 2). The high deposition rates are probably driven by biomass burning, soil emissions of NO_x and NH_3 as well as lightening production of NOx.

Table 2: Pre-industrial and contemporary N depositions (Tg N yr^{-1}) onto different hemisphere

Latitude	Pre- industrial	1990s
Northern hemisphere	4.03	26.65
Southern hemisphere	1.87	1.76
Tropical	7.64	15.55

Source: Adapted from Holland *et al.* 1999

While deposition of N to agricultural or croplands could serve as a source of nutrient, it could also adversely affect several processes in the soil. Nitrogen deposition can lead to reduction in biodiversity, soil acidification, and increased nitrous oxide emissions from soils. It also altered the balance of nitrification and mineralization/immobilization, and increased NO_3 leaching resulting in contamination of surface and underground water bodies. Increasing temperature in combination with aerobic conditions increases nitrification increases the mineral nitrogen in the soil. Episodic rainfall in arid and semiarid region increases the risk of nitrate loss due to leaching. Under anaerobic conditions, the rate of denitrification will also increase the loss of nitrogen as N_2O or N_2 which reduces the plant available nitrogen in soil. Depressing effects of temperature on symbiotic nitrogen bacteria also reduce the nitrogen fixation and increase the dependence of artificial nitrogen sources. The models simulating denitrification in soils indirectly address the impact of climate change by including adjustment functions for substrate availability (such as NO_3 concentration), soil temperature and moisture to calculate actual denitrification rate from the potential rate. The main problems with the models for simulating N regime of agricultural soils are that these do not explicitly include the effect of atmospheric N deposition and changes in ambient CO_2 concentration. So long term effect of climate change on nitrogen dynamics is complex and unpredictable.

Soil enzyme activity and microbial diversity

Factors influencing soil microbial activity exert control over soil enzyme production and nutrient availability (Sinsabaugh *et al.,* 1993). Warming and drought affect the activity of the soil enzymes involved in C and N mineralization in the mid and long term. In a recent study conducted in Mediterranean shrublands by Sardans *et al.* (2008) revealed that warming increased soil urease activity by 10% in the study period (1999-2005), and increased β-glucosidase activity 38%. Soil urease and β-glucosidase activities were positively correlated with soil temperatures in winter and negatively in summer. Drought reduced soil protease activity (9%) and did not affect β-glucosidase activity. They have observed that warming and drought have changed some soil enzyme activities related to P turn-over in some year seasons (Sardans *et al.* 2006). The effects of warming and drought on soil enzyme activities were due to a direct effect on soil temperature and soil water content and not to changes on soil organic matter quantity and nutritional quality (Sardans *et al.,* 2008).

Temperature, precipitation and vegetation changes considerably influence the microbial community and their activity in the soil. Particularly under condition of elevated CO_2 and changed temperature and moisture regimes, soil biodiversity and its function are expected to change. Elevated CO_2 elicited a 47% average increase in mycorrhizal abundance and that mycorrhizae were stimulated disproportionately more than roots (percentage colonization increased by more than 30%) (Treseder, 2004). Another meta-analysis reported that the mass of ectomycorrhizal (EM) fungi increased by 34% in CO_2- enriched environments, whilst that of AM (endomycorrhizal) fungi increased by 21% (Alberton *et al.,* 2005). Over all, changes in plant growth and rhizodepositions have an impact on soil microbial activity. However, reduced soil moisture creates unfavourable environment for microflora, microfauna as well as macrofuana like earthworms, termites, arthropods etc.

Conclusions

The climate system of our planet is changing due to emission of large amount of green house gases into the atmosphere through various anthropogenic activities.

The direct and indirect impact of climate change on soil-plant and human system is certain. Still we need to know many things about the impact of climate change. Knowledge of the response of plants to elevated atmospheric CO_2 in the given potential limitations in nutrients like nitrogen and phosphorus and how that affects soil organic matter dynamics is a critical need. There is also a great need for a better understanding of how soil organisms will respond to climate change because those organisms are incredibly important in a number of soil processes, including the carbon and nitrogen cycles. As these are highly complex and interconnected systems that make it difficult to isolate a single variable, such as temperature or precipitation patterns, to reach meaningful conclusions about how a change in that single variable affects the system being studied. However, we do know that there is a potential for some undesirable things to occur as a result of climate change. Therefore, it is critical that continued research into these areas be supported, with the particular goal of understanding the complex interactions that take place in the natural environment.

References

Ainsworth, E. A. and Long, S. P. (2005). What have we learned from 15 years of free-air CO2 enrichment (FACE) A meta-analytic review of the responses of photosynthesis, canopy properties and plant production to rising CO_2. *New Phytologist,* 165: 351-71.

Ainsworth, E. A. and Rogers, A. (2007). The response of photosynthesis and stomatal conductance to rising CO_2: mechanisms and environmental interactions. *Plant, Cell and Environment,* 30: 258-70.

Alberton, O., Kuyper, T. W. and Gorissen, A. (2005). Taking mycocentrism seriously: mycorrhizal fungal and plant responses to elevated CO_2. *New Phytologist,* 167: 859–68

Benbi, D. K. and Kaur, R. (2009). Modeling Soil Processes in Relation to Climate Change. *Journal of the Indian Society of Soil Science,* 57: 433-44.

Bobbink, R., Ashmore, M., Braun, S., Fluckiger, W. and van den Wyngaert, I J J. (2002). Empirical nitrogen critical loads for natural and semi-natural ecosystems: 2002 update, Available online *http:// www.iap.ch*. (2002).

Das, D. K. and Hati, K. M. (2010). Impact of climate change and global warming on soil processes and nutrient availability. pp 39-46. *Souvenir, 75th Annual Convention of the Indian Society of Soil Science*, 14-17 November 2010, Bhopal, India.

de Graaff, M. A., van Groenigen, K. J., Johan Six, Bruce Hungate and van Kessel C. 200). Interactions between plant growth and soil nutrient cycling under elevated CO_2: a meta-analysis. *Global Change Biology,* 12: 2077-91.

Emmett, B. A., Beier, C., Estiarte, M., Tietema, A., Kristensen, H. L., Williams, D., Penuelas, J., Schmidt, I. and Sowerby, A. (2004). The responses of soil process to climate change: Results from manipulation studies of shrub-lands across an environmental gradient. *Ecosystem,* 7: 625-37.

Ferris, R., R. H. Ellis, T. R. Wheeler and P. Hadley. (1998). Effect of High Temperature Stress at Anthesis on Grain Yield and Biomass of Field-grown Crops of Wheat. *Annals of Botany,* 82: 631-639.

Gifford, R. M. and Howden, M. (2001). Vegetation thickening in an ecological perspective: significance to national greenhouse gas inventories. *Environmental Science and Policy,* 4: 59-72.

Goulding, K. W. T., Bailey, N. J., Bradbury, N. J., Hargreaves, P., Howe, M., Murphy, D. V., Poulton, P. R. and Willison, T. W. (1998). Nitrogen deposition and its contribution to nitrogen cycling and associated soil processes. *New Phytologist,* 139: 49-58.

Hennessy, K., Lucas, C., Nicholls, N., Bathols, J., Suppiah, R. and Ricketts, J. (2005). *Climate change impacts on fire-weather in south-east Australia.* CSIRO, Melbourne.

Holland, E. A., Dentener, F. J., Braswell, B. H. and Sulzman, J. M. (1999). Contemporarory and pre-industrial global reactive nitrogen budgets. *Biogeochemistry,* 46: 7-43.

IPCC. (2007). *Climate Change 2007: The physical science basis. Summary for Policymakers.* Intergovernmental Panel on Climate Change. New Delhi.

IPCC. (2013). *Climate Change 2013: The Physical Science Basis. Summary for Policymakers.* Intergovernmental Panel on Climate Change. New Delhi.

Jeffrey, S. Dukes and Harold, A. Mooney. (1999). Does global change increase the success of biological invaders? *Trends in Ecology and Evolution (Amsterdam)* 14 (4): 135-9.

Kimball, B. A., Kobayashi, K. and Bindi, M. (2002). Responses of agricultural crops to free-air CO2 enrichment. *Advances in Agronomy, 77*: 293-368.

Kirschbaum, M. U. F. (2000). Will changes in soil organic carbon act as a positive or negative feedback on global warming? *Biogeochemistry,* 48: 21-51.

Lal, R. (2004). Soil carbon sequestration to mitigate climate change. *Geoderma,* 123: 1-22.

Lee, T. D., Barrott, S. H. and Reich, P. B. (2011). Photosynthetic responses of 13 grassland species across 11 years of free-air CO2 enrichment is modest, consistent and independent of N supply. *Global Change Biology*, doi: 10.111/j.1365-2486.2011.02435.x.

Long, P., Ainsworth, E.A. Leakey, A.D.B., Nösberger, J., Ort D.R. (2006). Food for Thought: Lower-Than-Expected Crop Yield Stimulation with Rising CO_2 Concentrations. *Science,* 312: 1918-21.

Natarajan, A., Rajendra, Hegde, Naidu, L. G. K, Raizada, A., Adhikari, R. N., Patil, S. L., Rajan, K., Dipak Sarkar. (2010). Soil and plant nutrient loss during the recent floods in North Karnataka: Implications and ameliorative measures. *Current Science,* 99: 1333-40.

NATCOM (2004). *India's Initial National Communication to United Nations Framework Convention on Climate Change.* Ministry of Environment and Forests, Government of India, New Delhi.

Rustad, LE., Campbell, JL., Marion, GM., Norby, RJ., Mitchell, MJ., Hartley, AE., Cornelissen, JHC. and Gurevitch, J. (2001). A meta-analysis of the response of soil respiration, net N mineralization, and above-ground plant growth to experimental ecosystem warming. *Oecologia,* 126: 543-62.

Samuel, B. StClair and Jonathan, P. Lynch. (2010). The opening of Pandora's Box: climate change impacts on soil fertility and crop nutrition in developing countries. *Plant and Soil,* 335: 101- 15.

Sardans, J., Penuelas, J. and Estiarte, M. (2006). Warming and drought alter soil phosphatase activity and soil P availability in a Mediterranean shrubland. *Plant and Soil,* 289: 227-38.

Sardans, J., Penuelas, J. and Estiarte, M. (2008). Changes in soil enzymes related to C and N cycle and in soil C and N content under prolonged warming and drought in a Mediterranean shrubland. *Applied Soil Ecology,* 39: 223-35.

Schutz, H., Seiler, W. and Conrad, R. (1990). Influence of soil temperature on methane emission from rice paddy fields. *Biogeochemistry,* 11: 77-95.

Sinsabaugh, R. L., Antibus, R. K, Linkins, A. E. and McClaugherty, C. A. (1993). Wood decomposition: nitrogen and phosphorus dynamics in relation to extracellular enzyme activity. *Ecology,* 74: 1586-93.

Smith, P., Martino, D., Cai, Z., Gwary, D., Janzen, H., Kumar, P., McCarl, B., Ogle, S., OMara, F., Rice, C., Scholes, B. and Sirotenko, O. (2007). Agriculture. *Climate change* 2007*: Mitigation. Contribution of Working Group III to the Fourth Assessment Report of the*

Intergovernmental Panel on Climate Change. pp. 497-540. (Eds) Metz, B., Davidson, O. R., Bosch, P. R., Dave, R. and Meyer, L. A. Cambridge University Press, Cambridge.

Sombroek, W. G. (1990). Soils on a warmer earth: the tropical regions. *Soils on a warmer earth: Effects of expected climatic change on soil processes, with emphasis on the tropics and subtropics*. pp. 157-74. (Eds) Scharpenseel H W, Schomaker M and Ayoub A. Developments in Soil Science 20, Elsevier, Amsterdam.

Steffen, W. and Canadell, P. (2005). *Carbon Dioxide Fertilisation and Climate Change Policy*. Australian Greenhouse Office, Department of Environment and Heritage, Canberra.

Stêpniewski, W., Stêpniewski, Z. and Ro¿ej, A. (2011). Gas exchange in soils. *Soil management: Building a stable base for agriculture*. pp. 117-44. (Eds) Hatfield J L and Sauer T J. SSSA, Madison.

Taub, D. R. and Wang, X. (2008). Why are nitrogen concentrations in plant tissues lower under elevated CO2? A critical examination of the hypotheses. *Journal of Integrative Plant Biology*, 50: 1365-74.

Thomson, A. M., Brown, R. A., Rosenberg, N. J., Izaurralde, R. C. and Benson, V. (2005). Climate change impacts for the conterminous USA: An integrated assessment. Part 3. Dryland production of grain and forage crops. *Climate Change*, 69: 43-65.

Treseder, K. K. (2004). A meta-analysis of mycorrhizal responses to nitrogen, phosphorus and atmospheric CO2 in field studies. *New Phytologist*, 164: 347-55.

Walther, G. R., Post, E., Convey, P., Menzel, A., Parmesan, C., Beebee, T. J., Fromentin, J. M., Hoegh- Guldberg, O. and Bairlein, F. (2002). Ecological responses to recent climate change. *Nature*, 416: 389-95.

Wheeler, T. R., Batts, G. R., Ellis, R. H., Hadley, P. and Morison, J. I. L. (1996). Growth and yield of winter wheat (*Triticum aestiaum*) crops in response to CO2 and temperature. *Journal of Agricultural Science, Cambridge* 127: 37-48.

Advances in Nutrient Dynamics in Soil - Plant System, pp 239-248

Editors: R. Elanchezhian, A.K. Biswas, K. Ramesh and A.K. Patra

21

Simulation Modeling for Improving Nitrogen Use Efficiency in Crops and Cropping Systems

M. Mohanty and N.K. Sinha

ICAR-Indian Institute of Soil Science, Nabi Bagh, Bhopal – 462038, India

Introduction

The term nitrogen use efficiency, (NUE) is widely used in research to cover problems in use of N in crop production systems. It describes N input/output ratio at different scales starting from farm levels to regional levels. An analysis of of NUE is helpful to analyze possible options for improving NUE in different crops and cropping systems. NUE at crop scale may be defined as the ratio between the yield and the N uptake by the crop. This ratio is sometimes called as the physiological NUE or utilization efficiency. High NUE can be achieved by high harvest index and a high productivity of dry matter per unit N uptake. The second component of NUE is the ratio of plant N uptake to the potentially available N. This ratio is called uptake efficiency or available N use efficiency. Enhancing the uptake efficiency of the crop is possible through enhancing the physiological uptake efficiency of the roots, enhancing root length density and matching the N supply and demand of a particular crop.

At the cropping system level a useful definition of NUE is the ratio of longterm yield to the amount of N applied to a field or plot or treatment over the longer periods. In addition to the components of NUE at the crop level, this definition of NUE includes two components of NUE: The first is the ratio of potentially available N for the crops to the total N supply. This ratio is called as the N loss ratio or available N efficiency. It envisages the N losses from the system through the process of leaching, denitrification and erosion. A second component is identified as the ratio of N supply to N fertilized. This ratio is affected by the changes in the soil organic carbon input to the system which is resulted in the process of N mineralization and immobilization and thus reflects the increase or decrease in the value of the ratio. Also, all N inputs to the system in addition to N fertilization, namely bilogical nitrogen fixation increases this ratio.

The components of NUE to crops and cropping systems are interrelated. If the uptake efficiency at crop level is low, the residual soil nitrate levels may be high. Thus, it increases the ratio of N supply to N fertilized, a component of NUE and thus, the losses from the system will rise. Similarly if the physiological efficiency at the crop level is low, the amount of N in crop residues and N mineralization in the system should increase which results in an increase in N supply to the following crop but also results in N losses. Therefore a low NUE at the crop level only determines a low NUE of the crop production system if this predetermines substantial N losses from the system. So, designing cropping systems with high NUE can therefore be achieved by using crops which make effective use of residual N in the soil and crop residues and other measures which reduce the N-loss from the system.

Simulation Modeling: Methodological Aspects to Determine NUE

The ratios presented above for different crops and cropping systems are useful in defining terms. It is the fact that NUE as per definition a static entity as a result of dynamic processes. The definitions, ratios and their analysis do not include the processes and the time scale involved. Thus, simulation modelling offers the possibility to overcome these shortcomings and is therefore used to analyse aspects of NUE in different crops and cropping systems.

Crop simulation models were first developed in the early 1960s, since then, simulation models are used to address research problems and to increase knowledge on crop growth, development and yield. Crop system simulators contain mathematical equations describing basic flow and conversion processes of carbon, water, and nitrogen balance that are integrated daily or hourly by the computer program to predict the time course of crop growth, nutrient uptake, and water use, as well as to predict final yield and other plant traits and outputs.

In due course, scientists have developed number of crop models that are capable of simulating the effects of weather, soil type, seed variety, fertilizer management and irrigation on crop growth and development. Results from such simulations can then be used to predict changes and detect trends in biophysical indicators such as crop yield, nutrient uptake, and nitrate leaching and soil carbon levels. Once set within the framework of a comprehensive information and decision support system, the crop models can facilitate the effective analysis of issues related to agricultural production, resource allocation, risk, environmental quality and land use.

Over the past 10 to 20 years, crop system model developers have succeeded in linking good crop C balance (N demand) with good soil water balance and good soil-crop N balance. The DSSAT V3.5 models (Hoogenboom et al., 1999; Jones *et al.,* 1998) were among the early models to succeed in this full linkage, but APSIM (Keating *et al.,* 2003; McCown, *et al.*, 1996) and other models are also at this stage of development. The DSSAT-CSM V4.0 model (Hoogenboom *et al.,* 2004; Jones *et al.,* 2003) was a further improvement, in its use of a land-unit module, which is the interface of crop-soil-weather, where the soil organic matter module used can be the CENTURY model (Gijsman, *et al.*, 2002) or the older Godwin soil organic matter model (Godwin and Jones, 1991; Godwin and Singh, 1998). A description of a few crop simulations models are given below.

Agricultural Production Systems sIMulator (APSIM)

APSIM was developed to simulate biophysical processes in farming systems, particularly as it relates to the economic and ecological outcomes of management practices in the face of climate risk. APSIM is structured around plant, soil and management modules. These modules include a diverse range of crops, pastures and trees, soil processes including water balance, N and P transformations, soil pH, erosion and a full range of management controls. APSIM resulted from a need for tools that can provide accurate predictions of crop production in relation to climate, genotype, soil and management factors, while addressing the long-term resource management issues.

Decision Support System for Agro-technology Transfer (DSSAT)

DSSAT is a software package integrating the effects of soil, crop phenotype, weather and management options that allows users to ask "what if" type questions and simulate results by conducting, in minutes on a desktop computer, experiments which would consume a significant part of an agronomist's career. It has been in use for more than 15 years by researchers in over 100 countries. The DSSAT simulates growth, development and yield of a crop growing on a uniform area of land under prescribed or simulated management as well as the changes in soil, water, carbon, and nitrogen that take place under the cropping system over time.

Cropping Systems Simulation Model (Crop Syst)

CropSyst (Cropping Systems Simulation Model) is a multi-year, multi-crop, daily time step crop growth simulation model, developed with emphasis on a friendly user interface, and with a link to GIS software and a weather generator (Stockle, 1996). The link to economic and risk analysis models is under development. The model's objective is to serve as an analytical tool to study the effect of cropping systems management on crop productivity and the environment. For this purpose,

CropSyst simulates the soil water budget, soil-plant nitrogen budget, crop phenology, crop canopy and root growth, biomass production, crop yield, residue production and decomposition, soil erosion by water, and pesticide fate. These are affected by weather, soil characteristics, crop characteristics, and cropping system management options including crop rotation (including fallow years), cultivar selection, irrigation, nitrogen fertilization, pesticide applications, soil and irrigation water salinity, tillage operations (over 80 options), and residue management.

INFOCROP

InfoCrop, a generic crop model, which simulates the effects of weather, soils, agronomic management (planting, nitrogen, residues and irrigation) and major pests on crop growth, yield, soil carbon, nitrogen and water, and greenhouse gas emissions. InfoCrop considers various processes such as, crop growth and development, soil water, nitrogen and carbon, and crop–pest interactions. Each process is described by a set of equations, in which the parameters vary depending upon the crop/ cultivar (Aggarwal *et al.,* 2006).

These are a few examples of models that can used for assessing the impact of climate change on crop productivity. There are many other models they can be used for similar purposes. The biogeochemical models like CENTURY, DNDC and RothC are used for carbon sequestration studies also. Are all the models suitable for climate change assessment? It is possible that the models those taking into account the effects of temperature, and CO_2 concentration on photosynthesis and radiation use efficiency are suitable. Most of the crop simulation models accommodate these effects though not for the field crops. The model need to be robust and has the wider applicability to all geographical situations can be a good tool for climate change assessments.

Thus, simulation models can complement conventional field experimentation in finding alternative nutrient management options for crops and cropping systems particularly for improving NUE and decision making processes. The APSIM model is one such model which has been used to simulate vegetative and reproductive development, growth, and yield as a function of crop characteristics, climatic factors, soil characteristics and crop management scenarios. It is able to simulate the effect of application of crop residues, green manures and FYM on N availability to soils and crops. A few examples of use of models in NUE in crops and cropping systems (soybean and wheat) are discussed hereunder.

Case Studies

The APSIM model has been parameterized and validated for the soybean (cv. JS 335) and wheat (cv. Sujata) crops for central Indian conditions. The parameterization of the model was carried out using data from a separate experiment on cv JS 335 and cv Sujata by collecting agronomic and physiological observations as per International Benchmark Sites Network for Agrotechnology Transfer (IBSNAT) format and as per model requirement. The validation of the soybean and wheat was carried out using an independent data set from a long-term experiment. A few case studies are given below.

Nitrogen requirement in wheat

The amount of N applied to soybean was fixed at 20 kg ha^{-1} while five N application rates were investigated for wheat: (1) N0 - No N (0 kg ha^{-1}), (2) N1 - 50 kg N ha^{-1} (3) N2 - 100 kg N ha^{-1}, (4) N3 - 120 kg N ha^{-1} and (5) N4 - 150 kg N ha^{-1}. Nitrogen was applied as a split application, 50% at time of sowing and 50% CRI stage. One presowing irrigation of 80 mm and five post sowing irrigations of 60 mm each at 20 days intervals were applied to wheat.

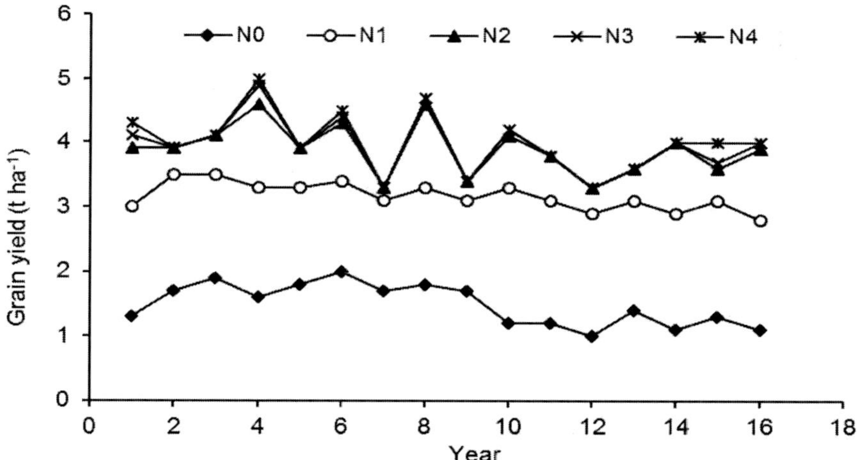

Fig. 1. Simulated wheat yield as influenced by N application rates

With sufficient irrigation (5 irrigations of 6 cm each at 20 days interval after CRI stage), application of 20 kg N ha^{-1} to soybean + 100 kg N ha^{-1} (N_2) to wheat constantly produced higher predicted yield over the 16 year period than 20 kg N ha^{-1} to soybean and 50 kg N ha^{-1} to wheat (N_1) (Fig. 1). The no N treatment (N_0) produced the lowest grain yield; (on average, 1.5 t ha^{-1} grain yield). The grain yield of wheat decreased with time in N_0 treatment. The simulated mean yield obtained from the application of treatment N_2 to wheat was 3.9 t ha^{-1} compared to 3.1 t ha^{-1} in treatment N_1. Increase in N application to wheat from 100 kg ha^{-1} to 120 and 150 kg ha^{-1} did not increase the yield significantly. The average yield obtained from the remaining treatments were 3.9 t ha^{-1} for treatment N_2, 4.0 t ha^{-1} for treatment N_3 and N_4.

Use of FYM for Efficient N Management in the Soybean-Wheat Cropping System

A total of six combinations of inorganic and organic sources of N were investigated. These combinations were designated as F_0 (No N applied), F_1 (20 and 100 kg urea-N ha^{-1} applied in soybean and wheat, respectively), F_2 (0 and 16 t FYM ha^{-1} applied in soybean and wheat, respectively), F_3 (16 and 0 t FYM ha^{-1} applied in soybean and wheat, respectively), F_4 (8 t ha^{-1} FYM and 50 kg urea-N /ha^{-1} applied in soybean and wheat, respectively), and F_5 (20 kg urea-N /ha^{-1} +5 t ha^{-1} FYM and 100 kg urea-N /ha^{-1} applied in soybean and wheat, respectively). FYM was applied and incorporated each year on 15 June which is 5 days before the sowing of soybean and wheat in treatments F_2, F_3 and F_4. The average grain yield of soybean was 1.5 t ha^{-1} and showed no response to either N fertilizer or FYM (Table 1).

The simulated wheat grain yield was very low when N was not applied (F_0). Application of 16t FYM ha^{-1} to soybean (F_3) produced a wheat yield of 3.6 t ha^{-1} which was 50% higher than the yield when the same amount of FYM was applied to the wheat crop itself (Table 1). The wheat yield obtained with only fertilizer N (100 kg N ha^{-1}) (F_1) was similar to that from the application of 8 t FYM ha^{-1} + 50 kg N ha^{-1} (F_4) and 20 kg N ha^{-1} + 5 t FYM ha^{-1} to soybean + 100 kg N ha^{-1} to wheat (F_5). The lowest wheat yield was obtained from the treatment F_0. The simulated wheat yield from 16 t FYM applied to soybean (F_3) was lower than that obtained from F_1, F_4 and F_5.

Table 1. Average yield of soybean and wheat as influenced by different N management regimes

Treatment	Simulated yield (t ha^{-1})	
	Soybean	Wheat
$F_{0=}$ 0 kg N ha^{-1} to soybean + 0 kg N ha^{-1} to wheat	1.5	1.5
$F_{1=}$ 20 kg N ha^{-1} to soybean + 100 kg N ha^{-1} to wheat	1.5	3.9
$F_{2=}$ 0 kg N ha^{-1} to soybean + 16 t FYM ha^{-1} to wheat	1.5	2.4
$F_{3=}$ 16 t FYM ha^{-1} to soybean + 0 kg N ha^{-1} to wheat	1.5	3.6
$F_{4=}$ 8 t FYM ha^{-1} to soybean + 50 kg N ha^{-1} to wheat	1.5	3.9
$F_{5=}$ 20 kg N ha^{-1} + 5 t FYM ha^{-1} to soybean + 100 kg N ha^{-1} to wheat	1.5	4.0

The simulated N balance over 16 years showed that the application of 16 t FYM ha^{-1} to wheat (F_2) led to the loss of 31% of applied N through leaching and denitrification pathways (Table 2). In contrast, application of the same amount of FYM to soybean (F_3) resulted in lower N losses (13%) and hence higher availability of N to following wheat crop. Integrated use of 8 t FYM ha^{-1} and 50 kg N ha^{-1} (F_4), application of 100 kg N ha^{-1} as urea-N (F_1), or an application of 16 t FYM ha^{-1} (F_3) all accounted lower loss of N than 0 kg N ha^{-1} to soybean + 16 t FYM ha^{-1} to wheat (F_2) (Table 2). The amount of N loss from the integrated use of 8 t FYM ha^{-1} to soybean and 50 kg N ha^{-1} to wheat (F_4) was lower than that obtained from inorganic application of 20 kg N ha^{-1} to soybean and 100 kg N ha^{-1} to wheat (F_1). The integrated application of FYM and fertilizer N produced the lowest N losses from the system amongst all the nutrient management systems studied here.

Table 2. Nitrogen balance for the soybean-wheat system as influenced by fertilizer regime, simulated over a period of 16 years

Treatment	F_1	F_2	F_3	F_4
Leaching loss (kg ha^{-1})	316	550	245	209
Denitrification loss (kg ha^{-1})	72	240	80	60
Total loss (kg ha^{-1})	388	790	325	269
N loss (kg ha^{-1} year^{-1})	24	49	20	17
Total N applied (kg ha^{-1})	1920	2560	2560	2240
% loss of applied N	20	31	13	12

Refer Table 1for F_1 to F_4 treatment

To understand why the N losses were lower when FYM is applied to the soybean rather than to wheat, it is helpful to consider the NO_3-N content of the soil profile. When FYM is applied to soybean the initial immobilization of N lowers the amount of NO_3-N in soil (Fig. 2) thereby protecting it from leaching and denitrification during the rainy season. Subsequent mineralization led to increasing NO_3 in soil with the concentrations being higher than when FYM was applied to wheat. The yields of wheat reflect the greater availability of NO_3 to the crop.

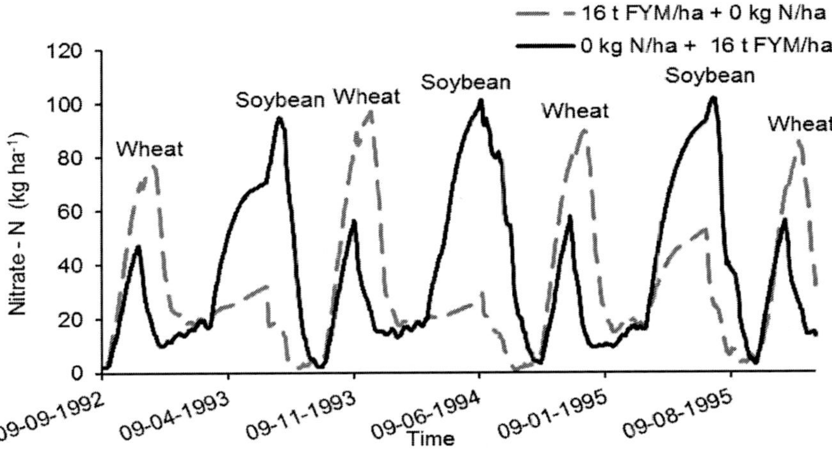

Fig. 2: Nitrate-N content in the profile (0-60 cm) as influenced by FYM application to soybean and wheat (16 t FYM ha^{-1} in each case). (*See colour version on page 379*)

The N losses via leaching and denitrification during four selected years (1991-96) were partitioned between the soybean growing seasons and the wheat season (Table 3). Clearly the losses are much higher during the soybean season which is during the monsoon. The amounts of N lost will depend on the NO_3 present at this time; as shown in Fig 2, NO_3 concentrations are highest during the monsoon when FYM is applied to the wheat crop.

Table 3. Total nitrogen loss between 1991 to 1996 as simulated by the model

Treatment	Soybean	Wheat	Total (kg ha^{-1})
Leaching loss (kg ha^{-1})			
F_2	162	27	189
F_3	72	8	80
De-nitrification (kg ha^{-1})			
F_2	68	2	70
F_3	16	5	21

F_2: 0 kg N ha^{-1} to soybean + 16 t FYM ha^{-1} to wheat; F_3: 16 t FYM ha^{-1} to soybean + 0 kg N ha^{-1} to wheat

Frequency of FYM Application and nitrogen management in wheat

In the above section, it was shown that application of FYM is most beneficial to wheat when applied at the time of sowing the soybean crop. The benefit of this application timing was reduced N losses and increased wheat grain yield. Here we explored the consequences of applying FYM less frequently. Application of FYM at 16 t ha^{-1} to soybean every year (A_1) produced a higher wheat yield than the same amount applied every three years (A_3) or five years (A_5) (Table 4). The minimum wheat yield obtained from the simulation study across all the treatments was 1.5 t ha^{-1}. A maximum of 4.4 t ha^{-1} was recorded in A_3 + 50 kg N ha^{-1} treatment (Table 4). The minimum yield obtained was from the 1st year of simulation in A_1, A_3 and A_5 treatments involving application of FYM alone. The yield gap between these treatments could be compensated by application of 50 kg N ha^{-1} to wheat every year in these treatments. There was also improvement in the average simulated wheat yield from 2.5 t ha^{-1} to 3.7 t ha^{-1} in A_3 and from 2.1 t ha^{-1} to 3.4 t ha^{-1} in treatment A_5 when 50 kg N ha^{-1} was applied to the system. The wheat yield was similar in the treatment A_1 and A_3 + 50 kg N ha^{-1} every year (Table 4).

Table 4. Simulated wheat grain yield as influenced by frequency of FYM application

Treatment	Grain yield (t ha^{-1})			
	Minimum	Maximum	Average	SD
A_1	1.5	4.3	3.6	0.8
A_3	1.5	3.5	2.5	0.6
A_5	1.5	3.3	2.1	0.6
A_3 + 50 kg N ha^{-1} every year*	2.9	4.4	3.7	0.4
A_5 + 50 kg N ha^{-1} every year*	2.9	4.0	3.4	0.4
SD = standard deviation				

A_1: 16 t FYM ha^{-1}+ 0 kg N ha^{-1} every year; A_3: 16t FYM + 0 kg N every three years; A_5: 16 t FYM ha^{-1} + 0 kg N ha^{-1} every five years * applied to wheat

The simulation study revealed that application of 16 t FYM ha^{-1} to soybean, or the application of 20 kg ha^{-1} of N to soybean and 100 kg ha^{-1} N to wheat, or combined application of 8 t ha^{-1} FYM to soybean + 50 kg N ha^{-1} to wheat can maintain the soybean and wheat yield in the long-term and are viable options of N management in the soybean-wheat system. Availability of FYM is always an issue with farmers who do not have enough cattle population to produce large amounts of FYM for the crops. However, yield could be maintained by the application of 50 kg N ha^{-1} to wheat every year. Thus, simulation models can be used in decision making process to enhance the NUE of different crops and cropping systems.

References

Carberry, P.S., McCown, R.L., Muchow, R.C., Dimes, J.P. and Probert, M.E. (1996). Simulation of a legume ley farming system in northern Australia using the Agricultural Production Systems Simulator. *Australian Journal of Experimental Agriculture*, 36: 1037-1048.

Crop systems dynamics. By Xinyou Yin and H. H. van Laar.

Crop-soil Simulation Models: Applications in Developing Countries. By Robin B. Matthews and William Stephens.

Keating, B.A., Carberry, P.S., Hammer, G.L., Probert, M.E., Robertson, M.J., Holzworth, D., Huth, N.I., Hargreaves, J.N.G., Meinke, H., Hochman, Z., McLean, G., Verburg, K., Snow, V., Dimes, J.P., Silburn, M., Wang, E., Brown, S., Bristow, K.L., Asseng, S., Chapman, S., McCown, R.L., Freebairn, D.M. and Smith, C.J. (2003). An overview of APSIM, a model designed for farming systems simulation. *European Journal of Agronomy*, 18: 267-288.

Kersebaum, K.C., Hecker, J.M., Mirschel, W. and Wegehenkel, M. (2007). Modelling water and nutrient dynamics in soil-crop systems: a comparison of simulation models applied on common data sets. In: Kersebaum, K.C., Hecker, J.M., Mirschel, W., Wegehenkel, M., (Eds.), Modelling water and nutrient dynamics in soil-crop systems, Springer, pp. 1-17.

Ludwig, F. and Asseng, S.(2006). Climate change impacts on wheat production in a Mediterranean environment in Western Australia. *Agricultural Systems*, 90: 159-179.

Luo, Q., Bellotti, W., Williams, M. and Bryan, B. (2005). Potential impact of climate change on wheat yield in South Australia. *Agricultural and Forest Meteorology*, 132: 273-285.

Mohanty, M., K Sammi Reddy, Muneshawar Singh, DLN Rao, R.C. Dalal, N. W. menzies, R. S. Chaudhary, Nishant K Sinha and A. Subba Rao (2013). Predicting nitrogen mineralized from crop residues, green manures and organic manures in relation to their quality using APSIM mode. *IISS-SPD Technical Bulletin* No.1, pp 53.

Mohanty, M., R.S. Chaudhary, Nishant K Sinha. K Sammi Reddy, P Dey and A Subba Rao (2013) Crop growth simulation modeling: theory and practices. IISS, Bhopal.

McCown, R.L., Hammer, G.L., Hargreaves, J.N.G., Holzworth, D. and Huth, N.I. (1995). APSIM - An agricultural production system simulation model for operational research. *Mathematics and Computers in Simulation* 39: 225-231.

Mohanty, M., Probert, M.E., Sammi Reddy, K., Dalal, R.C., Mishra, A.K., Subba Rao, A., Singh, M. and Menzies, N.W. (2012). Simulating soybean–wheat cropping system: APSIM model parameterization and validation. *Agricultural Ecosystem and Environ*ment, 152: 68-78.

Mohanty, M., Sammi Reddy, K., Probert, M.E., Dalal, R.C., Subba Rao, A. and Menzies, N.W. (2011). Modelling N mineralization from green manure and farmyard manure from a laboratory incubation study. *Ecological Modeling*, 222: 719-726.

Probert, M.E., Delve, R.J., Kimani, S.K. and Dimes, J.P. (2005). Modelling nitrogen mineralization from manures: representing quality aspects by varying C:N ratio of sub-pools. *Soil Biology and Biochemistry*, 37: 279-287.

Probert, M.E., Dimes, J.P., Keating, B.A., Dalal, R.C. and Strong, W.M. (1998). APSIM's water and nitrogen modules and simulation of the dynamics of water and nitrogen in fallow systems. *Agricultural Systems,* 56: 1-18.

Robertson, M.J., Carberry, P.S., Huth, N.I., Turpin, J.E., Probert, M.E., Poulton, P.L., Bell, M., Wright, G.C., Yeates, S.J. and Brinsmead, R.B. (2002). Simulation of growth and development of diverse legume species in APSIM. *Australian Journal of Agricultural Research*, 53: 429-446.

Whitbread, A.M., Robertson, M.J., Carberry, P.S. and Dimes, J.P. (2010). How farming systems simulation can aid the development of more sustainable smallholder farming systems in southern Africa. *European Journal of Agronomy,* 32: 51-58.

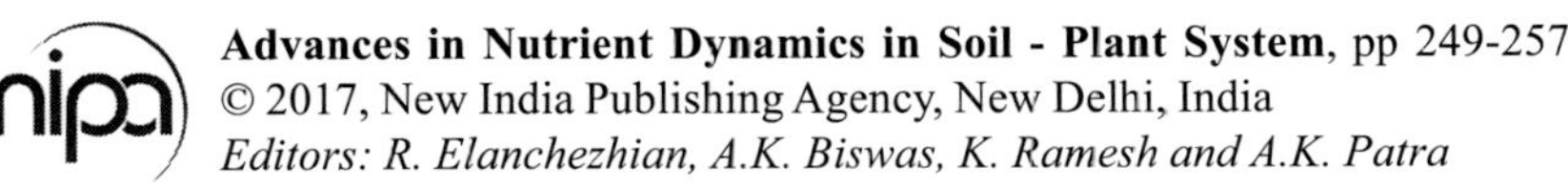
Advances in Nutrient Dynamics in Soil - Plant System, pp 249-257

Editors: R. Elanchezhian, A.K. Biswas, K. Ramesh and A.K. Patra

22

Soil Organic Carbon Dynamics with Respect to Climate Change

M.C. Manna

ICAR-Indian Institute of Soil Science, Nabi Bagh, Bhopal – 462038, India

Introduction

The soil is the largest terrestrial pool of organic carbon, about 1150 Pg (1Pg = 10^{15} g) compared with about 700 Pg in the atmosphere and 600 Pg in land biota (Lal and Kimble,1997). Organic carbon level of soils reaches a fixed equilibrium that is determined by a number of interacting factors such as precipitation, temperature, soil type, tillage, cropping systems, fertilizers, the type and quantity of crop residues returned to soil, and the method of residues management. In the temperate climate, it may take about 50 years for the organic carbon of soils to reach a new equilibrium level following a change in management. However, this period is much shorter in a semiarid and arid climate and also in tropical climate like India. Important factors controlling SOC levels include climate, hydrology, parent material, soil fertility, biological activity, vegetation patterns, land use and cultivation practices. SOC is sensitive to impact of human activities, *viz.* deforestation, biomass burning, land use changes and environmental pollution. Although the amount of SOC in Indian soils is relatively low, ranging from 0.1 to 1% and typically less than 0.5%, its influence on soil fertility and physical condition is of great significance. In this chapter we have discussed SOC potential in different soil orders of India, agro-ecological zones of India, influence of elevation, precipitation and temperature on organic carbon status under cultivated and forest lands of India.

SOC Potential in Different Soil Orders of India

Soil organic carbon stock of Indian soils is 10-12% of the tropical regions and about 3% of the total carbon mass of the world. The SOC stocks for India in terms of each soil order is estimated at 0-30 cm depths since such quantitative data reflect the kinds of soil with different amount of organic carbon (Table 1). In India, inceptisols contribute about 22% of the total SOC stock. Entisols contribute nearly 7% of the total SOC stock of Indian Soils. Vertisols are extensive in the central and southern part of India and contribute about 13% of the total SOC. Aridisols are in general poor in organic carbon due to their high rate of decomposition, low rate of plant growth and contribution to SOC. However, a few arid soils belonging to cold (Typic Camorthids) as well as hot (Typic Camorthids/ Natragids/ Calciortids) and arid ecosystem contribute about 37% of the total SOC stock mainly because of large area occupied by them. The Indian Mollisols contribute nearly less than 1% of the total SOC stocks due to the fact that only a small portion of geographical area of the country is covered by these soils. Most of Alfisols occur in sub-humid to humid regions of the country and contribute about 20% of the total SOC stocks. Oxisols occupying small area contributed less than 1% of the total SOC stock. Also poor accumulation of SOC in Oxisols is due to greater decomposition in tropical humid regions.

Table 1: Organic carbon stock in different soil orders of India

Soil orders	Organic carbon(0-30 cm depth) (Pg)	Percent of total carbon stock in India
Entisols	1.36	6.5
Inceptisols	4.67	22.2
Vertisols	2.62	12.5
Aridisols	7.67	36.5
Mollisols	0.12	0.6
Alfisols	4.22	20.0
Ultisols	0.14	0.8
Oxisols	0.19	0.9
Total	20.99	100

(Source: Velayutham *et al.,* (2000)

SOC Potential of Different Agro-ecological Zones of India

The maximum amount of SOC stocks is in surface of hot arid to semiarid regions covering agro-ecoregions of 2, 4, 6, 5, 7, and 8 followed by cold arid agro- ecoregions (1 and 3), and hot sub humid regions 9 to 15 (Table 2). The SOC stock in agro-ecoregions of 16 to 20 is comparatively less than those of arid and semiarid regions.

Table 2: SOC stock in different agro-ecoregions (AER) in India.

AER No.	Agro-Ecoregion	Area (M ha)	Organic C (Pg) 0-30	0-150
			---(cm depth)--	
1.	Cold arid ecoregion with shallow skeletal soils	15.2 (4.6)*	5.96	10.47
2.	Hot arid ecoregion with desert and saline soils in the western plains	31.9 (9.7)	2.51	10.43
3.	Hot arid ecoregion with red and black soils	4.9 (1.5)	1.21	4.21
4.	Hot arid ecoregion with desert and saline soils	32.2 (9.8)	3.88	15.24
5.	Hot semiarid ecoregion with medium and deep black soils	17.6 (5.4)	0.61	1.23
6.	Hot semiarid ecoregion with shallow and medium black soils	31.0 (9.4)	0.69	1.35
7.	Hot semiarid ecoregion with red and loamy soils	16.5 (5.0)	0.55	1.72
8.	Hot semiarid ecoregion with red and loamy soils	19.1 (5.8)	0.46	1.41
9.	Hot subhumid (dry) ecoregion with alluvium -derived soils	12.1 (3.7)	0.15	0.49
10.	Hot subhumid ecoregion with red and yellow soils	22.3 (6.8)	0.62	1.55
11.	Hot subhumid ecoregion with red and yellow soils	14.1 (4.3)	0.14	0.50
12.	Hot subhumid ecoregion with red and lateritic soils	26.8 (8.2)	0.67	1.70
13.	Hot sub-humid eco-region with alluvium-derived soils	11.1 (3.4)	0.17	0.79
14.	Hot sub-humid to humid with inclusion of per-humid eco-region with brown forest and podzolic soils	31.2 (6.4)	0.52	1.55
15.	Hot sub-humid (moist) to humid (inclusion of per-humid) eco-region with alluvium-derived soils	12.1 (3.7)	0.42	1.20
16.	Warm per-humid eco-region with brown and red hill soils	9.6 (2.9)	0.79	3.76
17.	Warm per-humid eco-region with red and lateritic soils	10.6 (3.2)	0.45	1.49
18.	Hot sub-humid to semiarid eco-region with coastal alluvium-derived soils	8.5 (2.6)	0.23	0.91
19.	Hot humid per-humid eco-region with red lateritic and alluvium-derived soils	11.1 (3.4)	0.84	2.82
20.	Hot humid to per-humid island eco-region with red loamy and sandy soils	0.8 (0.2)	0.12	0.37

*Percentage in parenthesis is proportion of total Indian area occupied by different agro-ecoregions
(Source: Sehgal et al., 1992)

Influence of Elevation, Precipitation and Temperature on Organic Carbon Status Under Cultivated and Forest Lands of India

Soils of India like most soils of the tropics have long been categorized as low in organic carbon and nitrogen although there are many variations like genetic, morphological, physical, chemical, and biological characteristics, associated with changing physiography, climate and vegetation. If we compare the isothermal and isohyets of India and North America, the organic carbon reserves of Indian soils, either virgin or cultivated, are higher, but they are lower than those of Central America (Jenny and Roychaudhuri,1960). The observed losses in soil organic carbon from managed ecosystems are greater in semiarid environments than in the humid low lands (Table 3). This indicates that a large portion of organic carbon under natural vegetation of arid and semiarid region is less recalcitrant than humid tropical soils. A study indicated that decline in total organic carbon in the agro-ecosystem was reduced two times faster than that of the soil carbon storage in the sub humid woodland forest and plantation. Long-term exhaustive practices of soil management and the climate on the organic carbon and nitrogen reserves of Indian soils indicate differentiation of organic C, total N and C/N ratio, a function of temperature, rainfall, and cultivation.

About 40% of the cultivated soils of the Indo-Gangetic alluvium are calcareous and, more precisely, they contain C as carbonates (Table 3). Initially, the geographic distribution of these calcareous bodies was conditioned by the flow patterns of the rivers, which traverse and erode calcareous strata in the Himalayan Mountains. But in due course of time, as precipitation increased the portion of calcareous soils declined by 20% at 127 to 152 cm rainfall. These large groups of cultivated, alluvial soils are richer in soil organic carbon than non-calcareous soils. This phenomenon may be meaningful if textural and climatic variables are simultaneously taken into account. In the drier section of the Indo-Gangetic Plains the areas of natural vegetation are tiny and far between. They consist of thorny volunteer shrubs on drifting sand dunes and on patches of temporarily abandoned land, and of clumps of wild bush grass. The native vegetation comprises brush, small *Acacias* and leguminous broad leaf trees, regularly pruned for cattle feed and intermingled with large specimens of *Euphoria* type bushes. The means of the combined vegetation types show a striking relation to elevation of northwest Himalayan soil. The soils above 1524 m are twice as rich in organic carbon when compared to soils below 1067 m. The surface layer (0-20 cm) is very dark and rich in carbon, and the subsoil is distinctly lower in organic carbon, but there exists no visible signs of fossilization.

Under natural vegetation, organic carbon may reach a near-steady state after 500 to 1000 years. Depletion of soil organic carbon under cultivated field was 23 to 48% of original value. It is documented that the agricultural soils of northwest India exclusive of the Himalayas have lost about one half to two thirds of their original organic carbon content. Northeast India consists of Tista-Brahmaputra plains, Assam Hill and valleys, and lesser Himalayan regions. Tista- Brahmaputra plains consist of relatively recent deposits which have been mapped as new alluvium and occasionally older, dissipated terraces which have weathered into reddish soils. The mean annual precipitation varies from 249 to 389, 129 to 1080 and 295 to 330cm under Tista- Brahmaputra plains, Assam Hill and valleys, and lesser Himalayan ranges, respectively. If the mean annual temperatures of Himalayan range are plotted against elevation, a nearly perfect straight line results with a negative gradient of 0.062°C per 30.5 m. The organic carbon values of these soils are two to three times higher than those of the cultivated soils of the Indo-Gangtic

Table 3: Influence of elevation, precipitation and temperature on organic carbon status under cultivated and forest lands of India

Location	Elevation (m)	Precipitation (cm)	Temperature (°C)	Carbon (%)	
				Cultivated	Native
A. Northwest India					
1. *Indo-Gangetic Plains*		25-51	25	0.3±0.033	0.59±0.211
		53-76	24-26	0.45±0.038	0.91±0.113
		79-102	23-25	0.55±0.037	-
		104-127	23-24	0.55±0.049	1.40±0.157
		130-152	23	0.35±0.059	1.24±0.297
2. *North-west Himalayan*					
Dehra Dun- Mussoorie	457-1067	216-224	23-20	1.44±0.145	1.81±0.270
	1067-1524	216-224	20-17	2.01	3.53
	1524-2134	216-224	17-14	3.37±0.365	3.99±0.346
Shimla	2195	155	13	2.91±0.386	4.48±0.258
3. *North-east India*					
Sriganganagar		25	25	0.33	-
Meerut		74	24	0.50	-
Biharigarh		122	24	0.51	-
Mohan		145	23	2.64	-
B. North-east India					
Tista- Bremhaputra Plain		249-389	24	1.37±0.119	2.32±0.160
Assam Hill & Valleys		129-1080	24-17	1.26±0.182	1.56±0.166
Himalayan Ranges					
	610-1311	300-315	21-17	3.18±0.221	4.82
	914-1158	218	19-18	2.23±0.219	-
	1524-2316	295-330	16-12	3.58±0.315	6.63±0.695

C. Southeast India					
Madurai- Kodaikanal	610	104	26	0.32±0.0025	1.73
Mountain Transect	945	117	24		3.23
	1280	130	22	2.31	
	1524	137	20		5.59
	1768	148	18		6.94
	2103	158	16	6.24	
D. West coast of India					
Dry coastal region		56-472	27	0.74±0.134	
Humid coastal region		108-223	27	1.89±0.272	1.86±0.212
E. Deccan Plateau and adjacent mountains					
Mysore-Bangalore area		79-86	25-23	0.52±0.036	1.68±0.230
Nagpure-Bellary area		51-124	27	0.55±0.124	1.09±0.170
Western Ghats- Nilgiri hills		130-917	24	1.25	2.59

(Source: Modified from Jenny and Raychaudhuri, 1960)

alluvium presumably because of higher rainfall (254-356 cm) and finer textures of the soils. For cultivated soils, the mean percentage value for carbon of Assam Hill and valleys is 1.26 ± 0.128.

In the regions of southeast India, the mean carbon content of cultivated soils of foot hills and plains is about 0.45%, and the values are nearly identical with those from the soils of the Indo-Gangetics plains having corresponding rainfall. The soils are covered with native vegetation consisting predominant of *Acacias, Euphorbia*, thorny shrubs, patches of poor stands of grass, and bare spots are largely unsuitable for agricultural production. The mean organic carbon content is 0.76 ± 0.076%. Organic carbon increased with increase in elevation under native vegetation and in cultivated fields of Madurai-Kodaikanal mountain sector.

The regions of the west coast of India, called Malabar and Kanara coast section, are laterite plateaus carrying bare iron stone crusts of panzer-like hardness and impenetrability. Along the entire dry coastal region, the range of annual rainfall is enormously increased from 56 to 472 cm and the mean annual temperature is about 27°C. The average organic carbon content is 0.74±0.134. The mean organic carbon per cent in non-paddy soils of humid coastal regions is about 0.92±0.155, which is significantly lower than in paddy soils (C=1.89±0.272). The organic carbon content in cultivated land is approximately half of the native vegetation. There is a decline in SOC concentration of cultivated soils by 30 to 60% compared with the antecedent level in undisturbed ecosystems even by 1960.

Factor Affecting SOC Storage

Soil organic carbon equilibrium is governed by a number of interacting factors such as temperature, moisture, texture, quality and quantity of organic matter application, soil type, soil tillage and cropping systems. Maintenance of soil organic carbon is an important tool for productivity and sustainability. Other important benefits of SOC in low-input agro-ecosystems are retention and storage of nutrients, increased buffering capacity, better soil aggregation, improved moisture retention, increased cation exchange capacity, and acting as a chelating agent. The addition of organic carbon improves soil structure, texture and tilth, activates a portion of inherent microorganisms, and reduces the toxic effects of pesticides.

a. Soil Type

Soil type is one of the important parameter that regulate soils organic carbon status of the soil. The major soil groups of India broadly fall into five groups *viz.,* alluvium derived soils (Inceptisol and Entisol: 74.3 million ha), black soil (Vertisol: 73.2 million ha), red, yellow and laterite soil (Alfisol and Oxisol: 87.6 million ha), and soils of desert regions (Aridisol : 28.7 million ha). The extent of clay aggregation is a direct controlling factor in organic carbon dynamics. Organic carbon content increases with clay content under desert, red, alluvial, laterite soil, saline and black soil, except mountain and forest soil which had the highest organic carbon at 34.5% clay possibly due to continuous deposition of unhumified organic carbon in these soils (Fig.1). Irrespective of climatic factor, increased amounts of sand, coarse loam, or gravelly sandy loam decrease the organic carbon content which may be due to less microbial proliferation and aggregation for carbon restoration.

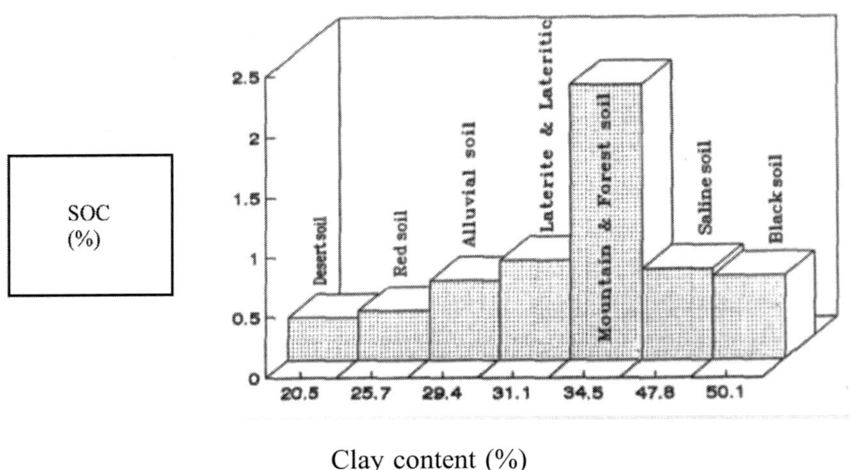

Fig.1: Relationship between clay and organic carbon content of soils (Modified from Ali *et al.,* 1966)

b. Rainfall and Temperature

Temperature and rainfall exert a significant influence on the decomposition of soil organic carbon and crop residues. A rise in the mean annual temperature reduces the level of SOC of cultivated soil in the humid region (Figure 2). Higher temperature activates the soil microbial population to a greater extent than plant growth. In temperate climate, the soils are several times richer in organic carbon than warmer climate. High rainfall and low temperature are conducive to accumulation of organic carbon in soils while high temperature and low rainfall decrease it.

Mean annual temperature (°C)

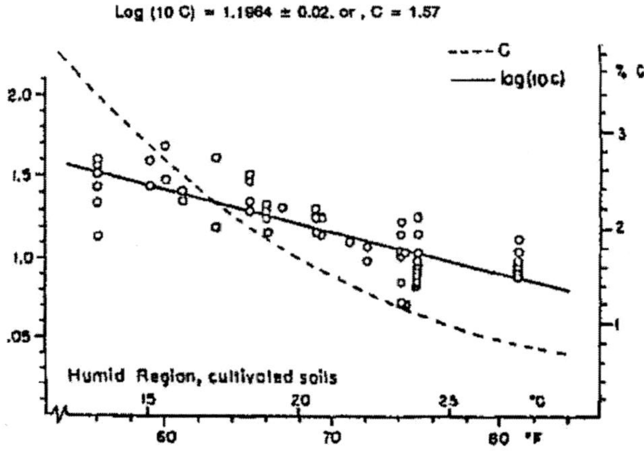

Fig. 2: Carbon –temperature function in the humid region (129-223cm); solid curve= logarithmic function; dashed curved =absolute function (Adapted from Jenny and Raychaudhuri,1960)

References

Ali, M. H., R.K. Chaterjee, and T.D. Biswas. (1966). Soil moisture tension relationship of some Indian soil. *Journal of Indian Society of Soil Science,* 14: 51-62.

Jenny, H. and S.P. Raychaudhuri. (1960). Effect of climate and cultivation on nitrogen and organic matter reserves in Indian soils. Indian Council of Agricultural Research, New Delhi, p 1-125.

Lal, R., Kimble, J. and Follett, R. (1997). In Soil properties and their management for carbon sequestration (Lal, R., *et al.*, Eds).USDA-NRCS-NSSC Lincoln, NE.

Sehgal, J.L. and Abrol, I.P. (1994). Soil degradation in India-status and impact. Oxford and IBH publishing Co.Pvt.Ltd. New Delhi, India.

Swarup, A., M.C. Manna and G.B. Singh. (2000). Impact of land use management practices on organic carbon dynamics in soils of India. In: Global Climate Change and Tropical Ecosystems (Lal *et al.,* eds). Advances in Soil Science (USA). pp. 261-281.

Velayutham,M., D.K.Pal. and T. Bhattacharyya (2000). Organic carbon stock of India. In: Global Climate Change and Tropical Ecosystem Lal R. et al. (eds.) Advances in Soil Sciences (USA)., pp.77-95.

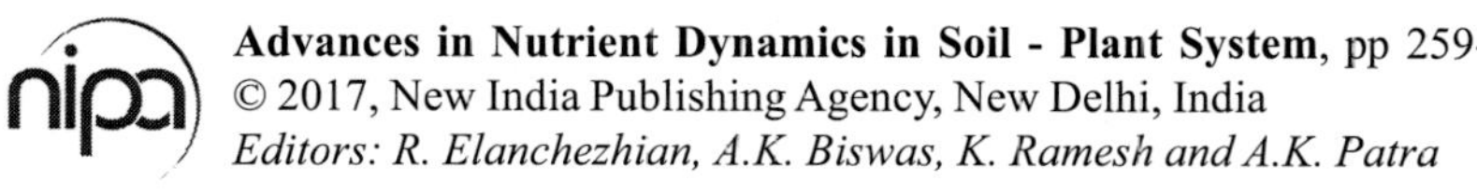
Advances in Nutrient Dynamics in Soil - Plant System, pp 259-269

Editors: R. Elanchezhian, A.K. Biswas, K. Ramesh and A.K. Patra

23

Soil Organic Carbon and Nitrogen Dynamics: A Key to Soil Health

Pramod Jha, Brijlal Lakaria and A.K. Biswas

ICAR-Indian Institute of Soil Science, Nabi Bagh, Bhopal – 462038, India

Introduction

Organic matter in soil is the key to soil health. SOM improves many physical, chemical, and biological characteristics of the soil, including water holding capacity, cation exchange capacity, pH, buffering capacity, and chelating of micronutrients. Furthermore, well decomposed SOM improves soil structure by increasing aggregation, enhances biological activities in the soil, slowly releases nutrients, and suppresses some diseases. A loss of SOM can lead to soil erosion, loss of fertility, compaction, and general land degradation.

Soil organic matter consists of partially disintegrated and decomposed plant and animal residues and other organic compounds synthesized by the soil microbes as the decay occurs (Brady, 1990). Soil humus (or humic material) is that part of SOM that is a heterogeneous mixture of organic compounds formed by degradation and synthesis. Soil humus makes up about 60-80% of SOM. The rest is less stable and partially decomposed organic residues. Humus is the stable fraction of SOM because it is relatively resistant to microbial attack.

Although the amount of SOC in Indian soils is relatively low, ranging from 0.1 to 1% and typically less than 0.5%, its influence on soil fertility and physical condition is of great significance. Intensive cropping and tillage systems have led to substantial decreases in the SOC through enhanced microbial decomposition and through wind and water erosion of inadequately protected soils, which is often accompanied by a decline in soil productivity.

The Role of Soil Organic Carbon

Total organic carbon influences many soil characteristics including colour, nutrient holding capacity (cation and anion exchange capacity), nutrient turnover and stability, which in turn influence water relations, aeration and workability. In soils with high clay content the contribution to cation exchange from the organic fraction is generally small compared to that from clay. In sandy soils, the relative contribution of the organic fraction is higher because there is less clay, even though the amount of total organic C is less. By providing a food source for micro-organisms, organic carbon can help improve soil stability by micro-organisms binding soil particles together into aggregates or 'peds'. Bacteria excretions, root exudates, fungal hyphae and plant roots can all contribute to better soil structure. Moist, hot and well-aerated conditions favour rapid decay of organic additions. If the rate of organic matter addition is greater than the rate of decomposition, the organic fraction in a soil will increase. Conversely, if the rate at which organic matter is added to soil is lower than the decomposition rate, the organic fraction will decline. At a steady state level, the rate of addition is equal to the rate of decomposition. Large organic additions can temporarily increase the organic fraction in a soil, but unless additions are maintained, the soil will revert to its steady state equilibrium, which is usually low.

Soil organic matter is a large reserve of potentially mineralizable nitrogen (N), sulfur (S), and other nutrients. Organic C is found to be highly correlated with organic N. The N requirements minus the N-supplying power of the soil based on SOM will indicate the N rate necessary to produce the yield goal for the selected crop. SOM in the form of humus enhances mineral breakdown and, in turn, nutrient availability. Organo-mineral complexes can also be formed with ions, particularly metallic ions such as Fe^{3+}, Cu^{2+}, Zn^{2+}, and Mn^{2+}, which will make them more available for plant uptake than the mineral form of these metals.

Increasing carbon content in the soil, through better management practices, produce a number of benefits in terms of soil biodiversity, soil fertility and soil water storage capacity and hence productivity. Soil carbon sequestration through the restoration of soil organic matter can further reverse land degradation and restore soil "health" through restoring soil biota and the array of associated ecological processes. In particular, through improved soil water storage and nutrient cycling, land use practices that sequester carbon will also contribute to stabilizing or enhancing food production and optimizing the use of synthetic fertilizer inputs, thereby reducing emissions of nitrous oxides from agricultural land.

Effect of Cultivation on Soil Carbon Status

A classical work on status of soil organic carbon of native and cultivated land of India was done by Jenny and Raychaudhary (1960). The samples were collected throughout India between the latitudes of 8 and 32 degrees N and analysed for total organic carbon under pristine/forest land use *vis a vis* cultivated lands. The mean annual temperatures (MAT) ranged from 10 to 30 °C while the mean precipitation ranged from near 0 to over 5 meters (Miller *et al.,* 2004). Soil sites were chosen with an emphasis on nearly-level, well-drained surfaces and, where possible, native vegetation. Each soil was sampled to a depth of 20 cm. Laboratory analyses were conducted in Berkeley, California for estimation of total organic C. The soil C sink capacity depends on the difference between the present SOC pool under agricultural land use compared to the SOC pool under an undisturbed or a natural ecosystem. Most soil loose the SOC pool upon conversion from natural to managed ecosystems. The observed losses in soil organic carbon from managed ecosystems are greater in semiarid environments than in the humid low lands (Table 1). This indicates that a large portion of organic carbon under natural vegetation of arid and semiarid region is less recalcitrant than humid tropical soils. A study indicated that decline in total organic carbon in the agro-ecosystem was reduced two times faster than that of the soil carbon storage in the sub humid woodland forest and plantations. In the drier section of the Indo-Gangetic Plains the decline in soil organic carbon was to the extent of 59.6% in cultivated land in comparison to native pristine land. This was closely followed in Deccan plateau where decline in soil organic carbon due to cultivation was to the tune of 57.0%. Invariably, cultivation leads to decline in soil organic carbon in the range of 29-60% from the initial value. It is documented that the agricultural soils of northwest India exclusive of the Himalayas have lost about one half to two thirds of their original organic carbon content. In the regions of southeast Indian the mean carbon content of cultivated soils of foot hills and plains is about 0.45%, and the values are nearly identical with those from the soils of the Indo-Genetics plains having corresponding rainfall. In general, SOC in cultivated soils are also inversely related to mean annual temperature and positively related to mean annual precipitation. In India, the total SOC loss, and decline in C/N ratio, increases with decreasing temperature, indicating that the cooler soil regions of India not only have a large potential for C storage, but a lesser N limitation than warmer climates (Miller *et al.,* 2004). This further corroborated that no one value (for relative C loss following cultivation) adequately describes the loss of SOC everywhere, and that accurate estimates of C loss will vary with climate and other factors.

Table 1. Depletion of SOC of cultivated compared with that in undisturbed soils

Region	SOC		Per cent reduction
	Cultivated	Native (g/kg)	
1. Northwest India			
Indo-gangetic plains	4.2 ±0.9	10.4±3.6	59.6
Northwest Himalays	24.3±8.7	34.5±11.6	29.6
2. Northeast India	23.2±10.4	38.3±23.3	39.4
3. Southeast India	29.6±30.1	43.7±23.4	32.3
4. West coast	13.2±8.1	18.6±2.1	29.1
5. Deccan Plateau	7.7±4.1	17.9±7.6	57.0

Adapted from Jenny and Raychaudhary, 1960

Strategies for soil carbon build up

One of the main options for greenhouse gas (GHG) mitigation identified by the IPCC is the sequestration of carbon in soils. Improving organic carbon content of terrestrial carbon pool by different agronomic measures like residue retention, application of organics, conservation agriculture and reducing soil erosion have been documented by several authors. Lal (2004) broadly classify soil carbon sequestration strategy in two categories.

1. Land use change

a) Restoration of degraded lands

A vast portion (120 m ha) of total geographical area of country is affected by various forms of land degradation. It offers an opportunity for storing carbon in soil by adoption of land restorative processes. The total potential of restoring degraded soils in India is 7 to 10 Tg C/y (Lal, 2004). Some of the important measures for land restoration and carbon sequestration strategies are green manuring, mulch farming/conservation tillage, afforestation/agroforestry, grazing management/ley farming, integrated nutrient management/manuring, diverse cropping systems etc.

b) Erosion control

Accelerated soil erosion depletes the SOC pool severely and rapidly. Soil conservation and water management, water harvesting and recycling, are important strategies of minimizing losses and restoring soil quality.

2. Soil/vegetation management

a) Residue management

Incorporating plant residues is one means by which we can add organic matter to soil. Removal of crop residues from field is known to hasten SOC decline especially when coupled with conventional tillage. Incorporation of crop residues favours immobilization because of wide C/N ratio. However, long-term use (repeated additions) are known to improve organic C content (Table 2).

Table 2: SOC (%) content as affected by cropping system and residue addition at the end of three years of cropping

Cropping system	(-) residue	(+) residue
Continuous	0.64	1.51
Intercropping	1.06	1.45
Rotation	0.65	1.74

Source: Anyanzwa *et al.,* (2010)

The extent of residue or crop cover left on the soil surface depends on the availability. In our country, there are great competing uses for crop residues as fuel or as thatching material or as feed. Therefore, crop residues are mostly disposed off from crop fields. In some situations, where they are available in abundance, crop residues are considered as waste material and disposed-off by burning such as in the rice-wheat growing areas of north India.

b) Integrated nutrient management

Balanced application of inorganic fertilizer and organic amendments greatly influence the accumulation of organic matter in soil and also influence the soil physical environment. Soil organic carbon (SOC) was significantly influenced by the fertilizer and organic manure applied over 28 years of cropping (Hati *et al.*, 2007). The results showed that the soil organic carbon (SOC) content in 100% NPK and 100% NPK + FYM treatments increased (Table 3), respectively, by 22.5 and 56.3% over the initial level (1.14 kg m^{-2}). Application of fertilizers in combination with manure resulted in greater accumulation and build up of SOC. This is because SOC is directly related to organic inputs.

Table 3: Effect of fertilizer and organic manure application under intensive cropping on soil organic carbon content (SOC) after 28 crop cycles

Treatment	SOC (kg m^{-2})
Control	1.12c
100%N	1.20c
50% NPK	1.30bc
100%NPK	1.39b
100%NPK+FYM	1.78a
Initial	1.14

Source: Hati *et al.,* (2007)

c) Improved cropping systems

Principal mechanisms of SOC sequestration with conservation tillage are increase in micro-aggregation and deep placement of SOC in the subsoil. Less tillage will influence the maintenance of C in un-decomposed residue and increase sequestered C in the soil. Incorporating plant residues is one means by which we can add organic matter to soil. Removal of crop residues from field is known to hasten SOC decline especially when coupled with conventional tillage. Crop rotations had significant influence on SOM content. Inclusion of legume in crop rotation resulted in build-up of SOM. It is interesting to note that even in semi-arid areas where the cropping systems are mainly focused on water conservation, SOC improvements are noticed.

d) Balanced fertilization

Long-term fertilizer experiments conducted over 30-35 years in different agro-ecoregions of India involving a number of cropping systems and soil types (Inceptisol, Vertisol, Mollisol and Alfisol) have shown a decline in SOC as a result of continuous application of fertilizer N alone. However, balanced use of NPK fertilizer either maintained or slightly enhanced the SOC over the initial values. Application of farmyard manure (FYM) and green manure improved SOC which was associated with increased crop productivity. The nutrient removal is far greater than the supply, it is therefore, extremely important to maintain SOC at a reasonably stable level, both in quality and quantity, by means of suitable addition of organic materials or crop residues. Some of the other measures outlined by Lal (2008) are mentioned in Table 4.

Table 4: Terrestrial carbon management options (Adapted from Lal, R. 2008)

Management of terrestrial C pool	Sequestration of C in terrestrial pool
Reducing emissions	*Sequestering emissions as SOC*
Eliminating ploughing	increasing humification efficiency
Conserving water and decreasing irrigation need	
Using integrated pest management to minimize the use of pesticides	Deep incorporation of SOC through establishing deep rooted plants, promoting bioturbation and transfer of DOC into the ground water
Biological nitrogen fixation to reduce fertilizer use	Asequestering emissions as SIC
Offsetting emissions	*forming secondary carbonates through biogenic processes*
Establishing biofuel plantations	Leaching of bicarbonates into the ground water
Biodigestion to produce CH_4 gas	
Bio-diesel and bioethanol production	
Enhancing use efficiency	
Precision farming	
fertilizer placement and formulations	
drip, sub-irrigation or furrow irrigation	

Nitrogen Dynamics in Soil

To meet the food needs of the burgeoning population, India will need to produce 300 million tonnes of food grains by 2020. At present more than 75% of the total food grains produced in the country are of rice and wheat. Use of nitrogenous fertilizers has contributed much to the remarkable increase in production of rice and wheat in India that has occurred during the past three decades. During the last half-decade or so while fertilizer N consumption is touching new heights, the production of both rice and wheat is showing a trend of plateauing. In fact, fertilizer N efficiency of foodgrain production expressed as partial factor productivity of N (PFP_N) has been decreasing exponentially since 1965. The PFP_N is an aggregate efficiency index that includes contributions to crop yield derived from uptake of indigenous soil N, fertilizer N uptake efficiency, and the efficiency with which N acquired by the plant is converted to grain yield. A decrease in PFP_N occurs as farmers move yields higher along a fixed N response function, unless other factors shift the response function up. In other words, an initial decline in PFP_N is an expected consequence of the adoption of N fertilizers by farmers and not necessarily bad within a system's context.

Applied N not taken up by the crop or immobilized in soil organic N pools-which include both microbial biomass and soil organic matter—is vulnerable to losses from volatilization, denitrification, and leaching. The overall NUE of a cropping system can therefore be increased by achieving greater uptake efficiency from applied N inputs, by reducing the amount of N lost from soil organic and inorganic N pools, or both. In many cropping systems, the size of the organic and inorganic N pools has reached steady-state or is changing very slowly, and the N inputs from biological N_2 fixation and atmospheric deposition are relatively constant (Cassman *et al.,* 2002).

Nitrogen is an element essential for all plant and animal life. The interlocking succession of nitrogen reactions occurring in the soil is known as the nitrogen cycle. Agriculture affects both nitrogen additions and subtractions to the soil. Additions include nitrogen fertilizers, crop residues, nitrogen fixation by legumes, and manures, irrigation water, rainwater, etc. Subtractions attributed to agriculture include crop removal (harvesting), plant uptake, and nitrogen leaching, emission losses, etc.

Under practical conditions, nutrient use efficiency (NUE) can be considered as the amount of nutrients taken up from the soil by plants and crops within a certain period of time compared with the amount of nutrients available from the soil or applied during that same period of time. Improving NUE in agriculture has been a concern for decades, and numerous new technologies have been developed in recent years to achieve this. The types of fertilizers and their management in agriculture will be at the forefront of measures to improve the global N balance in the short- and long-term. The most important task for the future is to further improve NUE or, more precisely, N-use efficiency, because a significant share of the added fertilizer N is lost during the year of application.

The applied nitrogenous fertilizer in soil is subject to several kinds of losses. The tricky situation with N fertilizers is that if we control one kind of loss it triggers another kind of loss with higher magnitude. Due to these circumstances, the apparent recovery of nitrogenous fertilizer is low and ranges from 30-50%. Under such circumstances, use of simulation models like DSSAT and DNDC may offer new opportunity for enhancing the yield and NUE of crops.

Nitrogen is one of the most important nutrient elements of plant. The soil reserve of nitrogen is not good enough to meet the plant nitrogen requirement. Therefore, supplementation of plant N requirement through chemical fertilizer is indispensable for getting a sustainable and economic yield of crops. However, application rate of N-fertilizer in a particular crop depends on available nitrogen content of soil. In India, soil available N is determined mostly by alkaline $KMnO_4$ method (Subbiah and Asija, 1956). Also availability of nitrogen in soil is generally predicted by soil carbon content (Walkley and Black C content). Accordingly, soil carbon content of <0.5% is regarded as low (<280 kg N ha^{-1}), medium 0.5-0.75% (280-560 kg N ha^{-1}) and high >1% (>560 N kg ha^{-1}). In this regard, Walkley-Black method is the most common, rapid and most widely used procedure, which requires minimum equipment, compared with other wet or dry combustion methods (Nelson and Sommers, 1982). The Walkley-Black method, however, gives variable recovery of soil organic C. A general standard conversion factor of 1.32 for incomplete oxidation of organic carbon is commonly used to convert Walkley-Black carbon to the total organic-C content, although true factors vary greatly between and within soils because of differences in the nature of organic matter, its distribution with soil depth and vegetation type (Grewal *et al.,* 1991). Two soils having the same Walkley-Black C content may have a different total carbon contents. The Walkley-Black method, which has been used for soil carbon stock gives only an approximation of soil organic carbon content. This is a common problem in global studies of soil carbon stocks (Schlesinger, 1977). In past, attempts were made to devise a relationship between TOC and Walkley-Black analysis. However, all workers came out with different conversion factors which ranged from 0.09-2.21 in different soils (Bremnerand Jenkinson, 1960; Kalembasa and Jenkinson, 1973; Orphanos, 1973; Richter *et al.,* 1973; Nelson and Sommers, 1975, Hussain, and Olson 2000; Karven *et al.,* 2000; Kamara *et al.,* 2007). The reason attributed for such wide variations in recovery of organic C with the Walkley-Black method was soil types, soil layers, resistance of crop residue against degradation, nature of organic compounds, type of management systems and climate. Hence, predicting N availability based on soil carbon index may not represent the true status of soil nitrogen supplying capacity.

Carbon stability governs N dynamics in soil

It has long been observed that there exists an intimate relationship between the soil N and carbon content. However, it was observed that two soils having a different textural composition but having same carbon content have different N dynamics in soil. It suggests that N dynamics in soil is not only governed by total C per se but also the relative allocation of carbons in different pools (i.e. mineralizable or resistant pool). Long-terrn fertilizer experiment conducted in three different locations of India (Jabalpur, Palampur and Ranchi) provides an opportunity to assess the impact of soil carbon pools on N dynamics of soil. Soil total nitrogen (TN) content increased in comparison to control under the treatments of NPK and NPK+FYM in all the sites (Fig. 1). Invariably, there was a buildup of the total N content of soil in all the 3 sites from the initial value. The total N content of soils across all sites ranged from 0.07 to 0.19 %. The highest content was recorded at Palampur (Alfisol) and Jabalpur (Vertisol) sites, and the least at Ranchi (Alfisol) site. In no site, total N content of NPK treatment was less than the control, although no significant difference in soil total N content was recorded between the 2 treatments in all the sites. The total N of NPK+FYM treatment plot was significantly higher than the rest of the treatments in all LTFE sites except in Ranchi (Alfisol). Similarly, the total N content increased to 0.14 per cent under permanent manurial trial (PMT) at Ranchi (Alfisol), whereas the NPK plot had 0.10 percent of total N. The total N content of the soil under the treatment of NPK and NPK+FYM increased to the extent of 9, 24, 14 and 33, 62, 16% over control at Jabalpur (Vertisol), Palampur (Alfisol) and Ranchi (Alfisol), respectively. Similarly, long-term adoption of an integrated approach (NPK+FYM) further increased the total N content of the soil by 22, 31 and 2% over NPK treatment in Jabalpur (Vertisol), Palampur (Alfisol) and Ranchi (Alfisol), respectively (Jha *et al.,* 2014). A close correlation was observed (r=0. 765, P=0. 01) between TOC and Nt (data not shown). As the TOC of soil increased, the total N content also showed a definite increase. The relationship of the different soil carbon pool to N availability is shown in figure 2 a and b. It was observed that the TOC content of soil bears a linear relationship with available nitrogen content of soil (R^2=0. 117, p=0. 01); however, acid-hydrolyzable carbon (Ca+Cs) showed excellent agreement with nitrogen availability (R^2=0. 642, p=0. 01).

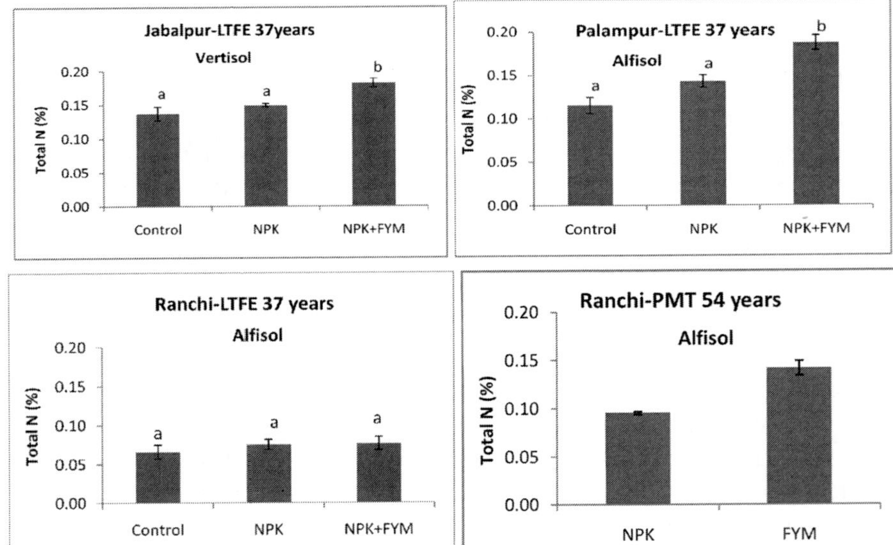

Fig. 1: Changes in total N content of soil under long-term chemical fertilization and manuring treatments (bar indicates standard error of mean, means followed by the same letter under a particular soil type are not different at 0.05 probability level using Student Newman Keuls test).

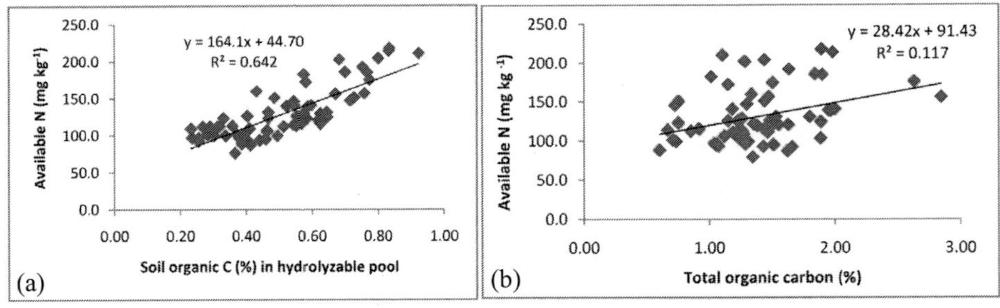

Fig.2a & b. Soil N availability as affected by soil carbon pools (*N=66*, inclusive of all sites).

References

Anyanzwa, H., Okalebo, J.R., Othieno, C.O., Bationo, A., Waswa, B.S. and Kihara, J. (2010). Effects of conservation tillage, crop residue and cropping systems on changes in soil organic matter and maize-legume production: a case study in Teso district. *Nutrient Cycling in Agroecosystems,* 88: 39-47.

Brady, N.C. (1990). The nature and properties of soils. 10th edition. 621 pp. Macmillan Publishing Co., New York, NY.

Bremner, J.M. and Jenkinson, D.S. (1960). Determination of organic C in soil. I. Oxidation by dichromate of organic matter in soil and plant materials. *Journal of Soil Science*, 11: 394-402.

Cassman, Kenneth G.; Dobermann, Achim R. and Walters, Daniel T. (2002). "Agroecosystems, Nitrogen-use Efficiency and Nitrogen Management". Agronomy & Horticulture — Faculty Publications. Paper 356.

Grewal, K. S., Buchan, G. D. and Sherlock, R.R. (1991). A comparison of three methods of organic carbon determination in some New Zealand soils. *Journal of Soil Science*, 42: 251-257.

Hati, K.M. Swarup, A, Dwivedi, A.K., Misra, A.K. and Bandyopadhyay, K.K. (2007). Changes in soil physical properties and organic carbon status at the topsoil horizon of a vertisol of central India after 28 years of continuous cropping, fertilization and manuring. *Agriculture, Ecosystems and Environment,* 119: 127–134.

Hussain, I. and Olson, K.R. (2000). Recovery rate of organic C in organic matter fractions of Grantsburg soils. *Communications in Soil Science and Plant Analysis*, 31: 995-1001.

Jenny, H. and S.P. Raychaudhuri. (1960). Effect of climate and cultivation on nitrogen and organic matter reserves in Indian soils. Indian Council of Agricultural Research, New Delhi, p 1-125.

Jha, P., Lalkaria, B.L., Biswas, A.K., Saha, R., Mahapatra, P., Agrawal, B.K., Sahi, D.K. Wanjari, R.H., Lal, R., Singh, M. and Rao, A.S. (2014). Effects of carbon input on soil carbon stability and nitrogen dynamics. *Agriculture, Ecosystems and Environment*, 189: 36-42. 10.1016/j.agee.2014.03.019

Kalembasa, S.J. and Jenkinson, D.S. (1973). A comparative study of titrimetric and gravimetric methods for determination of organic C in soil. *Journal of the Science of Food and Agriculture*, 24: 1085-1090.

Kamara, A., Rhodes, E.R. and Sawyer, P.A. (2007). Dry combustion carbon, walkley–black carbon, and loss on ignition for aggregate size fractions on a toposequence. *Communications in Soil Science and Plant Analysis*, 38: 2005-2012.

Kerven, G. L., Menzies, N. W., and Geyer, M. D. (2000). Analytical methods and quality assurance. *Communications in Soil Science and Plant Analysis*, 31: 1935-1939.

Lal, R. (2004). Soil carbon sequestration in India. *Climatic Change,* 65: 277–296.

Lal, R. (2008). Carbon sequestration. *Philosophical Transactions of the Royal Society of Biological Sciences,* 363: 815-830.

Miller, A.J., R. Amundson, Burke, I.C. and C., Yonker. (2004). The effect of climate and cultivation on soil organic C and N. *Biogeochemistry,* 67: 57-72.

Nelson, D.W. and Sommer, L.E. (1982). Total carbon, organic carbon, and organic matter. p.539-579. In A.L. Page (ed.) Methods of Soil Analysis. 2nd Ed. ASA Monogr. 9(2). Amer. Soc. Agron. Madison, WI.

Orphanos, P. (1973). On the determination of soil carbon. *Plant and Soil*, 39: 706-708.

Richter, M., Massen, G. and Mizuno, I. (1973). Total organic carbon and "oxidizable" organic carbon by the Walkley Black procedure in some soils of Argentine Pampa. *Agrochimica*, 17: 462-472.

Schlesinger, W.H. (1977). Carbon balance in terrestrial detritus. *Annual Review of Ecological Systems,* 8: 51-81.

Subbaiah, B.V. and Asija, G.L. (1956). A rapid procedure for the estimation of available nitrogen in soil. *Current Science*, 25: 259.

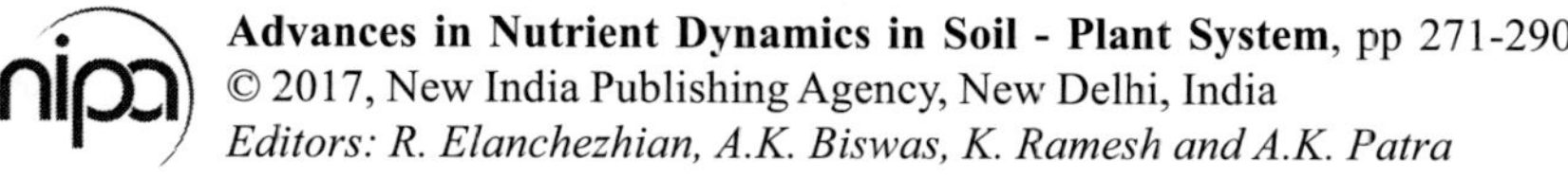
Advances in Nutrient Dynamics in Soil - Plant System, pp 271-290

Editors: R. Elanchezhian, A.K. Biswas, K. Ramesh and A.K. Patra

24

Impact of Conservation Agricultural Practices on Soil Organic Carbon and Nutrient Availability: An Overview

J. Somasundaram[1], Steven Reeves[2] K.M. Hati[1] and R.S.Chaudhary[1]

[1]ICAR-Indian Institute of Soil Science, Nabi Bagh, Bhopal – 462038, India
[2]Soil Processes, Department of Science, Information Technology and Innovation (DSITI), Ecosciences Precinct, Boggo Road, Dutton Park Brisbane, Queensland, Australia

Introduction

Our soil resource is generally taken for granted for many uses. Most people don't realize the importance of the resource and that it is fundamental for farming/agriculture. Without high quality soils, agriculture production cannot be achieved on a sustainable basis to meet ever growing demand. Thus, conservation agricultural practices are capable of improving soil health by increasing organic carbon, aggregation, improving infiltration and minimizing erosion losses.

Conservation agriculture (CA) practices involve key principles such cas minimum soil disturbances, providing a soil cover through crop residues or other cover crops, and crop rotations for achieving higher productivity. Conservation agriculture (CA) is a slower-evolving agricultural revolution that began at the same time as the Green Revolution and emerged as a new paradigm to achieve goals of sustainable agricultural production. It is a major step towards transition to sustainable agriculture. The key features of CA include: (i) minimum soil disturbance by adopting no tillage and minimum traffic for agricultural operations, (ii) leave and manage the crop residues on the soil surface, (iii) adopt spatial and temporal crop sequences / crop rotations to derive maximum benefits from inputs and minimize adverse environmental impacts (Abrol and Sanger, 2006; FAO, 2008;

Friedrich *et al.*, 2012; Somasundaram *et al.*, 2014a). In the conventional systems involving repeated intensive tillage, led to decline in soil organic matter through accelerated oxidation and burning of crop residues causing pollution, greenhouse gases emission and loss of valuable plant nutrients. Intensive seed-bed preparation with heavy machinery leads to declining soil fertility, biodiversity and accelerated soil erosion and land degradation. When the crop residues are retained on the soil surface in combination with no tillage, it initiates processes that lead to improved soil quality and overall resource enhancement. Therefore, conservation agriculture practices may lead to sustainable improvements in the efficient use of water and nutrients by improving nutrient balances and availability, infiltration and retention by soils reducing water losses due to evaporation, and improving the quality and availability of ground and surface water. In this chapter, an overview of soil organic carbon, nutrient availability and nutrient management strategies under conservation agricultural practices have been discussed.

Nutrient Management Strategy in CA

Worldwide, no-till (NT) farming systems have led to many benefits such as increased soil flora and fauna biodiversity, increased organic matter and improved soil structure and fertility (Thomas *et al.*, 2007a; Radford and Thornton, 2011). Many studies concerning no-till farming systems have also demonstrated advantages in economic, environmental and soil quality aspects over conventional tillage (CT). However, adoption of continuous NT has contributed to the build-up of herbicide resistant weed populations, increased incidence of soil- and stubble-borne diseases, and stratification of nutrients and organic carbon near the soil surface.

Unlike in conventional cultivation, nutrient management under conservation agriculture farming is a challenging issue. Application of manures and fertilizer nutrients in the amidst of crop residues is always a challenging task. Nutrient management strategies in CA systems need to be applied (as suggested by Kassam and Friedrich, 2009) based on the following four general aspects that:

(i) the biological processes of the soil are improved so that all the soil biota and microorganisms are under favourable conditions and that soil organic matter and soil porosity are built up and sustained;

(ii) there is adequate biomass production and biological nitrogen fixation for keeping soil energy and nutrient stocks sufficient to support higher levels of biological activity, and for covering the soil;

(iii) there is an adequate access to all nutrients by plant roots in the soil, from natural and synthetic sources, to meet crop demand; and

(iv) soil acidity is kept within an acceptable range for all key soil chemical and biological processes to function effectively and efficiently.

CA-Based Nutrient Management Practices

Integrated Soil Fertility Management (ISFM) and Integrated Natural Resources Management (INRM) approaches of various types and nomenclature have been in vogue in recent years in the scientific community. Focusing on soil fertility but without defining the tillage and cropping system, as often proposed by ISFM or INRM approaches, is only a partial answer to enhancing and maintaining soil health and productivity in support of sustainable production intensification, livelihood and the environment.

Generally, such approaches are focused more on "feeding the crop" and meeting crop nutrient needs in an input-output sense rather than managing soil health and productive capacity as is the case with CA systems. Also, most of the work under broad terms of ISFM or INRM over the past 15-20 years or so has been geared towards tillage-based systems of the first paradigm which have many unsustainable elements, regardless of farm size or the level of agricultural development. Unless the concepts of soil health and function are explicitly incorporated into ISFM or INRM approaches, sustainability goals and means will remain unconnected and sustainable production will be difficult to achieve, particularly by resource poor farmers (Kassam and Friedrich, 2009). Thus, CA systems have within them their own particular sets of ISFM or INRM processes and concepts that combine and optimize the use of organics with inorganic inputs integrating temporal and spatial dimensions with soil, nutrient, water, soil biota, biomass dimension, all geared to enhancing crop and system outputs and productivities but in an environmentally responsible manner. Over the past two decades or so, empirical evidence from the field has clearly shown that healthy agricultural soils constitute biologically active soil systems within landscapes in which both the soil resource and the landscape must operate with plants in an integrated manner to support the various desired goods and services namely food, fodder, livelihood, environmental services, etc provided by agricultural land use.

Moreover, CA principles and practices offer substantial benefits to all types of farmers in most agro-ecological and socio-economic situations. CA-based IFSM and INRM approaches to nutrient management and production intensification would be more effective for farmer-based innovation systems and learning processes such as those promoted through Farmer Field School networks/ Farm Science Centre.

Adopting CA-Based Nutrient Management Framework

CA has now emerged as a major "breakthrough" systems approach to crop and agriculture production with its change in paradigm that challenges the status quo. However, as a multi-principled concept, CA translates into knowledge-intensive practices whose exact form and adoption requires that farmers become intellectually engaged in the testing, learning and fine tuning of possible practices to meet their specific ecological and socio-economic conditions (Friedrich and Kassam 2009).

However, CA approach represents a highly biologically and bio-geophysical-integrated system of soil health and nutrient management for production that generates a high level of "internal" ecosystem services which reduces the levels of "external" subsidies and inputs needed. CA provides the means to work with natural ecological processes to harness greater biological productivities by combining the potentials of the endogenous biological processes with those of exogenous inputs. The evidence for the universal applicability of CA principles is now available across a range of ecologies and socio-economic situations covering large and small farm sizes worldwide, including resource poor farmers (Goddard *et al.*, 2007; FAO, 2008).

There are many different ecological and socio-economic starting situations in which CA has been and is being introduced. They all impose their particular constraints as to how fast the transformation towards CA systems can occur. In the seasonally dry tropical and sub-tropical ecologies, particularly with resource poor small farmers in drought prone zones, CA systems will take a longer time to establish, and step-wise approaches to the introduction of CA practices seem to show promise (Mazvimavi and Twomlow, 2006). These involve two components: the application of planting '*Zai-type*' basins which concentrate limited nutrients and water resources to the plant, and the precision application of small or micro doses of nitrogen-based fertilizer. In the case of degraded land in wet or dry ecologies, special soil amendments and nutrient management practices are required to establish the initial conditions for soil health improvement and efficient nutrient management for agricultural production (Landers, 2007). However, it is necessary to have a clear understanding about the CA system and should be followed holistically to sustain soil health and productivity. Moreover, efficient nutrient management interventions may be proposed which can contribute to the system effectiveness as a whole both in the short- and long-term.

Conservation Agriculture and Soil Organic Carbon

Conservation agricultural systems have been successfully developed for different regions of the world. These systems, however, have not been widely adopted by farmers for various reasons.

Through greater adoption of conservation agricultural systems, there is enormous potential to sequester soil organic carbon, which would:

(1) help mitigate greenhouse gas emissions contributing to global warming and

(2) increase soil productivity and avoid further environmental damage from the unsustainable use of inversion tillage systems, which threaten water quality, reduce soil biodiversity, and erode soil around the world.

Crop residues retained on the soil surface in CA (Fig. 1), in general, serve a number of beneficial functions, including soil surface protection from erosion, enhancing infiltration and cutting run-off rate, decreasing surface evaporation losses of water, moderating soil temperature and providing substrate for the activity of soil micro-organisms, and a source of SOC. Long-term implementation of conservation agricultural practices also increases organic matter levels in the soil. Lower soil temperatures and increased soil moisture regime contributes to slower rates of organic matter oxidation. An increase in organic matter is normally observed within the surface soil (0-10cm) which helps in better soil aggregation. Moreover, carbon turnover rate slows down when soil aggregation increases and soil organic carbon (SOC) is protected within stable aggregates (53-250 μm).

Fig 1: Residue retention under soybean-wheat system (left) and maize-gram system (right) in Vertisols of Central India (*See colour version on page 380*)

Minimum tillage practices, including no-till (NT) and reduced tillage (RT), have received attention due to their ability to both reduce soil erosion and increase C sequestration in agricultural surface soils (Cole *et al.*, 1997) by increasing aggregate stability. Alvarez (2005) reviewed the effect of nitrogen and no-tillage on soil organic carbon (SOC) from 137 sites and concluded that nitrogen fertilizer increased SOC but only when crop residues were retained. Furthermore, nitrogen fertilizer use in the tropics resulted in no SOC sequestration while in the temperate regions, there was a trend towards increasing SOC sequestration. After 22 years of no-till, Dalal (1992) found that soil total nitrogen decreased with the period of cropping irrespective of the tillage practices in subtropical cereal cropping in Australia. Similarly, Dalal *et al.* (2011) reported that tillage effects on soil organic carbon and soil total nitrogen were small following 40 years of no-tillage in Vertisols of the Queensland region. Crop residue and N fertilizer interactively increased SOC and total N stocks in 0-0.1m depth and cumulative stocks at 0-0.2 m and 0-0.3 m depth. It was evident that crop residue retention increased SOC and soil total N only when N fertilizers was applied.

In contrast to CA, conventional cultivation generally results in loss of soil C and nitrogen. However, CA has proven potential in converting many soils from sources to sinks of atmospheric C, sequestering carbon in soil as organic matter. In general, soil carbon sequestration during the first decade of adoption of best conservation agricultural practices is 1.8 tons CO_2 per hectare per year. On 5 billion hectares of agricultural land, this could represent one-third of the current annual global emission of CO_2 from the burning of fossil fuels (FAO, 2008). Lal *et al.* (1998) estimated that widespread adoption of conservation tillage on some 400 million ha of crop land by the year 2020 may lead to total C sequestration of 1500 to 4900 Mg.

Crop residue burning is a quick, labour-saving practice to get rid of residue that is viewed as a nuisance by farmers (Fig 2). However residue-burning has several adverse environmental and ecological impacts. The burning of dead plant material adds a considerable amount of CO_2 and particulate matter to the atmosphere and can reduce the return of much needed C and other nutrients to soil (Prasad *et al.*, 1999). Lack of soil surface cover due to burning or removal of the crop residues increases the loss of the mineral and organic matter–rich surface layer in run-off. In comparison to burning, residue retention increases soil carbon and nitrogen stocks, provides organic matter necessary for soil macro-aggregate formation and fosters cellulose–decomposing fungi and thereby carbon cycling.

Crop residues returned to the soil, on the other hand, help increase SOM levels, which facilitate greater infiltration and store greater water in the soil profile. Crop residues provide substrate to soil organisms which help in recycling of plant nutrients. Leaving crop residue on the field is another practice which could have an important impact on the global carbon cycle. The global annual production of crop residue is estimated to be about 3.4 billion Mg. If 15% of C contained in the residue can be converted to passive soil organic carbon (SOC) fraction, this may lead to C sequestration at the rate of 0.2×10^{15} g/yr (Lal, 1997). Similarly, restoring presently degraded soils, estimated at about 2 billion ha, and increasing SOC content by 0.01% /yr may lead to C sequestration at the rate of 3.0 Pg C/yr.

Fig. 2: Crop Residue Burnt by Farmers (*See colour version on page 380*)

Systems, based on high crop residue addition and no-tillage, tend to turn the soil into a net sink of carbon (Bot *et al.*, 2001).

In the USA, the total loss of carbon, from a plot of ploughed under wheat residues, was up to five times higher than from plots not ploughed, and the loss of carbon was equal to the quantity of carbon in the wheat residues which had remained in the field from the previous crop (CTIC, 1996a). Conservation tillage adoption on three-quarters of the land of USA would half respired CO_2 as compared to 1993, representing an accrual of almost 400 million tons (Bot *et al.*, 2001). Net soil C stock changes for US agricultural soils between 1982 and 1997 due to shifts towards CA are estimated to amount to 21.2 MMT C/year (Eve *et al.*, 2002). At an average rate of 0.51 t/ha/year, Brazil is sequestering about 12 million t of carbon on 23.6 million ha of no-tillage adoption. In Canada, at a CO_2 sequestration rate of 0.74 t/ha farmers practicing no-till would be sequestering about 9 million tons of CO_2 from the atmosphere each year, while at the same time enriching the soil in carbon (Bot *et al.*, 2001). It is estimated that wide dissemination of CA could offset as much as 16% of worldwide fossil fuel emissions (CTIC, 1996b).

A study conducted at IISS, Bhopal also reveals the effect of tillage systems on SOC was found to be significant only in the surface layer (0-15cm) and higher SOC values were observed under reduced tillage (RT) as compared to conventional tillage (CT) after three years of crop cycles (Fig 3). Furthermore, reduction in tillage operations coupled with residue retention helps in maintaining soil organic carbon (Somasundaram *et al.*, 2014b; Subba Rao and Somasundaram, 2013). Similarly, Bhattacharyya *et al.* (2012) reported that reduction in tillage intensity led to a significantly larger SOC accumulation in the surface soil layer (0–5 cm), but not in the 5- to 15-cm soil layer after 6 yr of cropping in a sandy clay loam soil (Typic Haplaquept) near Almora, India. The year-round NT management practice

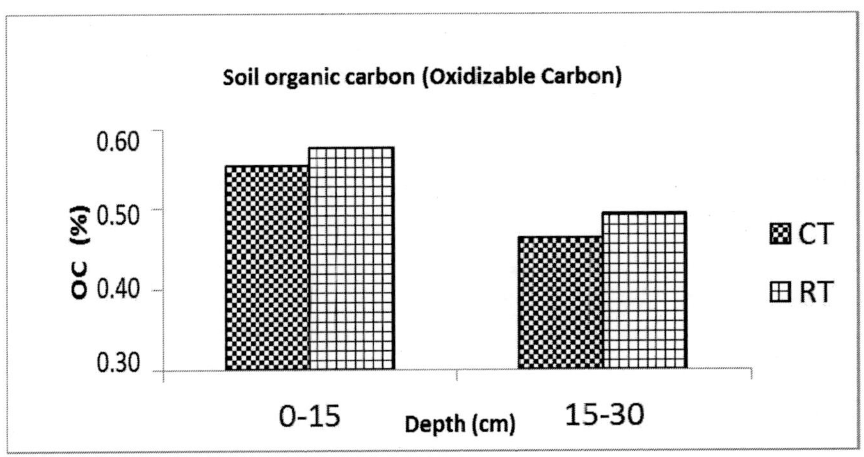

Fig. 3: Effect of different tillage on soil organic carbon

was very effective for SOC sequestration in a rainfed lentil–finger millet rotation system, where a net gain in SOC storage was about 0.37 Mg/ha/yr in the 0- to 15-cm soil layer.

Various Facets of Soil Carbon Sequestration under CA

The various components of CA like no-tillage, crop rotation and residue retention play a significant role in SOC sequestration which is discussed below.

Effect of Tillage on SOC

Reduced-tillage (RT) or no-tillage (NT) as a CA component may increase soil C compared with CT but these increases are often confined to near-surface layers (<10 cm). At deeper depths, soil C in CA maybe equal or even lower compared with CT. The potential of CA for storing C is not conclusive. It depends on antecedent soil C concentration, cropping system, management duration, soil texture, slope and climate (Govaerts *et al.*, 2009; Luo *et al.*, 2010). More data are available from temperate (i.e., USA) than from tropical regions. Across 100 comparisons, soil C stock in NT was lower in 7 cases, higher in 54 cases and equal in 39 cases compared with CT in the 0- to 30-cm soil depth after 5 years or more of NT implementation (Govaerts *et al.*, 2009). These studies were primarily from USA and Canada and some from Brazil, Mexico, Spain, Switzerland, Australia, and China. A meta-analysis found increased soil C in the topsoil (0-10 cm) on conversion of CT to NT but no significant difference over the soil profile upto 40 cm due to a redistribution of C in the profile (Luo *et al.*, 2010; Palm *et al.*, 2014). There are concerns that a majority of studies published prior to 2009 reported soil C on a fixed depth basis rather than equivalent soil mass basis (ESM) which can over estimate soil C sequestration for the treatment with higher bulk density, which is usually the NT treatment. Ellert and Bettany (1995) illustrated the importance of reporting soil C on an equivalent soil mass basis, rather than fixed soil depth (Fig 4). There is now general agreement that soil C stocks should be compared using ESM but many studies still do not use it because of methodological difficulties. The result of using fixed depth rather than ESM is that reports of changes in soil C stocks are confounded by management-induced changes in bulk density rather than outright changes in stock.

Conservation agricultural practices of soil surface retention of crop residues combined with reduced or no-tillage has the potential of conserving C and N in the agro-ecosystems as SOC and N. However, such sequestration of SOC and N may not be uniformly distributed in the soil profile. Long-term CA systems typically develop a highly stratified vertical distribution of SOC with time. It has been reported that significant increases in SOC with CA systems are usually confined to the surface 30 cm and even more typically to the surface 15 cm soil. Chang *et al.* (2012) through simulation modeling showed that NT systems can be regarded

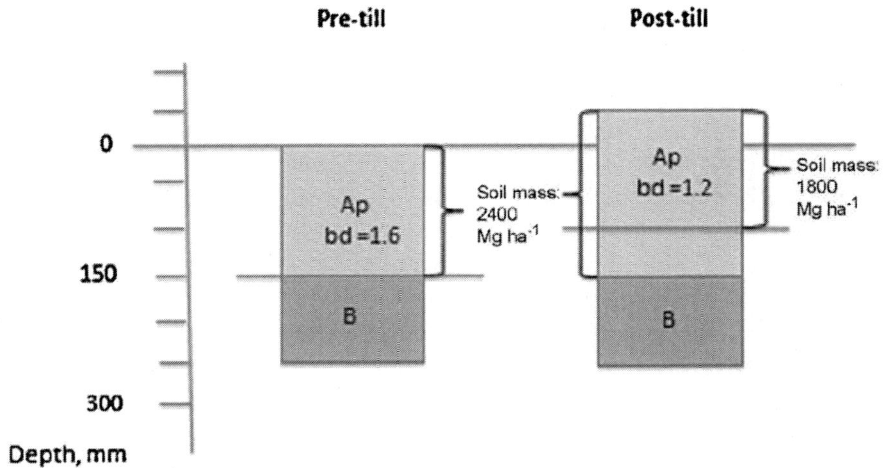

Fig. 4: Soil sampling between tilled and no tilled to a fixed depth or an equivalent soil mass basis (Adopted from Ellert and Bettany 1995; Palm *et al.* 2014).

as a sure management practices to enhance C sequestration, however, such sequestration dynamics are associated with vertical variability of soil texture, bulk density etc. NT generally, results in an increase in SOC in the 0 - 10 cm sampling layer, however, the effect of C sequestration may not be significantly different for the lower sample layers between CT and NT (Ramnarine, 2010). With continuous CA practices, the accumulation becomes more obvious in the surface layer as is evident from the increasing ratio of SOC in 0-15/15-30 cm. Franzluebbers (2010), on the basis of large numbers of reported information concluded that the ratio of SOC at 0-15 cm to that at 15-30 cm ranged from 2.4 at initiation to 3.1 at the end of 5 yr to 3.6 at the end of 12 yr. The stratification ratio of SOC was 0.9 and 3.8 under CT and NT respectively, at the end of 1 yr, 1.1 and 3.8 at the end of 2 yr, and 1.2 and 3.9 at the end of 3 yr in a tillage x cropping system experiment (Franzluebbers and Stuedmann 2008). Similar values for stratification ratio of SOC (0-15/15-30cm) were reported for four measurements during 30 yr of experimentation with moldboard plowing (1.7±0.2), NT management (3.4±0.1) and grass sod (4.1±0.2) (Diaz-Zorita and Grove, 2002). In a survey of 5 Southeastern states of United States, the stratification ratio of SOC (0-15/12.5-20 cm) averaged 1.4 with CT and reached a plateau of 2.8 within 10 yr of conservation tillage cropland, and a plateau of 4.2 with perennial pasture. The ratio is time dependent and increases with years of CA practices. Spargo *et al.* (2008) found that in three different soil types, the stratification ratio of SOC (0-2.5/7.5-15 cm) was linearly related to the number of years of continuous NT, initially 1.5 following CT and increasing to 3.6 with 14 yr of NT. Causarano *et al.* (2008) showed that the majority of the C stored under CA management in the Ultisols of Southeastern United States occurred within the surface 5 cm depth.

Although No-till, has been found to promote SOM stratification in top soil layers compared to CT (Thomas *et al.*, 2007b), when 0 to 0.3 m or deeper soil layers are considered, soil under both NT and CT practices usually contains similar amounts of SOC (Baker *et al.*, 2007). This appears to occur especially in subtropical and tropical regions (Alvarez, 2005), although only limited information is available on soil changes under long term CT and NT practices in these regions (Dalal, 1992; Mulvaney *et al.*, 2009). Other studies also indicated that SOC and TN stocks under NT were more stratified with greater accumulation near the soil surface (Gal *et al.*, 2007; Yong *et al.*, 2008). On the basis of published literature of 23 investigations with 47 site comparisons, Angers and Eriksem-Hamel (2008) showed that in most of the cases, NT had significantly greater SOC storage than full-inversion tillage (FIT) in the surface layer, whereas at 21 to 25 cm and 26 to 35 cm layers, the average SOC stocks were significantly greater under FIT than NT. Thus continuous NT practices may lead to profile stratification of SOC and TN pools, while the degree of stratification with depth expressed as a ratio, could indicate soil ecosystem function (Franzluebbers, 2002).

Both tillage and fertilizer management influences the SOC storage profile (Poirier *et al.*, 2009). In a long-term study on the interactive effect of tillage and mineral fertilizer management on soil C profile, Poirier *et al.* (2009) observed that NT enhanced the SOC contents in the soil surface layer, but moldboard plow resulted in greater SOC content near the bottom of the plow layer. Nitrogen and P fertilization with moldboard plow increased the estimated crop C inputs to the soil. Highest SOC stocks of 0 - 20 cm soil layer were observed in the NT treatment with highest N rates.

Implications of SOC Stratification

Such organic C enriched surface soil fosters productivity, regulates terrestrial water flow, cycles plant nutrients through diverse biological activity, and creates biologically active and diverse soil micro-organisms. This reconfigures the soil matrix into relatively stable structures with permanent bio-pores (channels), a process that is important in achieving high water infiltration. Stratification of SOC through CA systems proves a boon for infiltration of water in to the soil. For example, with doubling of SOC uniformly throughout the top 12 cm depth, infiltration was increased by 27% in a Typic Kanhaludult, but increased more than 200% when SOC was concentrated in the surface 3 cm of the soil (Franzluebbers, 2010). Therefore, SOC stratification is integrally linked to abatement of runoff and soil erosion. Rhoton *et al.* (2002) reported that soil loss was inversely related to the calculated SOC stratification ratio in field studies with small-plot rainfall simulations (4.6 – 5.5 m^2). A number of water-catchment and other field-plot investigations have also documented positive influences of conservation tillage and pasture management on soil stabilization and avoidance of water and nutrient run-off (Franzluebbers, 2008).

Effect of Crop Rotations on SOC

Crop rotations have less effect on soil C than tillage (West and Post, 2002). Crop rotations can affect soil C by increased biomass production and C inputs from the different crops in the system or through altering pest cycles, diversifying rooting patterns and rooting depth. Experimental designs have confounded crop rotations with tillage making it difficult to make conclusions about the effects of rotations alone. Crop rotation effects on soil C are often mixed (Corsi *et al.* 2012). High-residue producing crops may sequester more C than crops with low residue input. Intensification of cropping systems such as increased number of crops per year, double cropping, and addition of cover crops can result in increased soil C storage under NT (West and Post, 2002; Luo *et al.*, 2010). West and Post (2002) found interactions with crop rotations and tillage practice; in general, crop rotations sequestered more C than monocultures on conversion to NT, though there were notable exceptions with corn-soybean rotations with less soil C than mono-culture maize. It is generally recognized that the differential effects of rotations on soil C are simply related to the amounts of above and below ground biomass (residues and roots) produced and retained in the system (West and Post 2002). In Brazil, Boddey *et al.* (2010) attributed higher soil C storage in NT than CT to the inclusion of legume intercrops or cover crops in the rotations, and not due simply to higher production and residue inputs. They indicated slower decomposition of residues and lower mineral N in NT compared to CT resulting in higher root:shoot ratios and belowground C input with NT.

Effect of Residue Retention on SOC

Retention of crop residues is an essential component of CA for increasing or maintaining soil C. Factors that increase crop yields will increase the amount of residue available and potentially soil C storage. Fertility management may be the single most crucial factor to increase residue production and ultimately increase soil C storage, whether the system is no-tillage or conventional tillage or incorporates crop rotations (Giller *et al.*, 2009). This will be vital for increasing C inputs and soil C in low input-low productivity farming systems found in much of Sub Saharan Africa and parts of South Asia (Paul *et al.*, 2013; Theirfelder *et al.*, 2013b; Dube *et al.*, 2012; Ghimire *et al.*, 2012; Hillier *et al.*, 2012). As a rough comparison using average regional yields (Hazell and Wood, 2008) and a harvest index of 50% for maize, farms in the US generate 10 Mg ha^{-1} of maize residue while 3 and 1-2 Mg ha^{-1} are produced in South Asia and Sub Saharan Africa, respectively. A study by Paul *et al.* (2013) in Kenya illustrates the point that limited amounts of residue input may have little or no effect on increasing soil C. They found no difference in soil C concentration between CT and RT when both tillage systems received 4 Mg $ha^{-1}$of residue for six years. A similar lack of response to 4 Mg $ha^{-1}$of residue after four years of application was also seen in a subtropical area of Nepal (Ghimire *et al.*, 2012).

Conservation Agriculture and Nutrient Availability

Tillage, residue management and crop rotation have a significant impact on nutrient distribution and transformation in soils, usually related to the effects of conservation agriculture on SOC contents. Similar to the findings on SOC, distribution of nutrients in a soil under NT is different to that in tilled soil. Increased stratification of nutrients is generally observed, with enhanced conservation and availability (Franzluebbers and Hons, 1996). The altered nutrient availability under NT tillage compared to conventional tillage may be due to surface placement of crop residues in comparison with incorporation of crop residues with tillage (Ismail *et al.*, 1994). Slower decomposition of surface placed residues (Kushwaha *et al.*, 2000) may prevent rapid leaching of nutrients through the soil profile, which is more likely when residues are incorporated into the soil. However, the possible development of more continuous pores between the surface and the subsurface under no tillage may lead to more rapid passage of soluble nutrients deeper into the soil profile than when soil is tilled (Franzluebbers and Hons, 1996). Furthermore, the response of soil chemical fertility to tillage is site-specific and depends on soil type, cropping systems, climate, fertilizer application and management practices (Rahman *et al.* 2008).

The density of crop roots is usually greater near the soil surface under no tillage compared to conventional tillage (Qin *et al.*, 2004). This may be common under no tillage, as in the study of Mackay *et al.* (1987) where a much greater proportion of nutrients was taken up from near the soil surface under no tillage than under tilled culture, illustrated by a significantly higher P uptake from the 0–7.5 cm soil layer under no tillage than under conventional tillage. However, research on nutrient uptake by Hulugalle and Entwistle (1997) revealed that nutrient concentrations in plant tissues were not significantly affected by tillage or crop combinations. Although there are reports of straw burning increasing nutrient availability (Du Preez *et al.* 2001), burning crop residues is not considered sustainable given the well documented negative effects on physical soil quality, especially when it is combined with reduced tillage (Limon-Ortega *et al.*, 2002). Mohamed *et al.* (2007) observed only short-term effects of burning on N, P and Mg availability. As a consequence of the short-term increased nutrient availability limited nutrient uptake by plants after burning, leaching of N, Ca, K, and Mg increased significantly after burning (Mohamed *et al.*, 2007).

Nitrogen Availability

The presence of soil mineral N available for plant uptake is dependent on the rate of C mineralization. The literature concerning the impact of reduced tillage with residue retention on N mineralization is inconclusive. No tillage is generally associated with a lower N availability because of greater immobilization by the residues left on the soil surface (Bradford and Peterson, 2000). Some authors suggest that the net immobilization phase when no tillage is adopted, is transitory, and that in the long run, the higher, but temporary immobilization of N in no tillage systems reduces the opportunity for leaching and denitrification losses of mineral N (Follet and Schimel, 1989). According to Schoenau and Campbell (1996), a greater immobilization in conservation agriculture can enhance the conservation of soil and fertilizer N in the long run, with higher initial N fertilizer requirements decreasing over time because of reduced losses by erosion and the build-up of a larger pool of readily mineralizable organic N.

Tillage increases aggregate disruption, making organic matter more accessible to soil microorganisms and increasing mineral N release from active and physically protected N pools (Six *et al.*, 2002). Lichter *et al.* (2008) reported that permanent raised beds with residue retention resulted in more stable macro aggregates and increased protection of C and N in the micro aggregates within the macro aggregates compared to conventionally tilled raised beds. This increases susceptibility to leaching or denitrification if no growing crop is able to take advantage of these nutrients at the time of their release. Randall and Iragavarapu (1995) reported about 5% higher NO_3 -N losses with conventional tillage compared to no tillage. Jowkin and Schoenau (1998) reported that N availability was not greatly affected in the initial years after switching to NT in the brown soil zone in Canada. Larney *et al.* (1997) reported that, after eight years of tillage treatments, the content of N available for mineralization was greater in no-tilled soils than in conventionally tilled soil under continuous spring wheat. Wienhold and Halvorson (1999) found that nitrogen mineralization generally increased in the 0-5 cm soil layer, as the intensity of tillage decreased.

Govaerts *et al.* (2006) found after 26 cropping seasons in a high-yielding, high input irrigated production system that the N mineralization rate was higher in permanent raised beds with residue retention than in conventionally tilled raised beds with all residues incorporated, and also that N mineralization rate increased with increasing rate of inorganic N fertilizer application. The tillage system determines the placement of residues. Conventional tillage implies incorporation of crop residues while residues are left on the soil surface in the case of no tillage. These differences in the placement of residues contribute to the effect of tillage on N dynamics. Kushwaha *et al.* (2000) reported that incorporated crop residues decompose 1.5 times faster than surface placed residues. However, also the type of residues and the interactions with N management practices determine C and N mineralization.

Phosphorus Availability

Numerous studies have reported higher extractable phosphorus (P) levels in no tillage than in tilled soil, largely due to reduced mixing of the fertilizer P with the soil, leading to lower P-fixation. This is a benefit when P is a limiting nutrient, but may be a threat when P is an environmental problem because of the possibility of soluble P losses in runoff water (Duiker and Beegle, 2006). After 20 years of no tillage, extractable P was 42% greater at 0-5 cm, but 8-18% lower at 5-30 cm depth compared with conventional tillage in a silt loam (Ismail *et al.*, 1994). Also Matowo *et al.* (1999) found higher extractable P levels in no tillage compared to tilled soil in the topsoil. Accumulation of P at the surface of continuous no tillage is commonly observed. Concentrations of P were higher in the surface layers of all tillage systems as compared to deeper layers, but most strikingly in no tillage (Duiker and Beegle, 2006). When fertilizer P is applied on the soil surface, a part of P will be directly fixed by soil particles. When P is banded as a starter application below the soil surface, authors ascribed P stratification partly to recycled P by plants (Duiker and Beegle, 2006). They have also suggested that there may be less need for P starter fertilizer in long-term no tillage due to high available P levels in the topsoil where the seed is placed. Deeper placement of P in No tillage may be profitable if the surface soil dries out frequently during the growing season as suggested by Mackay *et al.* (1987). In that case, injected P may be more available to the crop. However, if mulch is present on the soil surface in no tillage the surface soil is likely to be moister than conventionally tilled soils and there will probably be no need for deep P placement, especially in humid areas.

Potassium Availability

No tillage conserves and increases availability of nutrients, such as potassium (K), near the soil surface where crop roots proliferate (Franzluebbers and Hons 1996). According to Govaerts *et al.* (2007), permanent raised beds had a concentration of K 1.65 times and 1.43 times higher in the 0-5 cm and 5-20 cm layer, respectively, than conventionally tilled raised beds, both with crop residue retention. In both tillage systems, K accumulated in the 0-5 cm layer, but this was more accentuated in permanent than in conventionally tilled raised beds. Other studies have found higher extractable K levels at the soil surface as tillage intensity decreases (Lal *et al.*, 1990). Du Preez *et al.* (2001) observed increased levels of K in no tillage compared to conventional tillage, but this effect declined with depth. Some authors have observed surface accumulation of available K irrespective of tillage practice (Duiker and Beegle, 2006). Follett and Peterson (1988) observed either higher or similar extractable K levels in no tillage compared to mould board tillage, while Roldan *et al.* (2007) found no effect of tillage or depth on available K concentrations.

Conclusions

Conservation agriculture (CA) practices improve soil aggregation compared to conventional tillage systems and No tillage without retention of sufficient crop residues in a wide variety of soils and agro-ecological conditions. It is apparent that a combination of reduced tillage with crop residue retention increases the SOC in the topsoil. Moreover, C-cycling is influenced by (CA), and similarly the N cycle is also altered. Adoption of (CA) systems with crop residue retention may result initially in N immobilization. However, rather than reducing N availability, CA may stimulate a gradual release of N in the long run and can reduce the susceptibility to leaching or denitrification, when no growing crop is able to take advantage of the nutrients at the time of their release. Also crop diversification, an important component of conservation agriculture, is an important strategy to govern N availability through rational sequences of crops with different C/N ratios. Tillage, residue management and crop rotation have a significant impact on macro and micro nutrients distribution and transformation in soils. The altered nutrient availability may be due to surface placement of crop residues in comparison with incorporation of crop residues with tillage. Conservation agriculture increases availability of nutrients near the soil surface where crop roots proliferate. Slower decomposition of surface placed residues prevents rapid leaching of nutrients through the soil profile. The response of soil chemical fertility to tillage is site-specific and depends on soil type, cropping systems, climate, fertilizer application and management practices. However, in general nutrient availability is related to the effects of conservation agriculture on SOC contents. The CEC and nutrient availability increase in the topsoil. Numerous studies have also reported higher extractable P levels in No tillage than in tilled soil largely due to reduced mixing of the fertilizer P with the soil, leading to lower P-fixation.

References

Abrol, I.P. and Sanger, S. (2006). Sustaining Indian Agriculture- Conservation Agriculture the way forward. *Current Science*, 91 (8): 1020-1024.

Alvarez, R. (2005). A review if nitrogen fertilizer and conservation tillage effects on soil carbon storage. *Soil Use and Management,* 21:38-52.

Angers, D. A. and Eriksen-Hamel, N. S. (2008). Full-Inversion Tillage and Organic Carbon Distribution in Soil Profiles: A Meta-Analysis. *Soil Science Society of America Journal,* 72 (5): 1370-1374. DOI: 10.2136/sssaj2007.0342

Baker, J.M., Ochsner, T.E., Venterea, R.T., and Griffis, T.J. (2007). Tillage and soil carbon sequestration - What do we really know? *Agriculture, Ecosystem and Environment,* 118: 1-5.

Bhattacharyya, R., Tuti, M.D., Kundu, S., Bisht, J.K. and Bhatt, J.C. (2012). Conservation Tillage Impacts on soil aggregation and carbon pools in a sandy clay loam soil of the Indian Himalayas. *Soil Science Society of America Journal,* 76(2): 617-627.

Blanco-Canqui, H. and Lal, R. (2007). Impacts of long-term wheat straw management on soil hydraulic properties under no-tillage. *Soil Science Society of America Journal,* 71: 1166-1173.

Boddey, R.M., Jantalia, A.C.P.J., Conceicao, P.C., Zanatta, J.A., Bayer, C., Mielniczuk, J.,Dieckow, J., Dos Santos, H.P., Denardin, J.E., Aita, C., Giacomini, S.J., Alves, B.J. and Urquiaga, S. (2010). Carbon accumulation at depth in Ferralsols under zero-till subtropical agriculture. *Global Change Biology,* 16: 784–795.

Bot, A.J., Amado, T.J.C., Mielniczuk, J. and Benites, J. (2001). Conservation agriculture as a tool to reduce emission of greenhouse gasses. I World Congress on Conservation Agriculture Madrid, 1-5 October, 2001.

Bradford, J.M. and Peterson, G.A. (2000). Conservation tillage. In *Handbook of soil science*, ed. M. E. Sumner, G247-G269. Boca Raton, FL, USA: CRC Press.

Causarano, H.J., Franzluebbers, Shaw, J.N., Reeves, D.W., Raper, R.L., and Wood, C.W. (2008). Soil organic carbon fractions and aggregation in the Southern Piedmont and Coastal Plain. *Soil Science Society of America Journal,* 72: 221-230.

Chang, K.H., Bartlett, P., and Wagner-riddle, C. (2012). Using DayCENT to simulate carbon dynamics in conventional and no-till agriculture. *Soil Science Society of America Journal,* 77: 941-950.

Cole, C.V., Duxbury, J., Freney, J., Heinemeyer, O., Minami, K., Mosier, A., Paustian, K., Rosenberg, N., Sampson, N., Sauerbeck, D., Zhao, Q., (1997). Global estimates of potential mitigation of greenhouse gas emissions by agriculture. *Nutrient Cycling in Agroecosystem,* 49: 221–228.

Corsi, S., Friedrich, T., Kassam, A., Pisante, M., Sà, J. D. M., (2012). Soil organic carbon accumulation and greenhouse gas emission reductions from conservation agriculture: a literature review, In: Corsi, S., Friedrich, T., Kassam, A., Pisante, M., Sà, J. D. M., (Eds.), Soil organic carbon accumulation and greenhouse gas emission reductions from conservation agriculture: a literature review, *Integrated Crop Management* Vol.16.

CTIC. Conservation Technology Information Centre. (1996a). Facilitating Conservation Farming Practices and Enhancing Environmental Sustainability with Agricultural Biotechnology 1 - 2 (ht tp://www. ctic .purdue . edu/media/pdf/Biotech Executive_Summary.pdf).

CTIC. Conservation Technology Information Centre (1996b). 17th Annual Crop Residue Management Survey Report. West Lafayette, In: Conservation Technology Information Center.

Dalal, R.C. (1992). Long-term trends in total nitrogen of Vertosols subjected to zero-tillage, nitrogen application and stubble retention. *Australian Journal of Soil Research,* 24: 281-291.

Dalal, R.C., Allen, D.E., Wang, W.J., Reeves, S. and Gibson, I. (2011). Organic carbon and total nitrogen stocks in a Vertisols following 40 years of no-tillage, crop residue retention and nitrogen fertilization. *Soil Tilland & Research,* 112:133-139.

Diaz-zorita, M., and Grove, J.H. (2002). Duration of tillage management affects carbon and phosphorus stratification in phosphatic Paleudalfs. *Soil Tillage & Research,* 66:165-174.

Dolan, M.S., Clapp, C.E., Allmaras, R.R., Baker, J.M., and Molina, J.A.E. (2006). Soil organic carbon and nitrogen in a Minnesota soil as related to tillage, residue and nitrogen management. *Soil Tillage & Research,* 89:221-231.

Dube, E., Chiduza, C. and Muchaonyerwa, P. (2012). Conservation agriculture effects on soil organic matter on a Haplic Cambisol after four years of maize–oat and maize–grazing vetch rotations in South Africa. *Soil Tillage & Research,* 123: 21–28.

Du Preez, C.C., Steyn, J.T., and Kotze, E. (2001). Long-term effects of wheat residue management on some fertility indicators of a semi-arid Plinthosol. *Soil Tillage & Research,* 63:25-33.

Duiker, S.W. and Beegle, D.B. (2006). Soil fertility distributions in long-term no-till, chisel/disk and moldboard plow/disk systems *Soil Tillage & Research,* 88:30-41.

Ellert, B.H. and Bettany, J.R. (1995). Calculation of organic matter and nutrients stored in soils under contrasting management regimes. *Canadian Journal of Soil Science,* 75:529-538.

FAO, (2008). Investing in Sustainable Agricultural Intensification: The Role of Conservation Agriculture – A Framework for Action' FAO Rome, August 2008 (available at www.fao.org/ag/ca/).

Follett, R.F. and Peterson, G.A. (1988). Surface Soil Nutrient Distribution As Affected by Wheat-Fallow Tillage Systems. *Soil Science Society of America Journal,* 52:141-147.

Franzluebbers, A.J. (2002). Soil organic matter stratification ratio as an indicator of soil quality. *Soil & Tillage Research,* 66:95-106.

Franzluebbers, A.J. (2010). Achieving soil organic carbon sequestration with conservation agricultural systems in the Southeastern United States. *Soil Science Society of America Journal,* 74:347-357.

Franzluebbers, A.J. and Hons, F.M. (1996). Soil-profile distribution of primary and secondary plant available nutrients under conventional and no tillage. *Soil & Tillage Research,* 39:229-239.

Franzluebbers, A.J., and Stuedemann, J.A. (2008). Early response of soil organic fractions to tillage and integrated crop-livestock production. *Soil Science Society of America Journal,* 72: 613-625.

Friedrich , T., Derpsch,R. and Kassam, A. (2012). Overview of the Global Spread of Conservation Agriculture, http://factsreports.revues.org/1941 Published 12 September 2012.

Friedrich, T. and Kassam, A.H. (2009). Adoption of Conservation Agriculture Technologies: Constraints and Opportunities. Invited paper, IV World Congress on Conservation Agriculture, 4-7 February (2009), New Delhi, India.

Gal, A., Vyn, T.J., Micheli, E., and McFree, W.W.(2007). Soil carbon and nitrogen accumulation with long-term no-till verses moldboard plowing over estimated with tilled-zone sampling depths. *Soil & Tillage Research,* 96: 42-51.

Giller, K.E., Witter, E., Corbeels, M. and Tittonell, P., (2009). Conservation agriculture and smallholder farming in Africa: The heretics' view. *Field Crop Research,* 114: 23–34.

Ghimire, R., Adhikari, K.R., Shah, S.C. and Dahal, K.R., (2012). Soil organic carbon sequestration as affected by tillage, crop residue, and nitrogen application in rice-wheat rotation system. *Paddy Water Environment,* 10: 95–102.

Goddard, T., Zoebisch, M., Gan, Y., Ellis, W., Watson, A. And Sombatpanit, S. (2007) (Eds.). No-Till Farming Systems. WASWC Special Publication No. 3, Bangkok, 544 pp.

Govaerts, B., Sayre, K.D., Ceballos-Ramirez, J.M., Luna-Guido, M.L., Limon-Ortega, A., Deckers, J., and Dendooven, L. (2006). Conventionally tilled and permanent raised beds with different crop residue management: Effects on soil C and N dynamics. *Plant and Soil,* 280: 143–155.

Govaerts, B., Sayre, K.D., Lichter, K., Dendooven, L., and Deckers, J. (2007). Influence of permanent raised bed planting and residue management on physical and chemical soil quality in rain fed maize/wheat systems. *Plant and Soil,* 291: 39-54.

Govaerts, B., Verhulst, N., Sayre, K.D., Dixon, J., and Dendooven, L. (2009). Conservation Agriculture and Soil Carbon Sequestration; Between Myth and Farmer Reality. *Critical Review in Plant Sciences,* 28: 97–122.

Hazell, P. and Wood, S., (2008). Drivers of changes in global agriculture. *Philosopical T. R. Soc. B.,* 363: 495–515.

Hillier, J., Brentrup, F., Wattenbach, M., Walter, C., Garcia-Suarez, T., Mila-i-Canals, L. and Smith, P., (2012). Which cropland greenhouse gas mitigation options give the greatest benefits in different world regions? Climate and soil-specific predictions from integrated empirical models. *Global Change Biology,* 18: 1880–1894.

Hulugalle, N.R. and Entwistle, P. (1997). Soil properties, nutrient uptake and crop growth in an irrigated Vertisol after nine years of minimum tillage. *Soil & Tillage Research,* 42:15-32.

Ismail, I., Blevins, R.L., and Frye, W.W. (1994). Long-Term No-Tillage Effects on Soil Properties and Continuous Corn Yields. *Soil Science Society of America Journal,* 58:193-198.

Jantalia, C.P., Resck, D.V.S., Alves, B.J.R., Zotarelli, L., Urquiaga, S., and Boddey, R.M. (2007). Tillage effect on C stocks of a clayey Oxisol under a soybean-based crop rotation in the Brazilian Cerrado region. *Soil & Tillage Research,* 95:97-109.

Jowkin, V. and Schoenau, J.J. (1998). Impact of tillage and landscape position on nitrogen availability and yield of spring wheat in the Brown soil zone in southwestern Saskatchewan. *Canadian Journal of Soil Science,* 78:563-572.

Kassam, A.H. and Friedrich, T. (2009). Perspectives on Nutrient Managment in Conservation Agriculture. Invited paper, 4th World Congress on Conservation Agriculture, 4-7 February 2009, New Delhi, India.

Kushwaha, C.P., Tripathi, S.K., and Singh, K.P. (2000). Variations in soil microbial biomass and N availability due to residue and tillage management in a dryland rice agro ecosystem. *Soil & Tillage Research,* 56:153-166.

Lal, R. (1997). Residue management, conservation tillage and soil restoration for mitigating greenhouse effect by CO -enrichment. *Soil & Tillage Research,* 43: 81-107.

Lal, R. (1998). Soil quality and sustainability. In: Methods for Assessment of Soil Degradation. (Eds: Lal, R.; Blum, W. H.; Valentine, C.; Stewart, B. A.) CRC Press, New York, 17-30.

Lal, R., Logan, T.J., and Fausey, N.R. (1990). Long-Term Tillage Effects on a Mollic Ochraqualf in North-West Ohio.3. Soil Nutrient Profile. *Soil & Tillage Research,* 15:371-382.

Landers, J. (2007). Tropical Crop-Livestock Systems in Conservation Agriculture: The Brazilian Experience. Integrated Crop Management Vol. 5. FAO, Rome.

Larney, F.J., Bremer, E., Janzen, H.H., Johnston, A.M., and Lindwall, C.W. (1997). Changes in total, mineralizable and light fraction soil organic matter with cropping and tillage intensities in semiarid southern Alberta, Canada. *Soil & Tillage Research,* 42:229-240.

Lichter, K., Govaerts, B., Six, J., Sayre, K.D., Deckers, J., and Dendooven, L. (2008). Aggregation and C and N contents of soil organic matter fractions in a permanent raised-bed planting system in the Highlands of Central Mexico. *Plant and Soil,* 305:237-252.

Limon-Ortega, A., Sayre, K.D., Drijber, R.A., and Francis, C.A. (2002). Soil attributes in a furrow-irrigated bed planting system in northwest Mexico. *Soil & Tillage Research,* 63:123-132.

Luo, Z., Wang, E., Sun, O.J., (2010). Can no-tillage stimulate carbon sequestration in agricultural soils? A meta-analysis of paired experiments. *Agriculture, Ecosystem and Environment,* 139, 224–231.

Mackay, A.D., Kladivko, E.J., Barber, S.A., and Griffith, D.R. (1987). Phosphorus and Potassium Uptake by Corn in Conservation Tillage Systems. *Soil Science Society of America Journal,* 51:970-974.

Matowo, P.R., Pierzynski, G.M., Whitney, D., and Lamond, R.E. (1999). Soil chemical properties as influenced by tillage and nitrogen source, placement, and rates after 10 years of continuous sorghum. *Soil & Tillage Research,* 50:11-19.

Mazvimavi, K. and Twomlow, S. (2006). Conservation Farming for Agricultural Relief and Development in Zimbabwe. In: No-Till Farming Systems (Goddard, T. *et al.*, Eds.), pp. 169-175. WASWC Special Publication No. 3, Bangkok.

Mohamed, A., Hardtle, W., Jirjahn, B., Niemeyer, T., and von Oheimb, G. (2007). Effects of prescribed burning on plant available nutrients in dry heathland ecosystems. *Plant Ecology,* 189:279-289.

Mulvaney, R.L., Khan, S.A., and Ellsworth, T.R. (2009). Synthetic nitrogen fertilizers deplete soil nitrogen: A global dilemma for sustainable cereal production. *Journal of Environmental Quality,* 38:2295-2314.

Palm, C., Blanco-Canqui, H., DeClerck, F. and Gatere, L. (2014). Conservation agriculture and ecosystem services: An overview. *Agriculture, Ecosystem and Environment,* 187: 87–105.

Paul, B.K., Vanlauwe, B., Ayuke, F., Gassner, A., Hoogmoed, M., Hurisso, T.T., Koala,S., Lelei, D., Ndabamenye, T., Six, J. and Pulleman, M.M. (2013). Medium-term impactof tillage and residue management on soil aggregate stability, soil carbon, andcrop productivity. *Agriculture, Ecosystem and Environment,* 164: 14–22.

Poirier, V., Angers, D.A., Rochette, P. *et al*. (2009). Interactive effects of tillage and mineral fertilization on soil carbon profile. *Soil Science Society of America Journal,* 73: 255-261.

Prasad, R., Gangaiah, B and Aipe, K.C. (1999). Effect of crop residue management in rice-wheat cropping system on growth and yield of crops and on soil fertility. *Experimental Agriculture*, 35:427-435.

Qin, R.J., Stamp, P., and Richner, W. (2004). Impact of tillage on root systems of winter wheat. *Agronomy Journal*, 96:1523-1530.

Radford, B.J. and Thornton, C.M. (2011). Effects of 27 years of reduced tillage practices on soil properties and crop performance in the semi-arid subtropics of Australia. *International Journal of Energy, Environment and Economics*, 19: 565–588.

Rahman, M.H., Okubo, A., Sugiyama, S., and Mayland, H.F. (2008). Physical, chemical and microbiological properties of an Andisol as related to land use and tillage practice. *Soil & Tillage Research*, 101:10-19.

Ramnarine, R. (2010). Soil tillage effects on the contributions of soil and plant carbon pools to CO_2 emissions using 13C natural abundance. Ph.D. diss. University of Guelph, Ontario.

Randall, G.W. and Iragavarapu, T.K. (1995). Impact of Long-Term Tillage Systems for Continuous Corn on Nitrate Leaching to Tile Drainage. *Journal of Environmental Quality*, 24:360-366.

Roldan, A., Salinas-Garcia, J.R., Alguacil, M.M., and Caravaca, F. (2007). Soil sustainability indicators following conservation tillage practices under subtropical maize and bean crops. *Soil & Tillage Research*, 93:273-282.

Schoenau, J.J. and Campbell, C.A. (1996). Impact of crop residues on nutrient availability in conservation tillage systems. *Canadian Journal of Plant Sciences*, 76:621-626.

Six, J., Conant, R.T., Paul, E.A., and Paustian, K. (2002). Stabilization mechanisms of soil organic matter: Implications for C-saturation of soils. *Plant and Soil*, 241:155-176.

Somasundaram, J., Chaudhary, R.S. Subba Rao, A., Hati, K.M., Sinha, N.K. and Vasanda Coumar, M.. (2014a). Conservation Agriculture for carbon sequestration and sustaining soil health. New India Publications Agency, New Delhi. P. 1-528. (ISBN: 978-93-83305-32-2).

Somasundaram, J., Chaudhary, R.S., Hati, K. M., Vassanda Coumar, M., Sinha, N.K., Jha, P., Ramesh, K., Neenu, S., Blaise, D., Saha, R., Ajay, Biswas, A.K., Maheswari, M., Rao, D.L.N., Subba Rao, A and Venkateswarlu, B. (2014b). Conservation Agriculture for Enhancing Soil Health and Crop Productivity. Published by ICAR-Indian Institute of Soil Science, Bhopal in collaboration with National Initiative on Climate Resilient Agriculture (NICRA), CRIDA, Hyderabad, 64 p.

Spargo, J.T., Alley, M.M., Follett, R.F., and Wallace, J.V. (2008). Soil carbon sequestration with continuous no-till management of grain cropping systems in Virginia Coastal Plain. *Soil & Tillage Research*, 100:133-140.

Subba Rao, A and Somasundaram, J. (2013). Conservation Agriculture for carbon sequestration and sustaining soil health, In: Agriculture Year Book 2013, Published by Agriculture Today (The National Agricultural Magazine) pp: 184-188.

Thierfelder, C., Mwila, M. and Rusinamhodzi, L., (2013). Conservation agriculture in eastern and southern provinces of Zambia: Long-term effects on soil quality and maize productivity. *Soil & Tillage Research*, 126: 246–258.

Thomas, G.A, Titmarsh, G.W, Freebairn, D.M, Radford, B.J (2007b). No-tillage and conservation farming practices in grain growing areas of Queensland—a review of 40 years of development. *Australian Journal of Experimental Agriculture*, 47: 887–898. doi:10.1071/EA06204.

Thomas, G.A., Dalal, R.C., and Stanley, J. (2007a). No-till effects on organic matter, pH, cation exchange capacity and nutrient distribution in a Luvisol in the semi-arid subtropics. *Soil & Tillage Research*, 94:295-304.

West, T.O. and Post, W.M. (2002). Soil organic carbon sequestration rates by tillage and crop rotation: global data analysis. *Soil Science Society of America Journal*, 66:1930-1946.

Wienhold, B.J. and Halvorson, A.D. (1999). Nitrogen mineralization responses to cropping, tillage, and nitrogen rate in the Northern Great Plains. *Soil Science Society of America Journal,* 63:192-196.

Yang, X.M., Drury, C.F., Reynolds, W.D., and Tan, C.S. (2008). Impact of long-term and recently imposed tillage practices on the vertical distribution of soil organic carbon. *Soil & Tillage Research,* 100:120-124.

Advances in Nutrient Dynamics in Soil - Plant System, pp 291-298

Editors: R. Elanchezhian, A.K. Biswas, K. Ramesh and A.K. Patra

25

Role of Biochar in Improved Crop Performance and Climate Change Mitigation

Brij Lal Lakaria, Pramod Jha and A.K. Biswas

ICAR-Indian Institute of Soil Science, Nabi Bagh, Bhopal – 462038, India

Introduction

Use of FYM, crop residues, leaf fall, root deposition etc. has been part and parcel in agriculture. These material undergoes microbial decay over the time slowly and the major portion of it finds way back to the atmosphere. Not only that, residue burning traditionally provides a fast way to clear the agricultural field of residual biomass, facilitating further land preparation and planting. However, in addition to the loss of valuable biomass and nutrients, biomass burning leads to release of toxic gases including GHGs. In this context, biochar offers a multidimensional opportunity to transform large scale agricultural waste streams from a financial and environmental liability to valuable assets that can enhance natural rates of carbon sequestration in the soil, reduce farm waste and improve the soil quality. We are aware that the atmospheric carbon dioxide concentration has been increasing over the years due to many reasons including industrialization and inappropriate agricultural practices such as deforestation and hopefully it would continue to increase in near future. Hence, attention of scientific community is to make soil a possible sink for atmospheric CO_2. When the carbon sequestration is thought with the perspective of reducing carbon dioxide emission to atmosphere at one hand and increasing the soil health on the other, there is a need to opt for technologies that can fulfill both these objectives. There is a growing interest in the use of charcoal/ biochar/ agrichar to sequester carbon in soil and improve soil fertility (Lehmann and Joseph, 2009). Biochar is not a new commodity available to us for carbon sequestration. It has been there for long but its importance has been realized very recently.

Definition and Preparation

When biomass is burnt in absence or minimal presence of oxygen it turns into black carbonic material called biochar. The process is termed 'slow pyrolysis', whereby organic material is heated under controlled temperatures that may vary between 300-500 °C. The product so obtained has a high carbon content (>60% C) and is thus a potential tool for long-term soil carbon storage as it is highly resistant to decay. Use of biochar in agriculture is infact an ancient agricultural practice that has been found to result in additional agronomic benefits. The conditions under which biochar is formed decides its quality, hence it is considered to be a highly variable material. It is presumed to be a stable material that can sustain in soil for thousands of years (Preston and Schmidt, 2006). The half-life of biochar varies from hundred to ten thousands of years (Zimmerman, 2010). Besides its resistance to decay, it has been found to attribute to increased crop production in many cases. The global production of biochar has been estimated to be between 50 and 270 Tg yr^{-1} with about 80 per cent of it remaining as residues in the soil (Kuhlbusch *et al.,* 1998; and Suman *et al.,* 1997).

Biochar is prepared by different methods varying from traditional to sophisticated ones. For instance, the Amerindians dug trenches or pits up to 6 feet deep and fill them with the cleared biomass. The soil that is excavated to create the hole, is then placed on top and the biomass is set on fire. Because of the high moisture content and lack of oxygen in the biomass, these fires smoke for hours and sometimes for days. The biomass thus undergo pyrolysis and become biochar or black charcoal of varying quality depending upon the conditions of preparation. Biochar can also be prepared at field level by making shallow pits and filling them with feed stock and igniting them from one side while keeping passage for smoke on the other side (Fig.1). It is also sometimes obtained as bi-product from the gasifiers, however, the recovery is low. In many countries biochar is being prepared on large scale through various processes. In Indian conditions, there is an immense scope for converting millions of tonnes of crop residues, which are not used as fodder, into biochar and use the same for enriching soil carbon.

Fig. 1: ITK in biochar preparation
(*Source*: http. biocharplus.blogspot.in)
(See colour version on page 381)

Slow Pyrolysis and Changes in Biomass

The organic resources that could be used for making biochar are numerous. These materils when subjected under slow pyrolysis undergoes a change and complex substances are formed. For instance, hemicellulose gasifies at 250°C – 300°C, cellulose splits into char and volatiles between 300°C and 450°C; Lignin splits into char and volatiles between 300°C and 750°C. Volatilization cools the remaining solid, but the gases burn and generate radiant heat. The amount of ash, resident nutrients and water in the biochar depends on the parent resource used for the same. For instance, mustard stalks and maize stover would pyrolyse more easily than the woody biomass.

Fig. 2: Raw material (a) and biochar (b) from subabool (*See colour version on page 381*)

The conversion of biomass carbon to biochar carbon leads to sequestration of about 50% of the initial carbon compared to the low amounts retained after burning (3%) and biological decomposition (<10-20% after 5-10 years), thereby yielding more stable soil carbon than burning or direct land application of biomass. This efficiency of carbon conversion of biomass to biochar is highly dependent on the type of feedstock, but is not significantly affected by the pyrolysis temperature (within 350-500°C common for pyrolysis).

Since the biomass can be pyrolysed at varying temperatures and duration for obtaining biochar. The amount of material obtained after pyrolysis must not be too little so that the process remain viable with respect to carbon sequestration. There is always decline in recovery of biomass after pyrolysis with increase in temperature and duration. Fig. 3 shows that while preparing subabool biochar there was a sharp decline in the recovery of biochar from subabool feedstock with increase in temperature. At 400°C temperature the recovery of biochar varied between 30.4 and 32.3 percent respectively across all the sizes of subabool. Similarly, the biochar produced from upper half containing the secondary and tertiary branches of pigeon pea recorded slightly less recovery of biochar as compared to more woody lower part. The recovery percentage was significantly less than that observed in case of subabool.

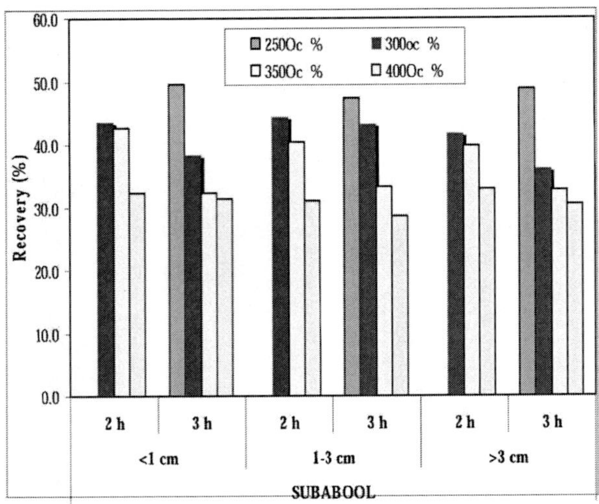

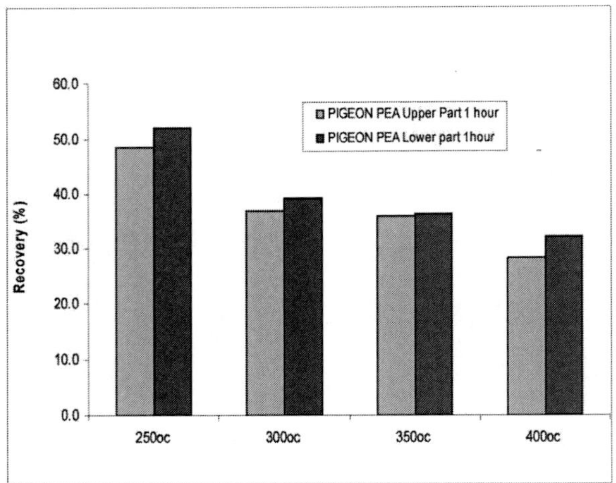

Fig. 3: Recovery of biochar from feedstock (*See colour version on page 382*)

Biochar is filled with tiny holes (Fig. 4) that hold water and provide habitat for micro-organisms. Soil amended with biochar has increased capacity to hold water and nutrients. Biochar is an excellent soil amendment for sequestering carbon (increasing SOC content) and water retention as well as providing a habitat for microbes. Biochar also adds some macro (P, K, N, Ca, Mg) and micronutrients (Cu, Zn, Fe, Mn) which are needed for sustainable agriculture. Black carbon may significantly affect nutrient retention and play a key role in a wide range of biogeochemical processes in the soil, especially for nutrient cycling. Chan *et al.* (2007) studied the influence of rate and type of biochar produced from poultry litter under different conditions on soil quality parameters. Dampster *et al.* (2012) prepared and characterized the bio-char made from trunk and large branches of *Eucalyptus marginata* at 600°C for 24 hours. They observed 75 per cent carbon and 3 g/kg total N in this biochar.

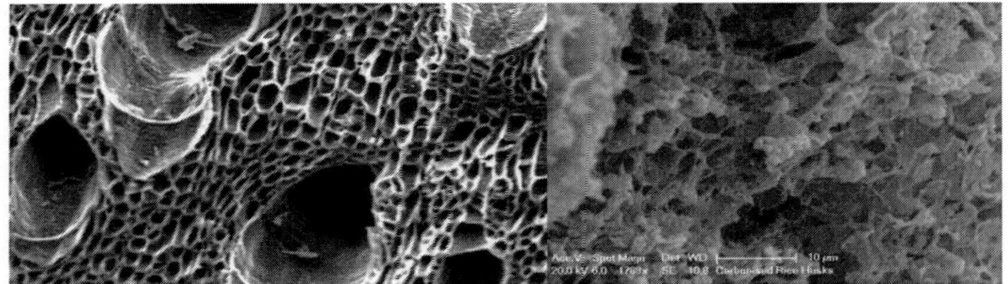

Fig. 4: Scanning electron microscopy of biochar (www.carboncommentary.com and UK Biochar Research Centre)

Biochar in Climate Change Scenario

Biochar can hold carbon in the soil for hundreds and even thousands of years. Biochar also improves soil fertility, stimulating plant growth, which then consumes more CO_2 in a feedback effect. The energy generated as part of biochar production can displace carbon positive energy from fossil fuels. Additional effects from adding biochar to soil can further reduce greenhouse gas emissions and enhance carbon storage in soil include: biochar reduces the need for fertilizer, resulting in reduced emissions from fertilizer production. It increases soil microbial life, resulting in more carbon storage in soil. It retains nitrogen and emissions of nitrous oxide (a potent greenhouse gas) may be reduced. Turning agricultural waste into biochar reduces methane (another potent greenhouse gas) generated by the natural decomposition of the waste. In addition to these advantages there are numerous benefits of sequestering biochar in soil from polical, economical, social and environmental viewpoints. Humus has this same ability to store nutrients but humus can take hundreds of years to form and there is no practical way through which we can speed up that process. Biochar on the other hand, can be made very quickly, making it like 'instant humus'. The indirect effect of biochar is that it prevents loss of nutrients from soil through leaching and improve release to plants slowly as per their need. It does not affect the soil C:N ratio. Microorganisms will release varying amounts of CO_2 depending on the availability of nitrogen in the soil. When the C/N ratio moves too far towards carbon, microorganisms will increase the amount of CO_2 they give off in their respiration in an attempt to restore their preferred C/N ratio.

Effects on Soil Properties and Crop Improvement

Our history has been well documented with practices like deforestation for fuel, cremation, celebrations in mass, slash and burn (Jhum cultivation) agriculture etc. that has already generated some amount of biochar. ***Terra preta*** was discovered in the 1950's by Dutch soil scientist Wim Sombroek in the Amazon rainforest. Terra preta still covers 10% of the Amazon Basin. Similar sites have been found in Ecuador, Peru, Benin and Liberia in West Africa. Deep in Amazonia, large tribes of

Amerindian lived and farmed some of the poorest soils on earth, Oxisols (tropical rain forest soils), for thousands of years. Their secret survival weapon was "slash and char" agriculture, turning about 50% of collected biomass into carbon soil. High fertility associated with the anthropogenic soils i.e. terra preta has been related to the high content of organic carbon in the form of char (Glaser *et al.*, 2001). The potentials might have been overlooked or unattended due to low amount and poor quality of biochar for want of any research information. In the 1970s and 80s, soil scientists started looking at these unique soils, but it was the late 1990s before they understood that it was a type of charcoal that enriched the soils. It has proved better effect on soil properties and crop yields. There is no side effect reported on micro-organisms and other fauna in soil. There are lots of positive influences reported on varying soil type on soil properties. Its preparation is of utmost importance as the preparation itself decides how much carbon the material is going to yield and of what resistance level. Biochar fertilizer is another product being considered of relevance to C sequestration. It is reported that black C can produce significant benefits when applied to agricultural soils in combination with some fertilizers. Apart from positive effects in both reducing emissions and increasing the sequestration of greenhouse gases, the production of biochar and its application to soil will deliver immediate benefits through improved soil fertility and increased crop production. Lehmann *et al.* (2003) in an Anthrosol showed significantly higher P, Ca, Mn and Zn availability than the Ferralsol and biomass production of both cowpea and rice increased by 38-45% without fertilization. The soil N contents were also higher in the Anthrosol but the wide C: N ratios due to high soil carbon content led to immobilization of N. Despite the generally high nutrient availability, nutrient leaching was minimal in the Anthrosol, providing an explanation for their sustainable fertility. Charcoal additions significantly increased plant growth and nutrition. Leaching of applied N fertilizer was significantly reduced by charcoal, and Ca and Mg leaching was delayed.

With application of biochar at very higher dose such as more than 50 t ha^{-1} significant changes in soil quality as reduction in tensile strength have been reported in literature. Chan *et al.* (2008) observed that poultry litter biochar, produced under two conditions, applied without N fertilizer, produced similar increases in dry matter yield of radish, which were detectable at the lowest application rate, 10 t ha^{-1}. The yield increase, compared with the un-amended control was 42 percent at 10 t ha^{-1} and 96 percent at 50 t ha^{-1} of biochar application which is attributed to the ability of these biochar to increase N availability. Significant additional yield increases, in excess of that due to N fertilizer alone, were observed when N fertilizer was applied together with the biochar, highlighting the other beneficial effects of these biochar.

With different combinations of biochar on two soil types (sandy loam and silt loam soils), in a green house study, cattle manure and N fertilizer in maize resulted in highest shoot dry weight due to improved nutrient retention from the biochar. At the same time, study revealed that higher N recovery can be improved by biochar application to sandy loam soil compared to silt loam soil suggesting soil textural effect in the effectiveness of biochar application for soil productivity.

Epilogue

It is crucial to maintain a threshold level of organic matter in the soil for maintaining physical, chemical and biological integrity of the soil and for sustained agricultural productivity. Efficient use of biomass by converting it as a useful source of soil amendment/nutrients is one way to manage soil health and fertility. The current availability of biomass in India is estimated at about 500 million tons/year. These residues are either partially utilized or unutilized due to various constraints. It is estimated that about 93 million tons of crop residues are burnt every year in India. Thus biochar holds promise to cater to needs of farmers in terms of soil health and crop yields improvement, environmentalists with regards to decreasing GHG emissions and carbon sequestration for much longer period. Also, intensification of forest management in specific areas is necessary to ensure sufficient wood and energy supplies to meet the current demand. Thus socialists and policy planners must accept the potential of biochar in agriculture and ensure that farmers are motivated so that the technology gets adopted may be through certain incentives if at all required.

References

Chan, K. Y., Van Zwieten, L., Meszaros, L., Downie, A. and Joseph.S. (2007). Agronomic values of green-waste biochar as a soil amendment. *Australian Journal of Soil Research,* 45: 629-634.

Chen, K.Y., Van Zwieten, L., Meszaros, I. A., Downie, C. and Joseph, S. (2008). Using poultry litter biochars as soil amendments. *Australian Journal of Soil Research,* 46:437-444.

Glaser, B., Haumaier, L., Guggenberger, G., Zech, W. (2001). The terra preta phenomenon: a model for sustainable agriculture in the humic tropics, *Die Naturwiss enschaften* 88: 37–41, doi:10.1007/s00114000193.

Jha, Pramod; Biswas, A.K., Lakaria, B.L. and Subba Rao, A. (2010). Biochar in agriculture – prospects and related implications. *Current Science,* 99(9): 1218-1225.

Kuhlbusch, T.A.J. (1998). Black carbon and the carbon cycle. *Science,* 280: 1903-1904.

Lehman J. and Joseph, S. (2009). Biochar for environmental management: an introduction In: Lehmann, J., Joseph, S. (eds) Biochar for environment management, science and technology, Earthscan, London, pp 1- 12.

Lehmann, J., Pereira da Silva Jr, J., Steiner, C., Nehls, T., Zech, W. and Glaser, B. (2003). Nutrient availability and leaching in an archaeological Anthrosol and a Ferralsol of the Central Amazon basin: fertilizer, manure and charcoal amendments. *Plant and Soil,* 249: 343-357.

Preston, C.M.; Schmidt, M.W.I. (2006). Black (pyrogenic) carbon: a synthesis of current knowledge and uncertainties with special consideration of boreal regions. *Biogeosciences,* 3:397-420.

Suman, D.O., Kulbusch, T.A.J. and Lim, B. (1997). Marine sediments : A reservoir for black carbon and their use as spatial and temporal records of combustion. In. Sediment records of biomass burning and global change (eds Clark *et al.,*). Global Environment Change, NATO, ASI Series Ist, Springer, Verlag, Berlin, 51:271-293.

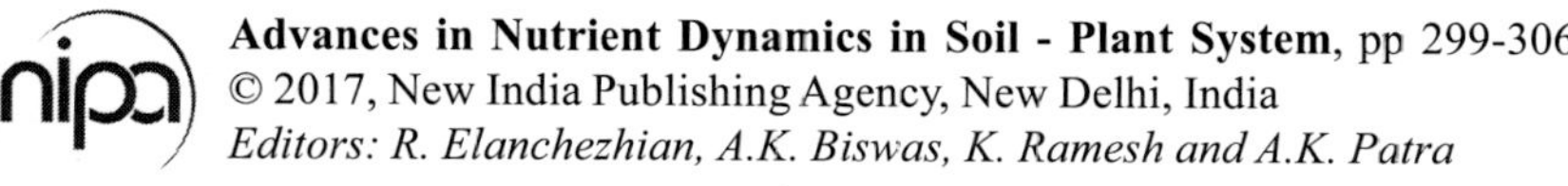
Advances in Nutrient Dynamics in Soil - Plant System, pp 299-306

Editors: R. Elanchezhian, A.K. Biswas, K. Ramesh and A.K. Patra

26

Recycling Urban Organic Wastes and Waste Water for Better Crop Production and Nutrient Use

J.K. Saha and M.V. Coumar

ICAR-Indian Institute of Soil Science, Nabi Bagh, Bhopal – 462038, India

According to Planning Commission Report (2014), India generates about 62 million tons of municipal solid wastes (MSW) every year from urban cities. There has been a significant increase in municipal solid waste generation in India in the last few decades. This is largely due to rapid population growth and economic development in the country (Annepu, 2012). Solid waste management (SWM) has become a major environmental issue because of serious environmental implications like global warming (through green house gases emission) and contamination of toxic pollutants (like heavy metals) in surface and groundwater bodies. Solid waste management is one of the essential services to be provided by municipal authorities in the country to keep urban areas clean. Out of total waste generated in India, only 12.45% waste is scientifically processed and rest is disposed in open dumps (CPCB Report, 2013). Although, incineration of solid wastes is being followed in many places for energy production, composting is considered most environment friendly method of SWM as it promotes recycling of nutrients in crop production.

Composting municipal solid waste involves managing conditions to accelerate the biological decomposition of its organic components. End product is an organically rich product with potential benefits for agricultural soils. The conditions for efficient biological decomposition of organic waste depend on optimum temperatures (50-65°C), moisture (50–60%), aeration (>10% O_2), pH (6.0–8.0), levels and carbon to nitrogen (25:1–35:1) ratios of the feedstock (Cooperband, 2002). If conditions deviate from these optimum conditions, the composting process is slowed and chemically unstable (immature) compost may be produced. When microorganisms degrade the organic materials under optimum oxygen levels, the process is called aerobic composting. In contrast, a different group of microorganisms can degrade the organic material under limited oxygen levels, where the process is called anaerobic composting. Aerobic composting is usually preferred over anaerobic composting because it is faster in biological oxidation and does not generate as many foul odors (i.e., ammonia, sulfur compounds, and organic acids).

Use of Municipal Solid Waste Composts in Agriculture and its Impact on Soil Resources

Agricultural lands can be utilized beneficially using municipal solid waste compost as an organic soil amendment. The organic matter present in many soils has gradually decreased over the past 100–200 years. Most agricultural cropping systems result in the depletion of organic matter. Soil organic matter acts as a sink and source of nutrients in the soil system because it has a high nutrient-holding capacity. It also acts as a large pool for the storage of nitrogen, phosphorus, and sulfur, and has the capacity to supply these and other nutrients for plant growth. Soil organic matter interacts with trace metals, often reducing their toxicity to plants. The physical benefits of organic matter on soil include improved soil structure, increased aeration, reduced bulk density, increased water-holding capacity, enhanced soil aggregation, and reduced soil erosion. The application of municipal solid waste compost to agricultural soil can be a means to return the organic matter to agricultural soil and in some cases reduce the cost of municipal solid waste disposal. Many studies have shown that compost may improve soil quality by increasing nutrient and water holding capacity, organic matter and cation exchange capacity, and providing different kinds of microelements to further improve crop yield and quality (Crecchio *et al.*, 2001; Soumare *et al.*, 2003; Termorshuizen *et al.*, 2004; Cherif *et al.*, 2009; Mylavarapu and Zinati, 2009).

In African country Ghana, it was observed that a product called Comilizer (@91 kg N/ha) derived from mixing composted municipal waste and ammonium sulphate fertilizer produced maize yield at par with split application of NPK (15-15-15) and ammonium sulphate at the rate of 150 kg N ha. Moreover N uptake and water use efficiency of maize plants grown with Comilizer was 11 and 12 per cent higher than with chemical fertilizer alone (Cofie *et al.,* 2010). Some organic wastes may have chemical, physical, and microbiological properties that would limit the extent to which they could be composted alone (Parr *et al.,* 1986). Hence detailed characterization of MSW compost is required for their efficient use in agriculture.

Physical soil properties: A primary benefit of MSW compost is the high organic matter content and low bulk density. Municipal solid waste compost has a high water holding capacity because of its organic matter content, which in turn improves the water holding capacity, aggregate stability and structure of the soil.

Biological soil properties: Soil ecology is increasingly being used to evaluate soil quality. It is thought that soil microbiological properties are most sensitive to changes in the soil environment. Biomass N, C, and S showed increases in the soil immediately after compost addition. In long-term experiments, it was found that multiple additions of MSW compost increased microbial biomass C, and this increase persisted years after application. Another measure of soil microbial health is the activity of soil enzymes involved in the transformation of the principal nutrients. Enzyme activities e.g., dehydrogenase, phosphodiesterase, alkaline phosphomonoesterase, arylsulphatase, deaminase, urease, and protease increases due to MSW compost application. MSW compost may stimulate the transformation of organic P to its inorganic and available form. Some enzyme activities were reported to decrease where MSW compost was applied for long. The decrease was attributed to the potential toxic effects exerted by trace elements in this particular compost.

Chemical soil properties: Increased soil pH of acidic soil is regarded as a major advantage when MSW compost is used. These increases were usually proportional to the application rate. The increase in the pH of soil may be due to the mineralization of carbon and the subsequent production of OH^- ions by ligand exchange as well as the introduction of basic cations, such as K^+, Ca^{2+}, and Mg^{2+}. Immature MSW compost tended to have a lower pH prior to thermophilic stage due to the intensive production of organic acids.

A survey of selected Indian MSW composts found that the EC of the composts were much higher than that of agricultural soils. Municipal solid waste composts applied at rates ranging from 40 to 120 Mg ha^{-1} were seen to proportionally increase the EC of soils to which they were applied. Most studies concluded that MSW compost increased the EC value in soils. In some cases, soil EC levels were excessive and inhibited plant growth. As with many other properties of MSW compost, the EC content of the MSW compost is likely related to the feedstock used in the compost and compost facility procedures.

The range of nitrogen concentrations that have been reported to be present in MSW compost is 0.26 to 1.71%. In some of the highly polluted cities of India like Delhi, Ahmadabad and Bangalore, N concentration of MSW ranged from 0.85 to 1.13% (Rawat *et al.,* 2013). The availability of nitrogen in MSW compost has been estimated at 10% in the first year after application with some reports of N release in the second year after application. While some studies showed that MSW compost increased soil N content, MSW compost is often reported to be less effective in supplying available N in the first year of application to the soil–plant system than inorganic mineral fertilizers. It is thought that N immobilization occurs in soils treated with compost because of increased soil microbial biomass. The type and ratio of feedstock and composting process should be the focus for increasing the inorganic N content of the compost when it is to be used as fertilizer. Mineralization of organic N in compost is dependent on many factors including C/N ratio of raw material, composting conditions, compost maturity, time of application, and compost quality (i.e., C/N ratio and C- and N-fractions). The composting process is equally as important as feedstock; compost made from the same feedstock but using different technologies could differ significantly. Optimum N transformations in MSW compost were found to occur at a temperature of 55.8 °C, moisture content of 60%, and an air flow rate of 10 L kg^{-1} h^{-1}. Researchers have also found aeration to play a large role in the inorganic N content of MSW compost. Low oxygen levels slow decomposition and increases the opportunity for adsorption of ammonia onto the solid materials leading to immobilization. The concentration of nitrogen in MSW compost has been seen to increase with composting time as carbon is utilized by microorganisms. Immature compost can cause N immobilization due to a high compost C/N ratio.

The range of phosphorus that has been found in MSW composts is 0.08 to 0.73% (mean 0.16%). Municipal solid waste compost has been reported to effectively supply P to soil with soil P concentration increasing with increasing application rates. Some reports observed that MSW compost provided equivalent amounts of P to soil as mineral fertilizers. A 10–50% of total P in MSW compost was available both the first and second year after application. Soil P availability was increased with the addition of MSW compost, however, soil P retention decreased with increasing compost application because of competition between organic ligands and phosphate for sites on metallic oxides as well as the formation of phosphohumic complexes which can increase P mobility. It has been concluded that MSW compost has a high capacity to supply P to plants given the compost is mature since the concentration of P in MSW compost tended to increase with composting time.

Long-term studies of MSW compost demonstrated that K was as available in MSW compost as in mineral K fertilizers. The range of K found in MSW compost produced in India is generally low and ranged between 0.12 to 1.31% (mean 0.44%). Of the total K in MSW compost, 36–48% was found to be plant available. Soil K concentrations are increased even when very low rates of MSW compost are used.

MSW compost in addition to plant nutrients and organic carbon, it also have some potentially hazardous heavy metals which was reported indifferent locations of the world. A variety of MSW composts manufactured in India were found to have high content of heavy metals, viz., Zn, Cu, Cd, Pb, Ni and Cr. Several reports have shown that application of trace element contaminated MSW in soil under long term resulted in heavy metal buid up in soil affecting soil and plant quality (Achiba *et al.,* 2009; Jordão *et al.,* 2006; Businelli *et al.,*2009). Different field crops and horticultural crops have been reported to take up heavy metals when soil was amended with MSW compost. Metal and trace metal availability from compost is thought to vary with compost maturity. As compost matures, the humic material in compost tends to increase and is capable of binding many metals thus decreasing their availability. The water-soluble fraction of Zn, Pb, Cu, and Cd were found to decrease and stabilize after the thermophilic stage of composting. The potential for excessive amounts of trace metals to contaminate the food chain through MSW compost additions is thought to depend on the source material used in the compost and the final concentration of the metals in the compost. Municipal solid waste compost tends to have higher concentrations of metals when sewage sludge is added with the feedstock, with the lowest metal concentrations found in MSW compost which has been made from source-separated waste. Furthermore, the earlier the sorting of waste occurs, such as at collection or before the composting process begins, the lower the heavy metal content in the finished product.

Nutrient Potential of Sewage water

Raw sewage water has a good nutrient potential in crop husbandary. Typical nutrients concentrations in sewage water are 20-85 mg N/l, 4-36 mg P/l, 7-20 mg K/l, 10-50 mg S/l besides other nutrients. Since the amount of N in sewage effluents is high, the C/N ratio for the organic components is low (typical values are 5), and a release of mineral N from organic substances is favored. Such sewage water, mainly in its raw (untreated) form in India, is used extensively as source of irrigation water in the nearby area of sewage canal mainly for fodder and vegetable production. It is observed that Phosphorus adsorbed in wastewater filters designed for P removal can be a valuable fertilizer when the filters have been exchanged with new media (Kvärnström *et al.,* 2004)

In general, yields of most crops are significantly higher in municipal sewage treated plots as compared to those in groundwater treated plots and crops in sewage water treated area respond less to NPK application compared to those in groundwater treated area. Experiments at different places also indicated significant nutrient supplying capacity of sewage water. In Vertisols of Bhopal, yield of wheat crop irrigated with untreated sewage water was almost equivalent to 50% of recommended dose of NPK fertilizers. Urban sewage water has been found to increase the nitrogen fixing bacteria, P solubilising organisms and pH in rhizospheres.

Wastewater irrigation improves the nutrient balance of the soil in respect of total nitrogen and available phosphorus. In contrast Na displaces Ca and diminishes the Ca saturation of the soil. In farmers' field (Vertisol) at Bhopal, a considerable increase in total and available P content has been observed due to long-term untreated sewage water irrigation.

Nine different samples of sewage sludges, composts and other representative organic waste were characterized for their use as agricultural soil amendments and it was observed that barring municipal sewage sludge other wastes can be safely used (Alavarenga *et al.,* 2015). They also recommended the composting of sewage sludge as a way to turn a less stabilized waste with lower heavy metal content and without sanitation problems.

Environmental concern: The health hazards associated with direct and indirect wastewater use are of two kinds: the rural health and safety problem for those working on the land or living on or near the land where the water is being used, and the risk that contaminated products from the wastewater use area may subsequently infect humans or animals through consumption or handling of the foodstuff or through secondary human contamination by consuming foodstuffs from animals that used the area (WHO, 1989). Also several workers have expressed environmental concern for the use of sewage water irrigation, *e.g.* nitrate pollution of ground water, production of green house gases N_2O through denitrification, salinity build up, soil and plant pollution by heavy metals, trace organics (like polycyclic aromatic hydrocarbons PAH, polychlorinated biphenyls PCB and other pesticides) and harmful micro-organisms. Application on sewage effluent having a high BOD value reduces the oxygen level in the soil, and consequently, denitrification may be enhanced. Dnitrification is microbially mediated loss of NO_3^- as N_2O and N_2 gases under anoxic conditions and considerable agricultural and environmental implications (N_2O is an important greenhouse gas).

Heavy metals concentration in domestic effluents is very low unless these are contaminated with industrial sewage effluents. At normal application rate, greater amounts of heavy metals are likely to be added to soil in one year through sludge than may be added in a century of effluent irrigation. Several reports have also indicated that heavy metal contents in sewage water are low and within the permissible limit. However, there are reports of heavy metal accumulation in soil due to long-term sewage water irrigation. Continuous application of sewage water for long period resulted migration of Cd, Co, Cr, Cu, Ni and Zn downward from cultivated layer; which was supposed to be due to increased mobility through complexation by dissolved organic C. Salts, in particular Na, have been found the most important inorganic contaminants for irrigated agriculture in many sewage farms. A high pH and accumulation of salts were common in sewage-irrigated soil.

Harmful pathogenic organisms (bacteria, viruses, protozoa and parasitic worms) are found in waste water and depending upon local conditions its concentration varies (Feachem *et al.,* 1983; Rose, 1986; Shuval *et al.,* 1986). The potential health hazard involved in the use of sewage effluent for irrigation of agricultural crops is the main effluent property used for its classification. The suitability of sewage effluent for agricultural irrigation is based on their microbiological quality. According to WHO guidelines, no crops which is eaten raw, should be grown with sewage water if it contains more than 100 coliform organisms in 100 ml.

Summary

Composting of municipal solid waste has potential as a beneficial recycling tool. It's safe use in agriculture, however, depends on the production of good quality compost, specifically, compost that is mature and sufficiently low in metals and salt content. The best method of reducing metal content and improving the quality of MSW compost is early source separation, perhaps requiring separation to occur before or at source of collection. Bioavailability should be addressed in the guideline limits, in addition to metal loading.

Sewage water has a good nutrient potential in crop husbandry as it contains considerable amount of N, P, K, and S besides other micronutrients. Since the amount of N in sewage effluents is high, the C/N ratio for the organic components is low and a release of mineral N from organic substances is favored. Such sewage water, mainly in its raw (untreated) form in India, is used extensively as source of irrigation water in the nearby area of sewage canal mainly for fodder and vegetable production. However, the low plant availability of nutrients in sewage sludge results in small yield increases even after many years of repeated sludge addition (Kirchmann *et al.,* 2016). Therefore, nutrient extraction from urban wastes instead of direct organic waste recycling is also a possible way forward. Moreover, pathogenic contamination to sewage farmers' as well as to consumers (particularly when vegetables consumed uncooked) is a major concern.

References

Achiba, W.B., Gabteni, N., Lakhdar, A., Laing, G.D., Verloo, M., Jedidi, N. and Gallali, T. (2009). Effects of 5-year application of municipal solid waste compost on the distribution and mobility of heavy metals in a Tunisian calcareous soil. *Agriculture, Ecosystem and Environment*, 130:156–163.

Alvarenga, P., Mourinha, C., Farto, M., Santos, T., Palma, P., Sengo, J., Morais, M.C. and Cunha-Queda, C. (2015). Sewage sludge, compost and other representative organic wastes as agricultural soil amendments: Benefits versus limiting factors. *Waste Management,* 40: 44-52.

Annepu, R. K. (2012). Sustainable solid waste management in India, Waste-to-Energy Research and Technology Council (WTERT). City of New York: Columbia University. Retrieved fromhttp://www.seas.columbia.edu/earth/wtert/sofos/Sustainable%20Solid%20 Waste %20Management%20 in%20India_Final.pdf.

Businelli D., Massaccesi L., Said-Pullicino D. and Gigliotti G. (2009). Long-term distribution, mobility and plant availability of compost-derived heavy metals in a landfill covering soil. *Science of Total Environment*, 407 (4): 1426–1435.

Cherif, H., Ayari, F., Ouzari, H., Marzorati, M., Brusetti, L., Jedidi, N., Hassen. A, and Daffonchio, D. (2009). Effects of municipal solid waste compost, farmyard manure and chemical fertilizers on wheat growth, soil composition and soil bacterial characteristics under Tunisian arid climate. *European Journal of Soil Biology*, 45:138–145.

Cofie, O., Veenhuizen, R.V., de Vreede, V. and Maesse, S. (2010). Waste Management for Nutrient Recovery: Options and challenges for urban agriculture. *Urban Agriculture magazine,* (23): 3-7.

Cooperband, L. (2002). The art and science of composting. Center for Integrated Agricultural Systems, University of Wisconsin-Madison.http://www.cias.wisc.edu/wpcontent/uploads/2008/07/artofcompost.pdf.

CPCB. (2013). Status report on municipal solid waste management. Retrieved from http://www.cpcb.nic.in/divisionsofheadoffice/pcp/MSW_Report.pdfhttp://pratham.org/images/paper_on_ragpickers.pdf

Crecchio, C., Curci, M., Mininni, R., Ricciuti, P. and Ruggiero., P. (2001). Short-term effects of municipal solid waste compost amendments on soil carbon and nitrogen content, some enzyme activities and genetic diversity. Biol. Fert. Soils 34:311-318.

Feachem, R.G., Bradley, D.J., Garelick, H. and Mara, D.D. (1983). Sanitation and Disease: Health Aspects of Excreta and Wastewater Management. John Wiley, Chicester.

Jordao, C.P., Nascentes, C.C., Cecon, P.R., Fontes, R.L.F. and Pereira, J.L. (2006). Heavy metal availability in soil amended with composted urban solid wastes. *Environmental Monitoring and Assessment*, 112, 309-326.

Kirchmann, H., Börjesson, G., Kätterer, T. and Cohen, Y. (2016). From agricultural use of sewage sludge to nutrient extraction: A soil science outlook. *Ambio*, Sep 2016.

Kvärnström, E., Morel, C. and Krogstad, T. (2004). Plant availability of phosphorus in filter substrates derived from small-scale wastewater treatment systems. *Ecological Engineering,* 22: 1-15.

Mylavarapu, R.S. and Zinati, G.M. (2009). Improvement of soil properties using compost for optimum parsley production in sandy soils. *Scientia Horticulturae*, 120:426–430.

Parr, J.F., Papendick, R.I. and Colacicco, D. (1986). Recycling of organic wastes for a sustainable agriculture. *Biological Agriculture and Horticulture,* 3: 115-130.

Planning Commission Report. (2014). Reports of the task force on waste to energy (Vol-I) (in the context of Integrated MSW management). Retrieved from http://planningcommission.nic.in/reports/genrep/rep_wte1205.pdf.

Rawat, M., Ramanathan, A.L. and Kuriakose, T. (2013). Characterisation of Municipal Solid Waste Compost (MSWC) from Selected Indian Cities—A Case Study for Its Sustainable Utilization. *Journal of Environmental Protection*, 4: 163-171.

Rose, J.B. 1986. Microbial aspects of wastewater reuse for irrigation. *CRC Critical Reviews in Environmental Control* 16(3): 231-256.

Shuval, H.I., Adin, A., Fattal, B., Rawitz, E. and Yekutiel, P. (1986a). Wastewater irrigation in developing countries: health effects and technical solutions. *Technical Paper Number 51.* World Bank, Washington DC. 324 p.

Soumare, M., Tach, F.M.G. and Verloo, M.G. (2003). Effect of municipal solid waste compost and mineral fertilization on plant growth in two tropical agricultural soils in Mali. *Bioresource and Techenology,* 88: 15-20.

Termorshuizen, A.J., Moolenaar, S.W., Veeken, A.H.M. and Blok, W.J. (2004). The value of compost. *Reviews in Environmental Science and Biotechnology*, 3:343–347.

WHO. (1989). Health guidelines for the use of wastewater in agriculture and aquaculture: Report of a WHO Scientific Group. *WHO Technical Report Series 778.* World Health Organization, Geneva. 74 p.

Advances in Nutrient Dynamics in Soil - Plant System, pp 307-323

Editors: R. Elanchezhian, A.K. Biswas, K. Ramesh and A.K. Patra

27

Analytical Techniques for Nutrient Estimation in Soil and Plants

I. Rashmi and S. Neenu

ICAR-Indian Institute of Soil Science, Nabibagh, Bhopal – 462038, India

Fertility of a soil can be assessed by analyzing various available nutrients present in the soil. Fertilizer recommendations for various crops and cropping sequences can be made on the basis of fertility status of a soil. Besides this, problematic soil can be ameliorated on the basis of soil test values. Among the various steps of soil testing programmes, soil sampling is the most vital step.

Soil Sampling

Principle

Soil testing is an essential component of soil resource management. Each sample collected must be a true representative of the area being sampled. Utility of the results obtained from the laboratory analysis depends on the sampling precision. Hence, collection of large number of samples is advisable so that sample of desired size can be obtained by sub-sampling. In general, sampling is done at the rate of one sample for every two hectare area. However, at-least one sample should be collected for a maximum area of five hectares. For soil survey work, samples are collected from a soil profile representative to the soil of the surrounding area.

Materials required

1. Spade or auger (screw or tube or post hole type)
2. *Khurpi*
3. Core sampler
4. Sampling bags
5. Plastic tray or bucket

Points to be considered

1. Collect the soil sample during fallow period.
2. In the standing crop, collect samples between rows.
3. Sampling at several locations in a *zig-zag* pattern ensures homogeneity.
4. Fields, which are similar in appearance, production and past-management practices, can be grouped into a single sampling unit.
5. Collect separate samples from fields that differ in colour, slope, drainage, past management practices like liming, gypsum application, fertilization, cropping system *etc.*
6. Avoid sampling in dead furrows, wet spots, areas near main bund, trees, manure heaps and irrigation channels.
7. For shallow rooted crops, collect samples up to 15 cm depth. For deep rooted crops, collect samples up to 30 cm depth. For tree crops, collect profile samples.
8. Always collect the soil sample in presence of the farm owner who knows the farm better.

Procedure

1. Divide the field into different homogenous units based on the visual observation and farmer's experience.
2. Remove the surface litter at the sampling spot.
3. Drive the auger to a plough depth of 15 cm and draw the soil sample.
4. Collect at least 10 to 15 samples from each sampling unit and place in a bucket or tray.
5. If auger is not available, make a 'V' shaped cut to a depth of 15 cm in the sampling spot using spade.
6. Remove thick slices of soil from top to bottom of exposed face of the 'V' shaped cut and place in a clean container.

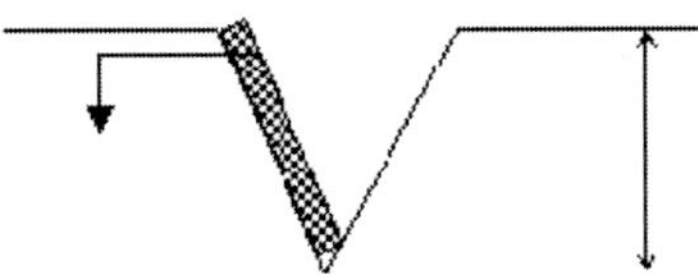

1. inch / 2.5 cm 6 inches (15 cm)

7. Mix the samples thoroughly and remove foreign materials like roots, stones, pebbles and gravels.
8. Reduce the bulk to about half to one kilogram by quartering or compartmentalization.

9. Quartering is done by dividing the thoroughly mixed sample into four equal parts. The two opposite quarters are discarded and the remaining two quarters are remixed and the process repeated until the desired sample size is obtained.
10. Compartmentalization is done by uniformly spreading the soil over a clean hard surface and dividing into smaller compartments by drawing lines along and across the length and breadth. From each compartment a pinch of soil is collected. This process is repeated till the desired quantity of sample is obtained.
11. Collect the sample in a clean cloth or polythene bag.
12. Label the bag with information like name of the farmer, location of the farm, survey number, previous crop grown, present crop, crop to be grown in the next season, date of collection, name of the sampler *etc.*

Collection of soil samples from a profile

1. After the profile has been exposed, clean one face of the pit carefully with a spade and note the succession and depth of each horizon.
2. Prick the surface with a knife or edge of the spade to show up structure, colour and compactness.
3. Collect samples starting from the bottom most horizon first by holding a large basin at the bottom limit of the horizon while the soil above is loosened by a khurpi.
4. Mix the sample and transfer to a polythene or cloth bag and label it.

Processing and storage

1. Assign the sample number and enter it in the laboratory soil sample register.
2. Dry the sample collected from the field in shade by spreading on a clean sheet of paper after breaking the large lumps, if present.
3. Spread the soil on a paper or polythene sheet on a hard surface and powder the sample by breaking the clods to its ultimate soil particle using a wooden mallet.
4. Sieve the soil material through 2 mm sieve.
5. Repeat powdering and sieving until only materials of >2 mm (no soil or clod) are left on the sieve.
6. Collect the material passing through the sieve and store in a clean glass or plastic container or polythene bag with proper labeling for laboratory analysis.
7. For the determination of organic matter it is desirable to grind a representative sub sample and sieve it through 0.2 mm sieve.
8. If the samples are meant for the analysis of micronutrients at-most care is needed in handling the sample to avoid contamination of iron, zinc and copper.

Brass sieves should be avoided and it is better to use stainless steel or polythene materials for collection, processing and storage of samples.

9. Air-drying of soils must be avoided if the samples are to be analyzed for NO_3-N and NH_4-N as well as for bacterial count.
10. Field moisture content must be estimated in un-dried sample or to be preserved in a sealed polythene bag immediately after collection.
11. Estimate the moisture content of sample before every analysis to express the results on dry weight basis.

(I) Estimation of pH of soils

Principle: Soil pH determined in 1:2.5 soil: water suspension by potentiometric method (Jackson, 1967).The pH value is a measure of hydrogen ion concentration of the soil water system and expresses the acidity and alkalinity of soil. pH is very important property of soil as it determines the nutrient availability, microbial activity and physical condition of the soil. The concept of pH was given by Sorensen in 1901. The pH of solution has been defined as negative logarithm of hydrogen ion activity which in very dilute solution is expressed as g ions L^{-1} or g mol L^{-1}.

In soil water system, some of the adsorbed hydrogen ions dissociate from the surface of the soil colloids into soil solution. These dissociated H+ give rise to soluble acidity or active acidity. pH is a sort of voltage measurement and to cover the entire range of 0 – 14, a potential measurement in the range of + 420 to - 420 mV is needed. A potential difference of 59.1 mV is developed for a difference of one pH unit. Instrument used for pH determination is a glass electrode pH meter with a calomel or reference electrode introducing salt bridge. Most digital pH meters now days have a single or combined electrode assembly.

Apparatus required: pH meter, beaker (100 ml), measuring cylinder, glass rod and a balance.

Reagents: Buffers of different pH values (4.0, 7.0 and 9.2) for standardization or calibrating the pH meter. Dissolve one tablet in 100 ml of solution.

Procedure

1. Weigh 10 g of air dried soil into a 100 ml beaker.
2. To this add 25 ml of distilled water.
3. Stir the contents with a glass rod intermittently for 30 minutes.
4. Switch on the pH meter and allow for warming.
5. Calibrate with the known buffers.
6. Immerse the electrodes in soil water suspension and record the pH reading.

(II) Estimation of EC of soils

Soil salinity, which is caused due to the total soluble salt content, is determined by measuring the electrical conductivity of soil. EC is a measure of the ability of soil solution to carry electric current by the migration of ions under the influence of an electric field. Like metallic conductors, solutions also obey ohm's law. Electrical conductivity was determined in 1:2.5 soil-water extract using conductivity bridge and expressed as dS/m (Jackson, 1973).

Principle: Pure water offers maximum resistance to the passage of electricity. If any soluble salts are present in water then they dissociate thus increasing the flow of current. The soil salinity in relation to plant growth is generally measured in terms of conductivity of the soil water suspension or saturation extract of the soil. The percentage of soluble salts in soils can then be computed from the following relationship

Percentage of soluble salts in soil = 0.064 x EC x SP/100,

Where SP the saturation percentage of soil and EC is in dS m^{-1}

Apparatus: Balance, beaker (100 ml), measuring cylinder, a glass rod, a conductivity meter.

Reagents: Potassium chloride (0.01 N): Weigh 0.7456 g of potassium chloride and make the volume to 1 L with distilled water. This solution gives an electrical conductivity of 1.41 dS m^{-1} at 25 ° C.

Procedure

1. Weigh 10 g of air dried soil into 100 ml beaker and add 25 ml of distilled water.
2. Carry out intermittent stirring with a glass rod for 30 minutes and then allow standing for obtaining clear supernatant solution.
3. Connect the salt bridge to the power supply, switch on the bridge and adjust it to room temperature with the help of temperature setting knob.
4. Calibrate the conductivity bridge with the help of standard KCl solution.
5. Dip the conductivity cell in the supernatant solution so that platinum electrodes are completely immersed in solution.

(III) Estimation of Organic Carbon

Principle: Organic matter estimation in the soil can be done by different methods. Loss of weight on ignition can be used as a direct measure of the organic matter contained in the soil. It can also be expressed as the content of organic carbon in the soil. It is generally assumed that on an average organic matter contains about 58% organic carbon. Organic matter/organic carbon can also be estimated by volumetric and colorimetric methods. However, the use of potassium dichromate ($K_2Cr_2O_7$) involved in these estimations is considered as a limitation because of its hazardous nature. Soil organic matter content can be used as an index of N availability (potential of a soil to supply N to plants) because the content of N in soil organic matter is relatively constant. The estimation of organic carbon is volumetric method given by Walkley and Black (1934).

Apparatus

a. Conical flask - 500 ml
b. Pipettes - 2, 10 and 20 ml
c. Burette - 50 ml

Reagents

1. Phosphoric acid – 85%
2. Sodium fluoride solution – 2%
3. Sulphuric acid – 96 % containing 1.25% Ag_2SO_4
4. Standard 0.1667M $K_2Cr_2O_7$: Dissolve 49.04 g of $K_2Cr_2O_7$ in water and dilute to 1 litre.
5. Standard 0.5M $FeSO_4$ solution: Dissolve 140 g Ferrous Sulphate in 800 ml water, add 20 ml concentrated H_2SO_4 and make up the volume to 1 litre.
6. Diphenylamine indicator: Dissolve 0.5 g reagent grade diphenylamine in 20 ml water and 100 ml concentrated H_2SO_4.

Procedure

1. Weigh 1.0 g of the prepared soil sample in 500 ml conical flask.
2. Add 10 ml of 0.1667M $K_2Cr_2O_7$ solution and 20 ml concentrated H_2SO_4 containing Ag_2SO_4.
3. Mix thoroughly and allow the reaction to complete for 30 minutes.
4. Dilute the reaction mixture with 200 ml water and 10 ml H_3PO_4.
5. Add 10 ml of NaF solution and 2 ml of diphenylamine indicator.
6. Titrate the solution with standard 0.5M $FeSO_4$ solution to a brilliant green colour.
7. A blank without sample is run simultaneously.

Calculation

Percent organic Carbon (X) =10 (S-T)/S x 0.003 x 100/Wt. of soil

Where,

S = ml $FeSO_4$ solution required for blank

T = ml $FeSO_4$ solution required for soil sample

3 = Eq W of C (weight of C is 12, valency is 4, hence Eq W is 12÷4 = 3.0)

0.003 = weight of C (1 000 ml 0.1667M $K_2Cr_2O_7$ = 3 g C. Thus, 1 ml 0.1667M $K_2Cr_2O_7$ = 0.003 g C)

Organic Carbon recovery is estimated to be about 77%. Therefore, actual amount of organic carbon (Y) will be:

Percent value of organic carbon obtained x 77/100 *Or* Percentage value of organic carbon x 1.3

Percent Organic matter = Y x 1.724 (organic matter contains 58 % organic carbon, hence 100/58 = 1.724)

Note: Published organic C to total organic matter conversion factor for surface soils vary from 1.724 to 2.0. A value of 1.724 is commonly used, although whenever possible the appropriate factor be determined experimentally for each type of soil.

(IV) Estimation of available nitrogen in soils

Principle: The easily mineralizable N is estimated by using alkaline $KMNO_4$ which oxidizes and hydrolyses the organic N present in the soil. The liberated ammonia is condensed and collected in boric acid and is titrated against standard H_2SO_4 using mixed indicator to determine the amount of ammonia liberated. This method has been widely adopted for the estimation of available nitrogen content in the soil due to its rapidity and reproducibility. The process of oxidative hydrolysis requires uniform heating temperatures for better results. This method was given by Subbaiah and Asija (1956).

Apparatus: Macrokjeldahl distillation unit

Reagents

1. $KMNO_4$ (0.32 %) – Dissolve 3.2 g of $KMNO_4$ salt in distilled water and make up the volume to 1 litre.
2. NaOH (2.5 %) – Dissolve 25 g of NaOH in distilled water making the volume to 1 litre.
3. Boric acid (H_3BO_3) – (2.5 %) Weigh 25 g of boric acid and dissolve in warm water and dilute to 1 litre.
4. Mixed indicator –0.5 g bromocresol green is mixed with 0.1 g methyl red in 100 ml of 95 % ethanol.
5. Standard H_2SO_4 (0.01 N) - Dissolve 0.3 ml of concentrated H_2SO_4 in 1 L solution and standardize against 0.01 N Na_2CO_3 using methyl red as indicator.

Procedure (With macro kjeldahl distillation unit)

1. Weigh 5g of soil sample into a 250 ml Kjeldahl flask.
2. Moisten the soil with about 10 ml of distilled water, wash down the soil adhering to the neck of flask, if any.
3. Add 50 ml of 0.32 % $KMnO_4$ solution.
4. Add a few glass beads or broken pieces of glass rod.
5. Add 2 to 3 ml of paraffin liquid, avoiding contact with upper part of the neck of the flask.
6. Measure 25 ml of 2.5 % boric acid containing mixed indicator in a 250 ml conical flask and place it under the receiver tube. Dip the receiver tube end in the boric acid.
7. Run tap water in condenser.
8. Add 50 ml of 2.5 % NaOH solution and immediately attach to the rubber stopper fitted in the alkali trap.
9. Switch the heaters on and continue distillation until 100 ml of distillate is collected.
10. First remove the conical flask containing distillate and then switch off the heater to avoid back suction.
11. Titrate the distillate against 0.01 N H_2SO_4 taken in burette until light red colour appears.
12. Run a blank without soil with each set of samples.
13. Carefully remove the kjeldahl flask after cooling and clean the flask.

Calculations

$$\text{Available N (Mineralizable N) in kg/ha} = \frac{(\text{TV-BV}) \times 0.02 \times 0.014 \times 2.24 \times 10^6}{\text{Wt of soil}}$$

(V) Estimation of available phosphorus in soils

Phosphorus occurs in soil both in organic and inorganic form. More than 50 % of P is present in inorganic form as Ca, Fe, and Al phosphates. Plant absorbs P in the form of the $H_2PO_4^-$ and HPO_4^{2-} (orthophosphates). For the estimation of available P, different extractants such as water, dilute acid, alkali and salt solution are used. The available P in neutral to slightly alkaline soil will be extracted with Olsen's extractant which is 0.5 M $NaHCO_3$ (pH 8.5). This method was given by Olsen and Wantanabe (1957).

Principle: Solubility of $Ca_3(PO_4)_2$ is controlled by the activity of Ca^{2+} in the solution and pH of the extractant. The bicarbonate of the extractant decreases the activity of Ca and prevents the release of phosphates from apatites and also releases the orthophosphates into the extract. The soluble phosphates form heteropoly complexes with molybdate forming ammonium phosphomolybdate which gives blue colour in the presence of ascorbic acid. The intensity of blue colour in the solution is proportional to the concentration of phosphates. The intensity of blue colour of both standards and sample solution will be measured in spectrophotometer at a wavelength of 660 nm which corresponds to red region. The concentration of 'P' in soil extract is calculated from the standard graph.

Reagents

1. Olsen's extractant i.e., $NaHCO_3$ (0.5 M) of pH – 8.5: Dissolve 84 g of $NaHCO_3$ in distilled water, adjust the pH to 8.5 and dilute to 2 L.
2. Reagent A : Dissolve 12 g ammonium molybdate in 200 ml of distilled water, dissolve 0.2908 g of antimony potassium tartarate in 100 ml distilled water and add the two solutions to 1000 ml of 2.5 M H_2SO_4 and dilute to 2 litres.
3. Reagent B: Dissolve 1.056 g of ascorbic acid in 200 ml of reagent A. This has to be prepared afresh.
4. H_2SO_4 (2.5 M): Dilute 140 ml of concentrated H_2SO_4 to 1 litre.
5. Standard stock P solution: Dissolve 0.439 g potassium dihydrogen orthophosphate (KH_2PO_4) in 1 litre solution. This contains 100 ppm P. Prepare 2 ppm P solution by diluting 50 times.

Preparation of Standard Curve

To prepare the standard curve; take 0, 1, 2, 3, 4 and 5 ml aliquots of 2 ppm P solution into a series of 25 ml volumetric flasks. To this add 5 ml of extracting solution followed by 1 drop of para nitrophenol indicator that gives yellow colour. Discolor the contents by adding dilute H_2SO_4 (2.5 M) drop wise to bring down the pH to around 5 at which pH, the formation of heteropoly complex of phosphomolybdates is perfect. Add 4 ml of reagent B and make up the volume to 25 ml with distilled water. Shake the contents and measure the intensity of blue colour in spectrophotometer at a wavelength 660 nm wave length. Plot the standard graph for P taking absorbance values on Y – axis and concentration of P on X – axis.

Procedure for sample preparation

1. Weigh 2 g of soil sample into a 150 ml conical flask.
2. To this add 40 ml of Olsen's extractant and a pinch of charcoal.
3. Shake the contents for 30 minutes and filter through whatman No. 42 filter paper.
4. Pipette out of 5 ml of extract into a 25 ml volumetric flask and add one drop of P-nitrophenol.
5. The contents are acidified with 2.5 M H_2SO_4

6. Add 4 ml of reagent B and make up the volume to 25 ml with distill water.
7. Shake the contents and measure the intensity of blue colour (absorbance) in a spectrophotometer after 10 minutes at 680 nm wave length.
8. Incorporate the absorbance value in the standard graph and obtain the concentration of P in the colored solution.

Calculations

$$\text{Available P in soil} = \frac{40 \times 25 \times \text{Reading} \times 2.24 \times 10^6}{2\text{g} \times 5\text{ml}}$$

(VI) Estimation of available potassium in soils

About 90-98 % of K in the soil is present in mineral forms such as feldspars, muscovite, biotite and illite. Potassium from these minerals is released slowly by weathering and usually is not of much significance to meet the crop requirements. The readily available K constitutes about 1-2 % of total K in mineral soils. It consists of soil solution and exchangeable K. The neutral normal CH_3COONH_4 solution which extracts both water soluble and exchangeable K is most commonly used for determination of available K in soil.

Principle: A known weight (5 g) of soil is taken and shaken with neutral normal ammonium acetate solution (25 ml). Ammonium exchanges with K+ on the soil colloids and the extract contains exchangeable and water soluble K, the amount of which is determined by using a flame photometer by emission spectroscopy.

Glassware and Apparatus required: Balance, mechanical shaker, conical flask, volumetric flasks, measuring cylinder, beaker, funnel and flame photometer.

Reagents

1. Neutral normal ammonium acetate: Dissolve 77.09 g of ammonium acetate in 800 ml of distilled water, adjust the pH to 7.0 with ammonia solution or acetic acid and dilute to 1 L with distilled water.
2. Standard solution of K (1000 ppm): Dissolve 1.907 g of KCl in distilled water and make the volume to 1 L.
3. Preparation of standard graph for K: From the stock solution (1000 ppm), different working standards are prepared as given below. Take aliquots of 0, 1, 2, 3, 4 and 5 ml of 1000 ppm K solution into a series of 100 ml volumetric flasks and make up to the mark to prepare 0, 10, 20, 30, 40 and 50 ppm K solutions respectively.

Procedure

1. Weigh 5 g of soil sample into a 250 ml conical flask.
2. To this add 25 ml of ammonium acetate (IN, pH = 7).

3. Shake the contents for 5 minutes and filter through whatman No. 1 filter paper.
4. The extract is aspirated to the flame and the frame photometer reading is noted down.

Calculations

$$\text{Available K in soil} = \frac{25 \times \text{Reading} \times 2.24 \times 10^6}{5g}$$

(VII) Estimation of available sulfur in soils

Sulphur is present in organic and inorganic forms. The amount depends upon the parent material, soil organic matter status and texture of the soil. Plants absorb sulphur in the form of sulphate ion (SO_4^{-2}). A large number of extractants like H_2O; monocalcium phosphate; $CaCl_2$; a mixture of ammonium acetate and acetic acid and NaCl have been used for extraction of available S. Among the different extractants $CaCl_2$ (0.15 %) was found to be the best extractant and determined by turbidometry as outlined by Black (1965) using spectrophotometer at 420 nm.

Principle: When the soil solution is shaken with $CaCl_2$ (0.15 %), the chloride ions displace the adsorbed sulphate during extraction. The filtrate is analysed for sulphur by turbidimetry method as outlined by Chesin and Yien (1950), in which turbidity produced due to the precipitation of SO_4^{-2} as $BaSO_4$ is measured on a spectrophotometer at a wavelength of 420 nm or corresponding to blue filter. The conditioning reagent is added to stabilize or suspend the $BaSO_4$ precipitate uniformly in the solution.

Glassware and Apparatus required: Balance, mechanical shaker, spectrophotometer, conical flask, volumetric flask, measuring cylinder, beaker, funnel, burette.

Reagent

a) $CaCl_2$ (0.15 %). Dissolve 1.5 g of $CaCl_2$ dihydrate in distilled water and make the volume to 1 litre.
b) Stabilizing agent or conditioning agent: Dissolve 75 g NaCl in 250 ml of distilled water in a 500 ml volumetric flask and add 30 ml of concentrated HCl followed by 100 ml ethanol and 50 ml glycerol with constant stirring. Make the volume to 500 ml.
c) $BaCl_2 . 2H_2O$
d) Standard sulphate solution: Dissolve 0.5434 g of AR grade K_2SO_4 in distilled water and dilute to 1 L. This is 100 ppm S solution.

Standard preparation: Pipette out 0, 0.5, 1.0, 1.5, 2.0, 2.5 ml of 100 ppm sulphur solution into 50 ml volumetric flask and to this add 5 ml of conditioning agent and a pinch of $BaCl_2$. Make the volume to the mark to prepare the working standard of 0, 1, 2, 3, 4 and 5 ppm respectively. After 10 minutes, the turbidity developed in the standards is measured in a spectrophotometer at a wave length of 420 nm.

Procedure for sample preparation

1. Take 5 g of soil into a 250 ml of conical flask.
2. To this add 25 ml of $CaCl_2$ (0.15 %) solution and shake for 30 min.
3. Filter through whatman No. 1 filter paper.
4. Pipette out 5 ml of extract into a 25 ml volumetric flask, add little amount of distilled water followed by 2.5 ml stabilizing agent and a pinch of $BaCl_2$. Shake the contents and make up the volume to the mark with distilled water.

Calculations

$$\text{Available sulphur (kg/ha)} = \frac{\text{ml of } CaCl_2 \text{ added to soil x Final volume x Reading (ppm) x 2.24}}{\text{Weight of soil x vol.of aliquot}}$$

Estimation of Total Nitrogen, Phosphorus and Potassium Content in Plant Samples

Plant analysis implies the determination of the content of nutrient elements in certain plant parts such as leaves, stems or in the whole plant. It is based on the concept that, the concentration of an essential element in a plant or plant part indicates the soil ability to supply that nutrient. Total plant analysis is quantitative in nature and is more reliable and useful. Total nutrient measurement involves digestion of a plant sample with a strong acid mixture. The volatile constituents disappear and non volatile mineral elements enter into solution. Heating is continued to until digest is reduced to a few ml of clear white residue.

Plant Sample Collection, Processing and Storage of Plant Samples

For correct interpretation it is essential, that sampling is done at the prescribed morphological stage of growth and from the correct plant parts. For micro nutrient studies in general, usually matured leaves just below the growing tip on main branches and stems are preferred. Sampling is normally recommended just prior to or at the time the plants begins its reproductive stage of growth. In case of macro nutrients like N, P and K, lower parts of plant are sampled. A true plant sample should be representative of the crop under field conditions. Plants infested with diseases and insects, which are under stress due to drought or excess water, those heavily coated with dust should be avoided.

The plant samples thus collected should be placed in paper bags, indication sample number. If the sampling sites are located far then place the plant parts in polyhene bags and keep in ice chest to minimize physiological activity and spoilage of samples due to high temperature and humidity. Plant samples should be repeatedly washed with distilled water in laboratory and then air dried first. Later plant samples are dried in oven at 60°C to 80°C for 24 to 36 hours. The dried oven plant samples should be ground in grinder and pass through 40mesh sieve. After grinding plant samples should be stored in air tight containers.

(a) Total Nitrogen in Plant Samples

Nitrogen plays important role in the synthesis of protein responsible for metabolic activities in plants. Nitrogen content of plant parts varies between 0.2 to 0.6 % of dried materials and green leaves contain higher N then stem, flowers and seeds. One common plant analysis is that of nitrogen (N) by Kjeldahl method. However, wet ashing with H_2SO_4 and H_2O_2 is also used for eliminating the use of selenium in the former method .

Principle: The plant samples are digested with sulphur salicylic acid mixture. Organic and nitrate nitrogen is converted to ammonium sulphate and the ammonia gas is distilled into boric acid and titrated with standard sulphuric acid. The nitrate content present in the sample form nitrocompounds by the reaction of salicyclic acid in acid medium.

Apparatus: Block-digester, Kjeldahl distillation unit, Conical flask, Pipettes, measuring cylinders etc.

Reagents

1) Sulfuric Acid (H_2SO_4)-salicylic acid: Dissolve 1g of salicylic acid in 30 ml of concentrated H_2SO_4
2) Digestion mixture: Mix 25g of K_2SO_4 with 5g of $CuSO_4.5H_2O$ and 0.5 g of metallic selenium powder by grinding in a mortar.
3) Sodium Hydroxide Solution (NaOH), 40%
4) Mixed indicator: 0.1g of bromocresol green + 0.07g of methyl red in 100ml of 95% ethanol
5) Boric Acid Solution (H_3BO_3), 2%: Dissolve 20g of H_3BO_3 in distill water and dilute the contents to 900ml. Add 20ml of bromocresol green + methyl red mixed indicator solution. Then add 0.1N NaOH drop wise till solution becomes reddish purple.Make up the volume to 1 litre.

Procedure

A. Digestion

1. Mix and spread finely ground (Cyclone mill) plant sample in a thin layer on a sheet of paper until it looks uniform.
2. Weigh 0.5 or 1g of dry plant material, and transfer quantitatively into a 100-ml digestion tube.
3. Add a few pumice boiling granules, and add about 3 g catalyst mixture
4. Add 10 ml concentrated sulfuric acid containing salicylic acid using a dispenser, and stir until mixed well.
5. Allow to stand overnight
6. Place tubes in a block-digester set at 100°C for 20 minutes and digest at low flame, and remove the tubes to wash down any material adhering to the neck of the tube with the same concentrated sulfuric acid. Thoroughly agitate the tube contents, and then place the tubes back on the block-digester set at 380°C for 2 hours after clearing.
7. After digestion is complete, remove tubes, cool, and bring to 100 ml volume with distilled water.

B. Distillation

1. Prior to distillation, shake the digestion tube to thoroughly mix its contents. And pipette 10 ml aliquot into a 100 ml distillation flask.
2. Add 10 to 15ml of 40% NaOH to make the contents alkaline. Distillation assembly is washed with small amount of distill water 2 to 3 times.
3. Before adding NaOH, boric acid mixed indicator solution should be kept ready at the receiving end of condenser outlet so that outlet is dipped in boric acid.
4. Carry out distillation by passing steam into distillation flask and the colour of boric acid mixed indicator solution changes from reddish purple to green and continue distillation for some more time to trap all the NH_3 released from distillation of sample.
5. After distillation, bluish green coloured ammonia trapped boric acid is titrated against 0.01N H_2SO_4 till colour changes to purple releasing boric acid with the formation of $(NH_4)_2SO_4$.
6. Run a blank without the plant material to check for contamination and to ensure precision.

$$\% \text{ N in plant sample} = \frac{\text{(S-B) X N of } H_2SO_4 \text{ x Volume of digest x 0.014 x 100}}{\text{Weight of plant sample x Aliquot taken for distillation}}$$

where,

S = Volume of standard H_2SO_4 used in sample titration

B = Volume of standard H_2SO_4 used in blank titration

(b) Total phosphorus in plant sample

Principle: The concentration of P in the plant samples is usually about one tenth of the N content. The concentration varies with plant parts and crop growth stage. The medium level of P in most crops is less than 0.3% on dry weight basis. Plant P is converted into orthophosphates during digestion. These orthophosphates react with molybdate and vanadate and give yellow coloured unreduced vando molybdo phosphoric heteropoly complex in nitric acid medium. The yellow colour is attributed to the substitution of oxyvanadium and oxymolybednum radicals for the oxygen of phosphate. The intensity of this yellow colour is directly proportional to the concentration of phosphates in the plant samples which can be read in spectrophotometer. Yellow colour developed in 30 minutes and is stable for 2 to 8 weeks.

Apparatus

Spectrophotometer or colorimeter, 410 nm wavelength, Block-digester, beakers, volumetric flasks, pipettes, and funnels.

Reagents

1. Perchloric Acid ($HClO_4$), 60%
2. Nitric acid (HNO_3)
3. Diacid mixture: HNO_3 : $HClO_4$ (9:4 ratio) mix 900ml of HNO_3 with 400ml of $HClO_4$
4. Ammonium Heptamolybdate-Ammonium Vanadate in Nitric Acid: Dissolve 22.5 g ammonium heptamolybdate [$(NH_4)_6Mo7O_{24}.4H_2O$] in 400 ml distill water (a). Dissolve 1.25 g ammonium metavanadate (NH_4VO_3) in 300 ml hot distill water (b). Add (b) to (a) in a 1L volumetric flask, and let the mixture cool to room temperature. Slowly add 250 ml concentrated nitric acid (HNO_3) to the mixture, cool the solution to room temperature, and dilute to 1L volume with distill water.
5. Phosphorus standard solution: Prepare 100 ppm of P standard stock solution by dissolving 0.2195g of KH_2PO_4 (AR grade) distilled water and make upto 500ml.

Working P standards: Prepare 0.5,1, 2, 3, 5, 7, 7, 10, 12, 15, 18 and 20 ppm P working standard solution by pipetting out 0.25, 0.5, 1.0, 1.5, 2.5, 3.5, 5.0, 6.0, 7.5, 9.0 and 10ml of stock P solution respectively into 50ml volumetric flasks separately. Add to each 10ml of vanadate molybdate reagent and make up the volume to 50ml with distilled water and shake. Allow to stand for 30 minutes and yellow colour will be developed.

Procedure for plant sample

1. Transfer 0.5 to 1.0 gm of plant sample into a 100ml conical flask or dry digestion tubes and wet the sample with 10ml of conc. HNO_3. Let it stand for atleast 2 hours or overnight. Pre digestion is required to avoid bumping or violent reaction on addition of diacid mixture to plant sample.
2. Then gently heat on a hot plate until the volume of the contents is reduced to about 4ml or even less and the material turns white sand.
3. Remove the flask from hot plate and allow it to cool. And add distilled water and make the volume upto 100 ml.

$$\% \text{ P in plant} = \frac{\text{ppm from graph X Volume of digest X Volume made after colour development X 100}}{\text{Weight of plant sample X Aliquot taken for colour development X } 10^6}$$

(c) Total potassium in plants

Principle: The plant digest containing alkali and alkaline earth metals fed to flame photometer, it vapourises to gaseous state. Atoms of some specific elements (Na or K) take energy from flame and get excited to the higher state. The excited state of the atom does not remain stable for longer time as it is unstable and reverts to its original state (ground level). While returning to the original level, the atoms loose there energy in the form of radiation. The characteristic colour and wavelength of radiation indicates the type of element (red or pink coloured radiation is due to K-776nm). The intensity of emitted radiation indicates the concentration of the particular element in test samples. The intensity of this fluorescent electromagnetic radiation can be measured by photosensitive detector or photocell of the flame photometer.

Reagent

1. Perchloric Acid ($HClO_4$), 60%
2. Nitric acid (HNO_3)
3. Diacid mixture: HNO_3 : $HClO_4$ (9:4 ratio) mix 900ml of HNO_3 with 400 ml of $HClO_4$

Working standards: prepare 1000 ppm K by dissolving 1.9069g KCl in 1 litre distilled water. Dilute a known concentration from 1000 ppm to get 100 ppm K. From this, prepare working standard solutions containing 1, 2, 3, 4, 5, 6, 7, 8 and 10 ppm K with suitable dilution using distilled water

Procedure

1. Transfer 0.5 to 1.0 gm of plant sample into a 100ml conical flask or dry digestion tubes and wet the sample with 10ml of conc. HNO_3. Let it stand for atleast 2 hours or overnight.
2. Pre digestion is required to avoid bumping or violent reaction on addition of diacid mixture to plant sample.
3. Then gently heat on a hot plate until the volume of the contents is reduced to about 4ml or even less and the material turns white sand. Remove the flask from hot plate and allow it to cool. And add distilled water and make the volume upto 100 ml.

Measurement

1. Transfer a known quantity of sample digest into a 50 ml volumetric flask make up the volume with distilled water.
2. Feed the diluted solution of plant digest and record the flame photometer reading.

Calculations

$$\text{\% K in plant sample} = \frac{\text{Sample reading (ppm) X Volume of digest X Dilution factor X 100}}{\text{Weight of plant sample}}$$

References

Black, C.A. (1965). *Method of Soil Analysis*. American Society of Agronomy Inc., Madison, Wisconsin, USA.

Jackson, M.L. (1967). *Soil Chemical Analysis*, Oxford IBH, Bombay.

Jackson, M.L. (1973). *Soil Chemical Analysis*, Prentice Hall of India Private Limited, New Delhi.

Olsen, S.R. and Watanabe, F.S. (1957). A method to determine phosphorus adsorption maxima of soils as measured by the Langmuir isotherm. *Soil Science Society of America Proceedings,* 21: 144-149.

Subbiah, B.V. and Asija, G.L. (1956). A rapid procedure for estimation of available nitrogen in soils. *Current Science,* 25: 259–260

Walkley, A. and Black. I.A. (1934). An examination of Degtjareff method for determining soil organic matter and a proposed modification of the chromic acid titration method. *Soil Science,* 37: 29-37.

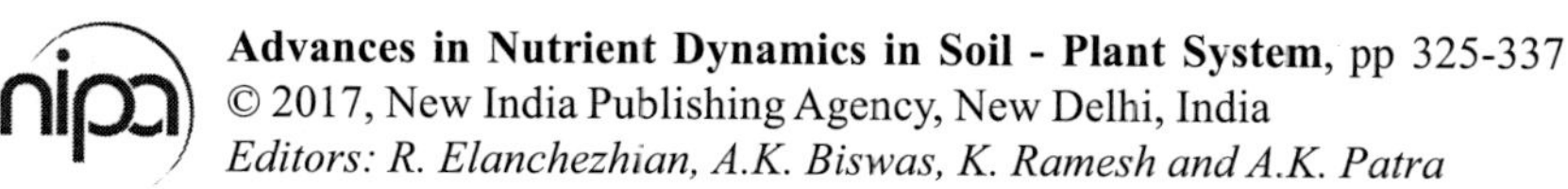
Advances in Nutrient Dynamics in Soil - Plant System, pp 325-337

Editors: R. Elanchezhian, A.K. Biswas, K. Ramesh and A.K. Patra

28

Techniques for Analysis of Soil Biological Parameters

J. K. Thakur and Asit Mandal

ICAR-Indian Institute of Soil Science, Nabibagh, Bhopal – 462038, India

Soil organisms play vital role in agri-ecosystem processes such as nutrient cycling, soil detoxification and organic matter decomposition. They also influence water infiltration by improving soil porosity through production of polysaccharides. Rhizosphere microorganisms help plant to acquire macro and micronutrients, produce phytohormones for better crop establishment. Soil microbes process organic matter to provide a balance of minerals and nutrients which are utilized by plants to achieve healthy and vigorous growth. Microbes also colonize on plants parts and offers protection by excluding many plant pathogens. They are the basic indicators of soil fertility and are responsible for maintaining a hospitable environment for crop growth.

Soil microorganisms are responsible for

- Transformation of raw elements from one chemical form to another. Important soil nutrients released by microbial activity are Nitrogen, Phosphorus, Sulfur, Iron and other micronutrients.
- Breaking down soil organic matter into a form useful to plants. This improves soil fertility by making nutrients available to the plants.
- Degradation of pesticides and other deleterious chemicals entered in the soil.
- Suppression of pathogenic microorganisms that cause diseases by producing antibiotics and other bioactive compounds.

Method to Study Soil Microbes

The techniques which give information on the conditions of organisms in soil are divided into four broad groups. They include; 1. Determination of form and arrangement of microorganisms in soil 2. Isolation and characterization of soil microorganisms 3. Detection of microbial activity in soil 4. Determination of microbial biomass. Depending upon goal of study we attempt to any of the above four broad groups using some of the methods described in the following section.

1. Direct Count

Small arthropods, earthworms, nematodes, bacteria, fungi etc can be estimated by direct count. Microscope with appropriate magnification and resolution capacity can be used to study organism of very small sizes. Direct counting by fluorescence microscopy can give 100– 1000 times more than the numbers obtained by plate counting (Johnsen *et al.,* 2001). Several stains specific to cellular molecules like proteins or nucleic acids have been used for microscopic study. They include fluorescein isothiocyanate (FITC), acridine orange (AO), ethidium bromide (EB) and europium chelate with a fluorescent brightener differential stain (DFS) for bacteria. Hyphae in agar films and on membrane filters can be stained by Phenol aniline blue (PAB), whereas metabolically active hyphae have been counted after their staining with fluorescent diacetate (FDA) (Bloem *et al.,* 1995, Nannipieri *et al.,* 2003). Stains for active bacterial cells include FDA or redox probes such as 2-(p-iodophenyl)-3-(p-nitrophenyl)-5-phenyl tetrazolium chloride (INT) or 5-cyano-2,3-ditoyl tetrazolium chloride (CTC). However, this procedure does not allow counting specific microbial species, and some of the stains used do not discriminate between living and dead microbial cells.

2. Culture Dependent Study

Traditionally microbial diversity has been quantified by counting techniques such as the plate count technique or the Most Probable Number (MPN) technique. The analysis of soil microbial communities using culture techniques employ variety of culture media designed to maximize the recovery of different microbial species. There are numerous examples where these techniques have revealed a diversity of microorganisms associated with various soil quality parameters such as disease suppression and organic matter decomposition (Hill *et. al.,* 2000). It has been estimated that less than 0.1% of the microorganisms found in typical agricultural soils are culturable using current culture media formulations (Torsvik *et al.,* 1990; Atlas and Bartha, 1998). Culture dependent study is inexpensive and can provide information on the active, heterotrophic component of the population. The limitations of culture dependent study include the difficulty in dislodging bacteria or spores from soil particles or biofilms, growth medium and growth condition selections. In addition, plate growth favours microorganisms with fast growth rates, numerically dominant species and those fungi that produce large numbers of spores.

3. Soil Enzyme Activity Study

Since enzymes catalyze all biochemical transformations, measurements of soil enzyme activities are useful indicators of biological activity. Soil enzyme activity measurements have been used as indices of land quality and soil health as well as to understand how human activity is changing biogeochemical cycles in ecosystems. Some of the enzymes and associated nutrient transformation processes are listed in table 1.

Table 1: Soil enzymes as indicators of nutrient cycling

Soil enzyme	Enzyme reaction	Indicator of microbial activity
Dehydrogenase	Electron transport system	C-cycling
β-glucosidase	Cellobiose hydrolysis	C-cycling
Cellulase	Cellulose hydrolysis	C-cycling
Phenol oxidase	Lignin hydrolysis	C-cycling
Urease	Urea hydrolysis	N-cycling
Amidase	N-mineralization	N-cycling
Phosphatase	Release of PO_4^-	P-cycling
Arylsulphatase	Release of SO_4^-	S-cycling

3.1 Fluorescein diacetate hydrolysis as a measure of soil biological activity

In soil more than 90% of the energy flow passes from the decomposition of organic matter by heterotrophic microbes (Schnurer and Rosswall, 1982). The decomposition of organic matter in soil is achieved by enzymes produced by several microbes inhabiting in the soil. These enzymes can also hydrolyse fluorescein diacetate (Green, *et al.*, 2006). Hence estimation of fluorescein diacetate hydrolysis activity of soil serves as indicator of active microbial community in the soil. Fluorescein diacetate (3' 6'-diacetyl-fluorescein) is a fluorescein conjugated to two acetate radicals. This colourless compound is hydrolysed by both free (exoenzymes) and membrane bound enzymes, releasing a coloured end product, fluorescein. This end product absorbs strongly in the visible wavelength (490 nm) and can be measured by spectrophotometry. The enzymes responsible for FDA hydrolysis are plentiful in the soil environment. Non-specific esterases, proteases and lipases also have been shown to hydrolyse FDA (Adam and Duncan, 2001).

3.2 Soil Microbial biomass carbon (SMBC)

As soil microorganisms play an important role in the retention and release of nutrients and energy; any attempt to assess nutrient and energy flow in soil systems must take into account the role of soil microbial biomass.

A more easily applicable, non-subjective and replicable method is described for total microbial biomass determination in soil samples at a particular point of time.

In the fumigation-extraction method, a direct measurement of C and other nutrients contained therein in microbial biomass is carried out. Overnight fumigation of chloroform is carried out to kill all the organisms in soil samples. The microbial biomass constituents released by $CHCl_3$ fumigation treatment can be extracted directly through chemical extractants. The readily oxidisable C contained in the extractant can be measured through standard chemical procedures. (Singh *et al.*, 2005). The method is based on assumptions that:

- Carbon in dead organisms is more rapidly mineralized than in the living organisms.
- Fumigation leads to a complete kill.
- Death of organisms in the non-fumigated soil is negligible compared to that in fumigated soil.
- The only effect of soil fumigation is to kill the living biomass.
- The fraction of dead biomass C mineralized over a given time period does not differ in different soils.

  (Jenkinson and powlson, 1976).

The proportion of microbial biomass carbon in total soil organic carbon is the microbial quotient (Insam and Domsch, 1988). It gives insight into the capability of a soil to support microbial growth and soils with high quality it is expected to maintained higher ratios of the microbial quotient. The metabolic quotient (qCO_2) is the ratio of basal respiration (BR) to the total microbial biomass carbon (Insam and Haselwandter, 1989). The qCO_2 indicates the efficiency by which soil microorganisms use C-resources in the soil. It is expected that stressed soils will provide higher qCO_2 values than less-stressed soils. (Brempong and Odei, 2013)

3.3 Dehydrogenase assay

Lenhard (1956) introduced the concept of determining the metabolic activity of microorganisms in soil and other habitats by measuring dehydrogenase activity which was further refined by Casida *et al.,* (1964) and is widely used for assessment of soil microbial activities. The technique involves the incubation of soil with 2,3,5-triphenyltetrazolium chloride (TTC) either in the presence or absence of added electron-donating substrates. Microbial dehydrogenase activity during this incubation results in reduction of the water soluble, colorless TTC to the water-insoluble, red 2,3,5-triphenyltetrazolium formazan. The latter is then extracted from the soil and read colorimetrically for quantitation (Casida, 1977).

3.4 Phosphatase assay

A large pool of soil phosphorus exists in organic form and can be available to the plant only if converted to inorganic form. Phosphatases play a key role in phosphorous mineralization, and P cycling. They are ubiquitous in soil and have been studied extensively because they catalyze the hydrolysis of organic phosphomonoester to inorganic phosphorous, making it available for plant uptake. According to their optimum pH, phosphatases are classified as acid (orthophosphoric monoester phosphohydrolase, pH 6.5) or alkaline (orthophosphoric monoester phosphohydrolase, pH 11). Acid phosphomonoesterase activity is more important in acid soils than is alkaline phosphomonoesterase activity and vice versa in alkaline soils. Phosphatases originate mainly from soil microorganisms, but in the rhizosphere and detritosphere they can also originate from plant cells (Nannipieri, *et al.,* 2011).

4. Physiological Profiling Based on Carbon Utilization Pattern

This technique is based on utilization of different carbon sources by soil microbes and it determines the catabolic versatility of microbial community of an ecosystem. Community-level physiological profiles have been facilitated by the use of a commercial taxonomic system, known as the BIOLOG® system, which is currently available and has been used extensively for the analysis of soil microbial communities. Utilization of carbon substrate from 95 different carbon substrate provided is detected by the reduction of a redox dye, which results in a color change that can be quantified spectrophotometrically. The pattern of substrates that are oxidized can be compared among different soil samples from a series of times or locations as an indication of differences in the physiological functions of microbial communities (Hill *et al.,* 2000). These community-level physiological profiles can be useful in assessing gross functional diversity.

5. Phospholipid Fatty Acids (PLFA) Analysis

Phospholipid fatty acids are potentially useful signature molecules due to their presence in all living cells and are found exclusively in cell membranes and not in other parts of the cell as storage products. Upon the cell death, membranes starts degrading fast and the component phospholipid fatty acids are metabolized rapidly hence phospholipids can serve as important indicators of active microbial biomass as opposed to non-living microbial biomass. Some Common signature phospholipids fatty acids of microbial community are given in Table 2.

Table 2: Common signature phospholipids fatty acids of microbial community (Hill *et al.*, 2000)

Common bacterial signatures	i15:0, a15:0, 15:0, 16:0, 16:1 ω 5, 16:1 ω 9, i17:0, a17:0, 17:0, 18:1 ω 7t, 18:1 ω 5, i19:0, a19:0
Aerobes	16:1 ω 7, 16:1 ω 7t, 18:1 ω 7t
Anaerobes	cy17:0, cy19:0
Sulfate-reducing bacteria	10Me16:0, i17:1 ω 7, 17:1 ω 6
Methane-oxidizing bacteria	16:1 ω 8c, 16:1 ω 8t, 16:1 ω 5c, 18:1 ω 8c, 18:1 ω 8t, 18:1 ω 6c
Cyanobacteria	18:2 ω 6
Protozoa	20:3 ω 6, 20:4 ω 6
Fungi	18:1 ω 9, 18:2 ω 6, 18:3 ω 6, 18:3 ω 3
Actinobacteria	10Me18:0
Microalgae	16:3 ω 3
Flavobacterium balustinum	i17:1 ω 7, Br 2OH-15:0

6. Genomic Study

Research toward the taxonomic resolution and characterizing of the soil metagenomes is currently underway. The diversity of the microorganisms in soil can be estimated using one of two approaches, cellular cultures or molecular biology techniques. Cellular methods of controlled growth of microorganisms under laboratory conditions have only yielded a fraction of taxa, and therefore only reveal a subset of the overall microbial community. Of the molecular techniques, conventional, vector-based cloning protocols have been used. However, the inherent cloning biases that hamper vector-based cloning procedures are unavoidable (Shokralla *et al.,* 2012). As a result, cloning methods have all but been replaced with newer molecular approaches. Prior to the use of next-generation sequencing, multiple molecular techniques were employed to address microbial diversity in natural ecosystems. While these techniques proved very useful for better ascertaining microbial communities of environmental samples, these techniques were costly, time consuming and could not be used to identify non-culturable microbes.

Usually microbial diversity in soil has been analyzed by methods such as viable and cultivable methods. The molecular techniques generally involve extraction of nucleic acid, directly or indirectly, from soil. They are independent of culture and according to their sensitivity can detect species, genera, families or even higher taxonomic groups. Few of the commonly used methods are described here;

6.1 G:C Ratio

GC ratio of an organism is percentage of guanine plus cytosine present in its DNA. This ratio can be determined by measuring the melting temperature of DNA or by chromatographic methods. The GC ratio of prokaryotes varies from as low as 20% to as high as 80%. GC ratio can be useful in drawing information about organism depending upon the situation. For example two organisms can have same GC ratio but they may not be taxonomically or phylogenetically same because for same base composition a number of arrangement is possible. Even if this analysis is considered to be low resolution, it can be used to indicate overall changes in microbial community structure, especially when the diversity is low.

6.2 DNA:DNA hybridization

GC ratio does not give any information about nucleotide sequence on DNA strand which limits its applicability. Base sequence is important because two organisms with same base sequence likely to contain many similar (if not identical) genes. Two DNAs would thus be expected to hybridize to one another in proportion in their gene sequences. Genomic hybridization measures the degree of sequence similarity in two DNAs and is useful for differentiating very closely related organisms where rRNA sequencing may fail to be definitive. In DNA Hybridization experiment, DNA isolated from one organism is made radioactive with ^{32}P or ^{3}H, cut into a relatively small size, heated to denature and mixed with an excess of DNA of second organism prepared in the same way but not radiolabeled. The DNA mixture is then cooled to allow it to reanneal and double stranded DNA is separated from remaining unhybridized DNA. There is no fixed convention as to how much hybridization between two DNAs is necessary to assign two organisms to the same taxonomic rank but hybridization value of 70% or greater are recommended for considering two isolates to be of same species and value of at least 25% for two organisms to reside in the same genus. DNA:DNA Hybridization is sensitive enough to delineate the subtle differences in the genus of the two organism and therefore can differentiate between two closely related organisms (Madigan and Martinko, 2006).

Ribotyping

Ribotyping measures the unique pattern that is generated from digestion of DNA of an organism by a restriction enzyme and the fragments are separated and probed with a ribosomal RNA probe. Because of difference in restriction enzyme cut site of two organisms in their ribosomal RNA sequence, different pattern will be generated for different organisms which are unique for that species with that restriction enzyme. In fact, ribotyping is so specific that it has been given the nick name "molecular fingerprinting" because of a unique series of bands appears for virtually any organisms. Ribotyping is both rapid and specific method of bacterial identification since it bypasses the sequencing and sequence alignment methods (Madigan and Martinko, 2006).

Fluorescent *in-situ* Hybridization (FISH)

Fluorescence *in situ* hybridization (FISH) enables in situ phylogenetic identification and enumeration of individual microbial cells by whole cell hybridization with oligonucleotide probes. A large number of molecular probes targeting 16S rRNA genes have been reported at various taxonomic levels. The FISH probes are generally 18–30 nucleotides long and contain a fluorescent dye at the 5' end that allows detection of probe bound to cellular rRNA by epifluorescence microscopy. In addition, the intensity of fluorescent signals is correlated to cellular rRNA contents and growth rates, which provide insight into the metabolic state of the cells. FISH can be combined with flow cytometry for a high resolution automated analysis of mixed microbial populations. The FISH method was used to follow the dynamics of bacterial populations in agricultural soils treated with s-triazine herbicides. A variety of molecular probes were used to target specific phylogenetic groups of bacteria such as α, β, γ and δ subdivisions of Proteobacteria and Planctomycetes (Rastogi and Saini, 2011).

Ribosomal Intergenic Spacer Analysis (RISA)

Ribosomal intergenic spacer analysis (RISA) involves PCR amplification of a portion of the intergenic spacer region (ISR) present between the small (16S) and large (23S) ribosomal subunits. The ISR contains significant heterogeneity in both length and nucleotide sequence. By using primers annealingto conserved regions in the 16S and 23S rRNA genes, RISA profiles can be generated from most of the dominant bacteria existing in an environmental sample. RISA provides a community-specific profile, with each band corresponding to at least one organism in the original community. The automated version of RISA is known as ARISA and involves use of a fluorescence-labeled forward primer, and ISR fragments are detected automatically by a laser detector. ARISA allows simultaneous analysis of many samples; however, the technique has been shown to overestimate microbial richness and diversity. Ranjard *et al.,* (2001) evaluated ARISA to characterize the bacterial communities from four types of soil differing in geographic origins, vegetation cover, and physicochemical properties. ARISA profiles generated from these soils were distinct and contained several diagnostic peaks with respect to size and intensity. Their results demonstrated that ARISA is a very effective and sensitive method for detecting differences between complex bacterial communities at various spatial scales (between- and within- site variability) (Rastogi and Saini, 2011).

Multilocus Sequence Typing (MLST)

The methods described above have limitation of targeting only single gene (SSU rRNA) which restricts its resolution in differentiating the species at strain level. Multilocus sequence typing (MLST) involves sequencing fragments of six to seven "housekeeping genes" from an organism and comparing these with the same gene set from different strains of the same organism. MLST has sufficient resolving power to differentiate even very closely related strains but is not suitable above the species level (Madigan and Martinko, 2006).

RAPD (Random Amplified Polymorphic DNA)

The RAPD analysis is a commonly used molecular marker in genetic diversity studies. Knowledge of the DNA sequence for the targeted gene is not required for RAPD study as the primers will bind randomly in the given genomic DNA. The primers are allowed to anneal randomly at multiple sites on the genomic DNA under low annealing temperature, typically ≤35°C. RAPD primers are decamer (10 nucleotide length) DNA fragments from PCR amplification of random segments of genomic DNA with single primer of arbitrary nucleotide sequence and which are able to differentiate between genetically distinct individuals, although not necessarily in a reproducible way. The principle is that, a single, short oligonucleotideprimer, which binds to many different loci, is used to amplify random sequencesfrom a complex DNA template. This means that the amplified fragment generatedby PCR depends on the length and size of both the primer and the target genome. The assumption is made that a given DNA sequence (complementary to that of the primer) will occur in the genome, on opposite DNA strands, in opposite orientation within a distance that is readily amplifiable by PCR. These amplified products are usually separated on agarose gels (1.5-2.0%) and visualised by ethidiumbromide staining (Senthil Kumar and Gurusubramanian, 2011). The RAPD band generated for an organism is compared from the other organisms' RAPD band prepared in same way.

Terminal Restriction Fragment Length Polymorphism (T-RFLP)

Terminal restriction fragment length polymorphism (T-RFLP) analysis is a popular high-throughput fingerprinting technique used to monitor changes in the structure and composition of microbial communities (Schütte *et al.,* 2008). With the T-RFLP method, after community DNA is isolated from a field sample, DNA coding for 16S rRNA is specifically amplified by PCR using primer pairs designed from the conserved region of the gene. The PCR product is then digested by a restriction enzyme and the length profile of terminal restriction fragment (TRF) labeled by the fluorescent dye is detected for the identification of species composition of the bacterial communities. The ratios of each PCR amplicon, estimated by measuring fluorescence emission intensity, could indicate the relative abundance of bacterial species. This method can be used to estimate the relative abundance of dominant bacterial species (relatively quantitative) and to study the pattern shift (in terms of dominant bacterial groups and their relative abundance) of bacterial community structures.

Length Heterogeneity PCR

Length heterogeneity PCR (LH-PCR) analysis is similar to the T-RFLP method except that the latter detects amplicon length variations that are produced after restriction digestion, whereas in LH-PCR different microorganisms are discriminated based on natural length polymorphisms that occur due to mutation within genes (Mills *et al.,* 2007). Amplicon LH-PCR interrogates the hyper variable regions present in 16S rRNA genes and produces a characteristic profile. LH-PCR utilizes a fluorescent dye-labeled forward primer, and a fluorescent internal size standard is run with each sample to measure the amplicon lengths in base pairs. The intensity (height) or area under the peak in the electropherogram is proportional to the relative abundance of that particular amplicon. One advantage of using LH-PCR over the T-RFLP is that the former does not require any restriction digestion and therefore PCR products can be directly analyzed by a fluorescent detector. The limitations of LH-PCR technique include inability to resolve complex amplicon peaks and underestimation of diversity, as phylogenetically distinct taxons may produce same length amplicons (Mills *et al.,* 2007). LH-PCR was used in combination with fatty acid methyl ester (FAME) analysis to investigate the microbial communities in soil samples that differed in terms of type and/or crop management practices (Ritchie *et al.,* 2000). LH-PCR results strongly correlated with FAME analysis and were highly reproducible, and successfully discriminated different soil samples.The most abundant bacterial community members, based on cloned LH-PCR products, were members of the β-Proteobacteria, Cytophaga–Flexibacter–Bacteriodes and the high G + C content Gram-positive bacterial group.

Metagenomic approach

The collective genomes of a total community in a given habitat are called metagenome. The term metagenomics was first used by Jo Handelsman *et al.,* (1998) in the University of Wisconsin. Metagenomics is an emerging field in which the power of genomic analysis is applied to entire communities of microbes, bypassing the need to isolate and culture individual microbial species. It is also referred to as "community genomics" or "environmental genomics". Metagenomics enables us to study microorganisms by deciphering their genetic information from DNA that is extracted directly from communities of environmental microorganisms, thus bypass the need for culturing, isolation and purification of culture. This discipline builds on the successes of culture-independent 16S rRNA surveys of environmental samples. It offers insights into the evolutionary history as well as previously unrecognized physiological abilities of microbial communities specialized to live in a given environmental niche. The resultant wealth of genes and molecular structures deciphered from uncultured microorganisms has tremendous potential in the development of novel biocatalysts for industrial and medical applications.

Summary of molecular methods for soil microbial diversity studies and their applications (Lynch *et al.*, 2004)

Method	Type of information and resolution	Application in soil microbial analysis
Mole % G+C composition	Genetic community profile, overall community composition. Low resolution	Comparative analysis of overall changes in community composition
PCR-DGGE /TGGE sequcncing of individualbands	Genetic fingerprinting of communities, affiliation of Predominant community members. Intermediate resolution	Comparative analysis of community structure, spatial and temporal changes in community composition
PCR—T-RFLP	Community composition, relative abundance of numerically dominant community members. Intermediate resolution	Comparative analysis of distribution of microbial populations, monitoring changes in community composition
PCR—ARDRA	Genetic fingerprinting of simple communities, populations or phylogenetic groups Discrimination at lower taxonomic (species) levels High resolution	• Comparative analysis of microbial population dynamics. • Diversity within phylogenetic or functional groups of microorganisms
PCR—RISA	Genetic fingerprinting of populations or phylogenetic groups. Simultaneously analysis of different microbial groups. Discrimination at species or group level High resolution	• Comparative analysis of microbial population dynamics. • Diversity within phylogenetic or functional groups of microorganisms
FISH	Detection and specific counting of metabolic active microorganisms. Intermediate resolution	Comparative analysis of community structure. Detectionand identification of active cells. Direct phylogenetic information on community members

Approaches to Metagenomic Analysis (Handelsman, 2004)

- Isolation and purification of DNA from soil samples, often without prior cultivation of the organism present.
- Selection of suitable vectors and host system for cloning of the DNA obtained.
- Analysis of the cloned fragments using either sequence-based or functional screening approaches

References

Adam, G. and Duncan, H. (2001). Development of a sensitive and rapid method for the measurement of total microbial activity using fluorescein diacetate (FDA) in a range of soils. *Soil Biology & Biochemistry*, 33: 943-951.

Atlas, R.M. and Bartha, R., (1998). Microbial Ecology: Fundamentals and Applications. Benjamin/ Cummings, Redwood City, CA, 694 pp.

Bloem, J., Bolhuis, P. R., Veninga, M. R. and Wieringa, J. (1995). Microscopic methods for counting bacteria and fungi in soil. In: Methods in Applied Soil Microbiology and Biochemistry (eds K. Alef & P. Nannipieri), pp. 162–173. Academic Press, London.

Brempong, S. A. and Odei, E. O. (2013). Microbial biomass carbon determination by chloroform fumigation and substrate induced respiration methods. *Agricultural Science Research Journal* 3(12): 372- 378.

Casida, L.E. Jr. (1977). Microbial Metabolic Activity in Soil as Measured by Dehydrogenase Determinations. *Applied Environmental Microbiology,* 34(6): 630-636.

Casida, L.E., Jr., Klein, D.A. and Santoro, T. (1964). Soil dehydrogenase activity. *Soil Science*, 98:371-376.

Green, V.S., Stott, D.E. and Diack, M. (2006). Assay for fluorescein diacetate hydrolytic activity: Optimization for soil samples. *Soil Biology & Biochemistry,* 38: 693–701.

Handelsman, J. (2004). Metagenomics: Application of genomics to uncultured microorganisms. *Microbiology and Molecular Biology Reviews,* 68 (4):669–685.

Hill, G T., Mitkowski, N.A., Aldrich-Wolfe, L., Emele, L.R., Jurkonie , D.D., Ficke, A., Maldonado-Ramirez, S., Lynch, S.T. and Nelson. E. B. (2000). Methods for assessing the composition and diversity of soil microbial communities. *Applied Soil Ecology,* 15: 25–36.

Insam, H. and Domsch, K. H. (1988). Relationship between soil organic carbon and microbial biomass on chronosequences of reclaimed sites. *Microbial Ecology,* 15:177-188.

Insam, H. and Haselwandter, K. (1989). Metabolic quotient of the soil microflora in relation to plant succession. *Oecologia,* 79:174- 178.

Jenkinson, D.S. and Powlson, D.S. (1976). The effects of biocidal treatment on metabolism in soil. A method for measuring soil biomass. *Soil Biology and and Biochemistry,* 8:209-213.

Johnsen, K., Jacobsen, C.S., Torsvik, V. and Sorensen, J. (2001). Pesticide effects on bacterial diversity in agricultural soils – a review. *Biology and Fertility of Soils*, 33: 443–453.

Lenhard, G. (1956). The dehydrogenase activity in soil as a measure of the activity of soil microorganisms. *Z. Pflanzenernaehr. Dueng. Bodenkd.* 73:1-11.

Lynch, J.M., Benedetti, A., Insam, H., Nuti, M. P., Smalla, K., Torsvik, V. and Nannipieri, P. (2004). Microbial diversity in soil: ecological theories, the contribution of molecular techniques and the impact of transgenic plants and transgenic microorganisms. *Biology and Fertility of Soils,* 40: 363–385.

Madigan, M. T. and Martinko, J. M. (2006). Brock Biology of Microorganisms. Eleventh Edition. Pearson Prentice Hall.

Mills, D. K., Entry, J. A. and Gillevet, P. M. (2007). Assessing microbial community diversity using amplicon length heterogeneity polymerase chain reaction. *Soil Science Society of America Journal,* 71:572–578.

Nannipieri, P., Ascher, J., Ceccherini, M. T., Landi, L., Pietramellara G. and Renella, G. (2003). Microbial diversity and soil functions. *European Journal of Soil Science*, 54: 655–670.

Nannipieri, P., Giagnoni, L., Landi, L. and Renella, G. (2011). Role of Phosphatase Enzymes in Soil. E.K. Bunemann *et al.,* (eds.), Phosphorus in Action, Soil Biology 26, *PP,215-243.* DOI 10.1007/978-3-642-15271-9_9. Springer.

Ranjard, L., Poly, F., Lata, J. C., Mougel, C., Thioulouse, J. and Nazaret, S. (2001). Characterization of bacterial and fungal soil communities by automated ribosomal intergenic spacer analysis fingerprints: biological and methodological variability. *Appl. Environ. Microbiol.* 67:4479–4487.

Rastogi, G and Sani, R. K.(2011). Molecular Techniques to Assess Microbial Community Structure, Function and Dynamics in the Environment. In, *Microbes and Microbial Technology:Agricultural and Environmental Applications*, Ahmad *et al.,* (eds.), PP 29-57. DOI 10.1007/978-1-4419-7931-5_2, Springer Science.

Ritchie, N. J., Schutter, M. E., Dick, R. P. and Myrold, D. D. (2000). Use of length heterogeneity PCR and fatty acid methyl ester profiles to characterize microbial communities in soil. *Applied Environmental Microbiology,* 66:1668–1675.

Schnurer, J. and Rosswall, T. (1982). Fluorescein Diacetate Hydrolysis as a Measure of Total Microbial Activity in Soil and Litter. *Applied Environmental Microbiology,* 43(6): 1256-1261.

Schütte, U.M.E., Abdo, Z., Bent, S.J., Shyu. C., Williams, C.J., Pierson, J.D. and Forney, L.J. (2008). Advances in the use of terminal restriction fragment length polymorphism (T-RFLP) analysis of 16S rRNA genes to characterize microbial communities. *Applied Microbiology and Biotechnology,* 80:365–380.

Senthil Kumar, N. and Gurusubramanian, G. (2011). Random amplified polymorphic DNA (RAPD) markers and its applications. *Science Vision,* 11 (3): 116-124.

Singh, D., Chhonkar, P. K. and Dwivedi, B. S. (2005). Manual on Soil, Plant and Water Analysis. 200p. Westville Publishing House, New Delhi.

Thakur, J. K., Sahu, A., Singh, U. B., Mandal, A. and Manna, M. C. (2015). Molecular Techniques in Soil Biodiversity Study. Pp 505-524. In Microbial Diversity: A Boon for Agricultural Sustainability. Sinha *et al.,* (eds.), Biotech Books, New Delhi.

Torsvik, V., Salte, K., Sorheim, R. and Goksoyr, J. (1990). Comparison of phenotypic diversity and DNA heterogeneity in a population of soil bacteria. *Appled Environmental Microbiology,* 56: 776-781.

Advances in Nutrient Dynamics in Soil - Plant System, pp 339-346

Editors: R. Elanchezhian, A.K. Biswas, K. Ramesh and A.K. Patra

29

Inductively Coupled Plasma- Emission Spectroscopy (ICP-ES)

M. V. Coumar and S. Rajendiran

ICAR-Indian Institute of Soil Science, Nabibagh, Bhopal – 462038, India

A new analytical technique called inductively coupled plasma emission spectroscopy has been used for simultaneous multi element analysis of biological materials and soils (Hou and Jones 2000; Wikipedia, 2010; Bradford and Cook, 1997). This technique offers advantages over AAS and other multi-element methods because matrix problems are eliminated or minimized through use of the high temperature argon plasma. Apart from multi- element capability at all concentration levels, plasmas are noted for relative freedom from chemical and ionization interferences that are common with AAS, and detection limits are equal to or better than AAS, depending on the element to be analyzed. Elements such as Al, P, S and B, which are either poorly measured at low concentrations or not possible by AAS, are readily determined with higher sensitivity by ICP.

The ICP-ES technique has been applied to the analysis of a large variety of agricultural and food materials. Types of samples include soils, fertilizers, plant materials, feedstuffs, foods, animal tissues, and body fluids. Analyses of these materials are required to determine levels of essential nutrients as well as levels of toxic elements in the materials (Bradford and Cook, 1997). Most agricultural and food materials are generally not in the form of dilute aqueous solutions nor are they readily soluble in distilled water. Therefore, analyses of these materials by ICP-ES often require rigorous sample preparation procedures to be carried out prior to analysis. Fortunately for the analyst, the use of modern microwave sample digestion techniques is helping to simplify the sample preparation steps for agricultural and food materials as well as many other sample types.

Principle

ICP is based on the observation of atomic emission spectra when samples in the form of an aerosol, thermally generated vapour or powder are injected into an inductively coupled plasma atomization and excitation source. By definition, plasma refers to a hot gas in which a significant fraction of their atoms or molecules is ionized. Plasmas are electrically conducted and have been referred to as electrical flames, as no combustion takes place.

The ICP is produced by passing initially ionized argon gas through a quartz torch located within an induction coil (Cu coil) which is connected to a radio frequency (RF) generator. The ratio frequency generator produces 1.5 to 3 KW power at a frequency of 27.1 MHz. An oscillating magnetic field is formed within the quartz torch in response to the radio frequency energy passing through the coil. Electrons and ions passing through the oscillating electromagnetic field flow at high acceleration rates within the quartz torch space. As argon gas enters the magnetic field associated with the induction coil, its atoms collide with the accelerated ions and electrons resulting in the ionization of the argon gas. These collisions give rise to heating, which produces plasma with temperature ranging from 6000 to 10000 °K. The resultant plasma is contained within the torch by means of argon flow (Ghosh *et al.,* 2013).

The method of presenting the sample to the plasma is similar to that used in flame atomic absorption. The liquid sample is aspirated into the plasma through a nebulizer system by using argon carrier gas at a rate of about 1 litre argon/minute. The prevailing high temperature in the plasma leads to complete vaporization, atomization and excitation of the element to be analyzed. The exited neutral atoms or ions of the sample emit radiation of characteristics wavelengths where the intensity of the emitted radiation is measured by the spectrophotometer component of the ICP- AES instrument.

Instrumentation

In inductively coupled plasma emission spectrometry, the sample in the form of liquid is normally introduced into instrument which is subsequently converted to aerosol through the process called nebulization with the help of nebulizer. Once the liquid is converted to aerosol, it is further transported to the plasma torch where sufficient energy is available for vaporization, atomization and ionization process. When these excited atoms and ions return to their ground state or to lower excitation states they will emit electromagnetic radiation in the ultra-violet/ visible range of the spectrum. Each excited element emits specific wavelengths that has a typical emission spectrum. The intensity of the radiation is proportional to the element concentration. A representation of the layout of a typical ICP-ES instrument is shown in Figure 1.

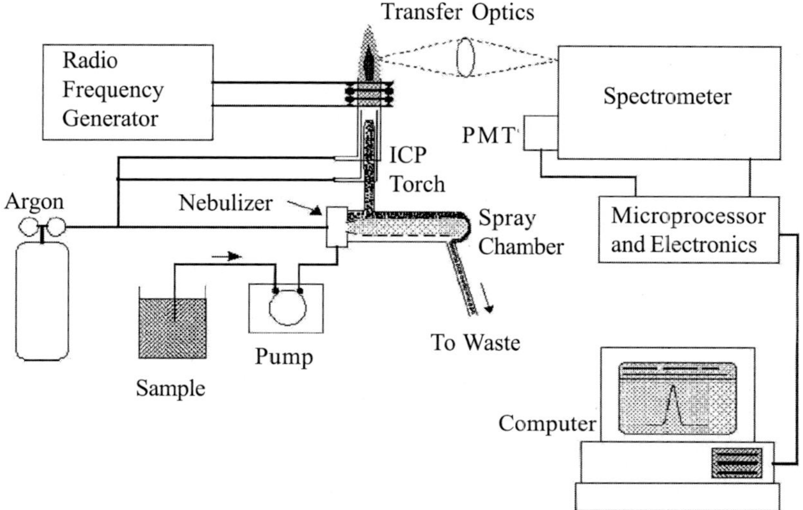

Fig.1: Component of ICP-ES
(Adapted from the schematic diagram of Ghosh *et al.,* 2013)

Nebulizers

Nebulizers are devices mostly used in ICP for sample introduction into the instrument and subsequent transportation to plasma. Nebulizer converts sample state from liquid to aerosol. The formation of aerosol normally depends on neubulizer type, solvent matrix and gas flow rate. The aerosol undergoes the process of desolvation, vaporization and atomization in the presence of high temperature near plasma region. Basically two types of nebulizers *i.e.* pneumatic and ultrasonic nebulizers are used usefully in ICP-OES. The ideal sample introduction system would be one that delivers the entire sample to the plasma in a form that the plasma could reproducibly undergo the process as mentioned above. Since only small droplets are useful in the ICP, the ability to produce small droplets for a wide variety of samples largely is determined by suitable nebulizer.

Spray chambers

Spray chamber is placed between nebulizer and plasma torch. Only very few droplets in the aerosol produced by nebulizer are transported to the plasma torch. Hence, the main purpose of spray chamber is to remove larger droplets having more than 10 μm in diameter in the aerosol and allowing only smaller droplets further inside plasma torch. A secondary purpose of the spray chamber is to smooth out pulses (to dampen pulses) that may occur during nebulization while pumping the sample solution. Thermally stabilized spray chambers are sometimes adopted to decrease the amount of liquid introduced into the plasma, thus providing stability especially when organic solvents are involved. Most of the spray chamber designed for ICP generally allows less than 10 μm in diameter droplets further into plasma torch which constitutes only 1 to 5% of the sample that is introduced by the nebulizer. The remaining droplet having size greater than 10 μm in diameter are drained out with the help of spray chamber which constitutes 95 to 99% of the droplet injected by the nebulizer. The material used for spray chamber must be corrosion-resistant materials and heat resistant because samples containing corrosive solvent like hydrofluoric acid, used for sample injection may damage spray chambers.

Plasma Torch

The torch unit of an ICP is used to create and sustain plasma. The main purpose of the torch is to (1) desolvation-evaporate the solvent from the analyte injected by the spray chamber, (2) atomization- break the ionic bonds and form gaseous state atoms, and (3) ionization- atom ionization and excitation. The torch contains three concentric quartz glass tubes, for samples and argon gas used to aspirate it pass through the center tube. The spacing between the two outer tubes is kept narrow so that the plasma generating gas (argon) passes through the middle tube, and the argon passing through the outer tube is used to cool the quartz torch. The outside chamber is designed to make the gas spiral tangentially around the chamber as it proceeds upward so as to keep the quartz tube cool. The chamber between the outer flow and the inner flow sends gas directly under the plasma toroid. This flow keeps the plasma discharge away from the intermediate and injector tubes and makes sample aerosol introduction into the plasma easier. In normal operation of the torch, this flow, called the auxiliary flow (intermediate flow) is usually introduced to reduce carbon formation on the tip of the injector tube when organic samples are being analyzed. However, it may also improve performance with aqueous samples as well. With some torch and sample introduction configurations, the intermediate flow may be as high as 2 or 3 L/min or not used at all.

Radio Frequency Generators

The radio frequency (RF) generator is the devices that provides the power to the RF coil at oscillating frequencies and sustain the plasma discharge. Plasma is created when the argon gas is seeded with a spark from a Tesla coil (basically an automatic gas grill lighter). The spark ionizes some of the argon, and the cations and electrons produced from that accelerate towards the RF coil. Due to their kinetic energy and collisions with other atoms a large amount of heat is generated, enough to generate and sustain a plasma at temperatures up to 10000 ^{0} K. Plasma RF power also affects the plasma temperature i.e. the greater the power intake, the higher the plasma temperature. The net effect of power on analyte sensitivity depends on the ratio of analyte signal to background noise. This power, typically ranging from about 700 to 1500 watts, is transferred to the plasma gas through a load coil surrounding the top of the torch. The load coil, which acts as an antenna to transfer the RF power to the plasma, is usually made from copper tubing and is cooled by water or gas during operation. In terms of an electrical analogy, the term "inductively coupled" in ICP is a result of the coupling of the induction coil and the electrons. The copper induction coil serves as the "primary winding" of the radio-frequency transformer and the "secondary winding" is the oscillating electrons and cations in the plasma; the two "windings" are thus coupled together because the second winding depends on the presence of the first.

Optics

The spectrometer isolates analytical wavelengths from the emitted light plasma. The majority of wavelengths lie within the region 160 to 860 nm. Separation of light into its component wavelengths is normally achieved using a diffraction grating. There are three types of diffraction grating: ruled, holographic and echelle grating. Many spectrometers are either flushed with nitrogen or argon or maintained under vacuum to remove any oxygen.

Detectors

Once the proper emission line has been isolated by the spectrometer, the detector and its associated electronics are used to measure the intensity of the emission line. By far the most widely used detector for ICP is the photomultiplier tube (PMT). New generation detectors have recently been introduced. These are solid-state detectors, but are also referred to as charged - transfer devices. There are two sub-classifications: charge – coupled devices (CCDs) and charge-injection devices (CIDs). A CID consists of a two dimensional array of detector elements when coupled to an Echelle spectrometer, is capable of simultaneous line analysis over the range 170 -800 nm (Earle *et al.,* 1993; Zander *et al.,* 1999).

Readout devices and data processing

The detector produces an electrical signal which is processed by an electronic circuit before being measured by a read-out device. When single element detection was the goal of instrumentation, a simple chart recorder or digital display was sufficient for data collection. However, with the advances in ICP-ES technology today, computers are a necessity, not only to control the automatic sampler and also for performing the task of sample logging, construction of calibration curves which facilitates rapid and efficient handling of data.

Preparation of soil and plant samples

Digestion of soil and plant samples for total elemental analysis by ICP-ES is similar to that used for various emission instruments. Universal/multi-element soil extractants are used for the extraction of soil samples. Acidified AB-DTPA and Mehlich No.1 extracts have been analysed by ICP-ES.

Preparation of standard solution

Procedure for the preparation of stock standard solution containing 1000 mg/litre of an element from pure metal wire or suitable compounds of the element is similar to that for AAS. However, multi element working standard solutions (secondary standards) should be made in such a way that these contain maximum number of the elements compatible with stability considerations and match with the sample solutions in kind and strength of acids. In soil and plant analysis, one set of secondary standards is required for each multi element extracting solution, and one for each soil and plant digest.

Wavelength Selection

Four main factors govern wavelength selection in ICP-ES: analyte concentration, overall sample composition, potential spectral interferences, and the spectral range of the instrument (Hou and Jones, 2000). In the case of natural waters, the major analytes are Na, K, Ca, Mg, Si and total S, and their concentrations typically range from 0.1 to 10^4 mg L^{-1}. With the possible exception of the determination of potassium in very dilute waters, appropriate spectral lines can be selected which will allow all these constituents to be determined directly in most samples, without the need for dilution or pre-concentration. In the case of Ca and Mg, it is necessary to avoid their most sensitive lines in order to achieve calibration ranges which can be extended to high enough concentrations. For minor and trace elements, present at concentrations ranging from <1 to 10^3 μg L^{-1} the most sensitive emission lines are usually chosen, subject always to any constraints imposed by spectral interferences from major components of the sample. Several elements, including Sr, Ba, Li, B, Fe and Mn, can be determined directly in natural waters at typical environmental concentrations. However, the limits of detection attainable in ICP-ES are insufficient to allow many important transition and rare earth elements to be determined without pre-concentration.

Precautions

- Filter soil extracts, and soil and plant tissue digests with Whatman No.42 filter paper to prevent clogging of the nebulizer.
- To prevent clogging of the nebulizer tip, either use high salt nebulizer (Babington type) or standards and samples having very low salt content.
- Avoid mixing of chemicals that cause precipitation during the preparation of multi-element working standard solutions.

Limitation in ICP

ICP technique is applicable to the determination of a large number of elements. The detection limits for these elements are generally in the μg/L (ppb) range. As in many techniques, the detection limit is regarded as the lowest concentration at which the analyst can be relatively certain that an element is present in a sample. Measurements made at or near the detection limit, however, are not considered to be quantitative. For purposes of rough quantitation (±10%), it is recommended that an element's concentration should be at least five times higher than the detection limit. For accurate quantitation (± 2%), the concentration should be greater than 100 times the detection limit. While most of the over 70 elements that can be determined by ICP have low detection limits, it is worthwhile to discuss the elements that are usually not determined at trace levels by ICP. These elements fall into three basic categories. The first category includes those elements that are naturally entrained into the plasma from sources other than the original sample. For example, in an argon ICP, it would be hopeless to try to determine traces of argon in a sample. A similar limitation might be encountered because of the CO_2 contamination often found in argon gas. When water is used as a solvent, H and O would be inappropriate elements and C if organic solvents were used. Entrainment of air into the plasma makes H, N, O and C determinations quite difficult.

Another category of elements generally not determined at trace levels by ICP includes those elements whose atoms have very high excitation energy requirements such as the halogens, Cl, Br and I. Though these elements may be determined, the detection limits are quite poor compared to most ICP elements. The remaining category includes the man-made elements which are typically so radioactive or short-lived that gamma ray spectrometry is preferable for their determination.

Interference

Interference is anything that causes the signal from an analyte in a sample to be different from the signal for the same concentration of that analyte in a calibration solution. Despite the fact that the presence of an interference can be devastating to the accuracy of a determination, there is no analytical technique that is completely free from interferences. However, modern trace elemental analysis instruments have been designed to minimize the interferences.

When ICP was first introduced to the analytical community, the claim was often made that the technique was nearly free from interferences. This claim was made because the classic chemical interferences that were found in flame atomic absorption spectrometry were not observed in ICP. Soon after analysts began measuring trace element concentrations in a wide variety of samples, however, the reality of the existence of some interference became apparent. The interferences that we know about today in the ICP are spectral in origin. Other interferences are often the result of high concentrations of certain elements or compounds in the sample matrix and are not too severe for most samples.

The best way to guard against inaccurate results due to unexpected interferences with ICP is an adequate quality control program. The components of quality control will vary with the sample type, the degree of precision and accuracy required, and the penalty anticipated if errors exceed acceptable levels. The most generally applicable quality control procedure is to analyze samples of known composition along with the unknowns. These reference materials should match the sample matrix and the concentration range of the analyte elements.

Reference materials representing many different types of matrices with numerous certified trace element concentrations can be purchased from the U. S. National Institute of Standards and Technology (formerly the National Bureau of Standards).

Summary

Optical spectrometry has made considerable advances since the first flame-based instruments were introduced in the 1960s. Recent advances in plasma technology, monochromator layout, and computer based detector systems such as charge transfer devices have given a new meaning to "state-of-the-art" technology. Inductively coupled plasma optical emission spectroscopy (ICP-OES) has been used for simultaneous multi element analysis of biological materials and soils. Today flame atomic absorption spectrometer (FAAS) instruments are only used in situations where only one or two elements are analyzed, whereas, ICP is applicable to the determination of a large number of elements simultaneously with detection limits in parts per billion (ppb) range. While ICP-OES detection limits are significantly better than flame- based techniques, ICP-MS (mass spectrometry) yields even better detection limits and can distinguish between different isotopes.

References

Hou, X. and Jones, B.T. (2000). Inductively Coupled Plasma/Optical Emission Spectrometry. In: Encyclopedia of Analytical Chemistry, R.A. Meyers (Ed.) pp. 9468 – 9485.

Wikipedia, the free encyclopedia (2010). Inductively coupled plasma mass spectrometry. Accessed at http://en.wikipedia.org/wiki/1nductively coupled plasma mass spectrometry 7/11/2010.

Bradford, T and Cook, M.N. (1997). Inductively Coupled Plasma (ICP) Accessed 6/10/2010 at http://www.cee.vt.edu/ewr/environmental/teach/smprimer/icp/icp.html.

Boss, C.B and Fredeen, K. J. (1997). Concept, Instrumentation and Techniques in Inductively Coupled Plasma Optical Emission Spectrometry, 2nd edition, Perkin-Elmer, Norwalk, CT.

Ghosh, S., Prasanna, V.L., Sowjanya, B., Srivani, P., Alagaraja, M. and Banji, D. (2013). Inductively Coupled Plasma –Optical Emission Spectroscopy: A Review. *Asian Journal of Pharmaceutical Analysis*, 3(1): 24-33.

Earle, C.W., Baker, M.E., Bonner Denton, M. and Pomeroy, R.S. (1993). 'Imaging Applications for Chemical Analysis Utilizing Charge Coupled Device Array Detectors', *Trends in Analytical Chemistry*, 12 (10): 395 – 403.

Zander, A.T., Chien, R.L., Cooper, C.B. and Wilson, P.V. (1999). An Image Mapped Detector for Simultaneous ICP-AES, *Analytical Chemistry*, 71(16): 3332-3340.

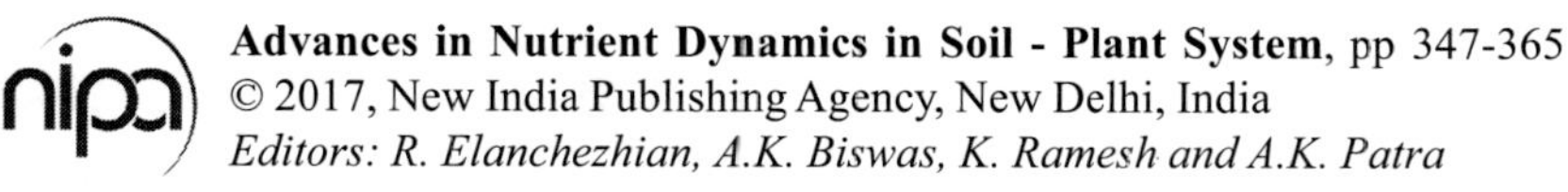
Advances in Nutrient Dynamics in Soil - Plant System, pp 347-365
© 2017, New India Publishing Agency, New Delhi, India
Editors: R. Elanchezhian, A.K. Biswas, K. Ramesh and A.K. Patra

30

Application of Radioisotopes in Soil Fertility and Plant Nutrition Studies

M. V. Coumar, A.K. Patra, S. Rajendiran, J. Somasundaram
B.L. Lakaria, J.K. Saha and S. Kundu

ICAR-Indian Institute of Soil Science, Nabi Bagh, Bhopal – 462038, India

Introduction

After the Second World War, isotopes have become available in sufficient quantities for use in different fields of scientific research. There has been tremendous increase in the number and scope of applications of these new tools and agricultural research is no exception. During the last quarter of a century, the use of isotopes and radiations has become one of the important tools in the solution of practical problems in the area of soil-plant nutritional studies throughout the world. In India, these studies were initiated as early as 1955 with the establishment of the Radiotracer Laboratory at the Indian Agricultural Research Institute, New Delhi. The use of these nuclear tools got further fillip in our country with the establishment of the Nuclear Research Laboratory at the Indian Agricultural Research Institute, New Delhi in 1968 as a UNDP Project on Nuclear Research in Agriculture, operated by the Joint FAO/IAEA division of the International Atomic Energy Agency. Also, the agriculture division of Bhabha Atomic Research Centre, Mumbai and the special radioisotope laboratories established by several agricultural universities had made significant contributions to the use of isotopes and radiations in agricultural research.

Radioisotopes have played an important role in improving productivity in agriculture in a sustainable manner. The use of radioisotope techniques in research on soils and plants has made substantial contributions to our understanding of the behavior and movement of plant nutrients in soil, their uptake by roots, their essential functions in plants, and the relationship of soil water to these processes. Radioisotope techniques offer unique advantage in that one can distinguish between uptake of the nutrient from the fertilizers and that from the soil. Fertilizers tagged with radioisotopes provide a means to find out the amount of fertilizer taken up by crops and the portion that is lost. Radioisotopes have also proven exceedingly valuable in assessing the relative economic advantage to be obtained by leaf application of liquid fertilizers (foliar feeding) as compared to soil or irrigation water application of fertilizers.

Radioisotopes applications in soil physics research are made in various fields such as erosion control, land drainage, water conservation, watershed management, soil tilth and tillage. Soil moisture content can be measured accurately and rapidly in the field using either a neutron scattering method or a gamma radiation source such as cobalt 60 or cesium 137. Precise soil density measurements also can be made using these gamma ray sources. Water labeled with radioactive isotope of hydrogen is being used as a tracer for studies of the efficiency of irrigation systems. Radioactive isotopes have also made progress in the studies of root distribution in soil. Thus, the use of radioisotopes and radiation in the life sciences particularly in the field of agricultural science has provided us with new techniques by which we can unravel many of the scientific complexities of the various basic processes and their application for enhancing productivity.

Radioactive Isotopes as Tracer

Nearly 100 years ago, scientists discovered that there are several elements in the earth crust with different atomic configuration. The differences in atomic configuration of a particular element are termed as isotopes. Isotopes are nuclei having same proton numbers but different number of neutrons and hence they are elements with same atomic number but different mass numbers. Isotopes have similar chemical properties, but differ in their physical properties. Later it was found that among the isotopes there exit stable and radioactive isotopes. A radioactive isotope acts as an indicator or tracer that makes it possible to follow the course of a chemical or biological process. The advantage of radioisotopic tracer use in agricultural research is that it behaves in the similar way as that their stable counterparts do, however it can be readily traced. Mostly the usefulness of radioisotopes as a tracer normally depends on their physical properties of the particular element. The three most important physical properties of radioisotopes which determine its use are half-life, mode of decay and decay energy. For example, if the half-life of a nuclide is very short, any compound labelled with it will be difficult to prepare, use and measure within the time of decay. Further, the mode and energy of decay (α, β and γ) will determine how the nuclide will be measured. For measurement, a small quantity of a radioisotope is introduced into the substance to be studied and its path is traced by means of a Geiger-Muller (G.M) counter. The radioactive isotopes emits particle which are captured in photomultiplier tubes and measured, whereas the stable isotopes are separated from each other by passing a gas containing them through a strong magnetic field, which deflects them differentially according to their mass.

Radioisotopes have played an important role in improving productivity in agriculture in a sustainable manner. Further, the use of such techniques in research on soils and plants has also made substantial contributions to our understanding of the behavior and movement of plant nutrients in soil, their uptake by roots, their essential functions in plants, and the relationship of soil water to these processes. Therefore, the following sections in this book chapter will briefly discuss the role of tracer techniques in plant nutrition and soil fertility research.

A. Predicting nutrient movement in soil and its uptake by crop plants using radio isotopic tracer techniques

The most successful use of isotopes in soil chemistry research has been in studying the chemical properties of soil, especially in estimating the amount of nutrient elements available to plants grown on a particular soil. Estimating the amount of nutrient availability in soil in proportion to that from the applied fertilizer material is the most widely established application of radioisotopes in soil-plant nutrition studies. Amount of available nutrient from the total pool in soil is known as 'A" value, a factor which is of greater significance in determining the amount of fertilizer nutrient that should be applied to a particular soil. If certain crop species are grown on a specific soil supplied with known amount of radioactive phosphorus, a measurement of the ^{31}P (stable) and ^{32}P (radioactive) taken up by the plants will show how much phosphate has come from the applied radioactive fertilizer and how much from the soil that occurs in soil naturally. Since the amount of ^{31}P and ^{32}P in the fertilizer added to the soil is already known, the ratio of the specific activities of plant and fertilizer phosphate will indicate the amount of phosphorus in the soil of equal availability as that in the fertilizer.

Several researches dealt with the chemical constituents and properties of soil in relation to plant requirements. Research attempts were also made to describe the process of phosphate dissolution from the solid phase into the soil solution and the subsequent uptake of phosphate from the solution by the plant (McAuliffe *et al.,* 1947, Larsen, 1952, Frossard and Sinaj, 1997). Likewise, research enumerated the process of sulphate retention in soil as influenced by different soil properties (Wen *et al.,* 1999). Further, radioisotopes has been used extensively to investigate the processes involved in the uptake and translocation of nutrient elements by plants and the results of such investigations were reported by number of authors.

B. Radioisotopes application in Nutrient use efficiency studies

Sustainability of agriculture largely depends on our capacity to manage soils for better production and productivity. Fertilizers are one of the essential inputs for maintaining or increasing the soil fertility level in intensive agricultural systems. The efficient use of fertilizer is of great importance because fertilizers are not only expensive, but for many countries they mean a great expenditure in foreign currency. In Sri Lanka the benefit of adopting the findings of research based on isotopic studies on fertilizer use efficiency resulted not only in saving of fertilizer cost but also increased crop yield with reduction in production cost. Due to the substantial increases in the cost of the fertilizers and their limited supplies to resource-poor farmers, enhanced nutrient management was pursued through maximizing the efficiency of nutrient uptake from various inorganic and organic sources. The purpose of applying fertilizers is primarily to supply the crop with essential plant nutrients to ensure normal plant growth and environmental protection (FAO, 1983b; 1984). It is therefore essential that a maximum amount of the applied fertilizer finds its way into the plant and that the minimum is lost due to poor placing, fertilizers source selection, bad timing, etc. Nuclear based technologies can help to optimize fertilizer recommendation for specific crop and location including proper selection of fertilizer, dose, timing and placement of fertilizer. The only direct means of measuring nutrient uptake from the applied fertilizer is through the use of isotopes. Extensive work has been conducted using N- fertilizers labeled with the stable isotope ^{15}N and P- fertilizers labeled with the radioactive isotope ^{32}P (Broeshart, 1974, Fried, 1978, IAEA, 1980, 1983, FAO, 1980, Zapata and Hera, 1995).

In isotopic- aided fertilizer experiments, a labeled fertilizer is added to the soil and the amount of fertilizer nutrient that a plant has taken up is determined. In this way, different fertilizer practices (placement, timing, sources, etc.) can be studied. The first parameter to be determined when studying the fertilizer uptake by a crop by means of the isotope techniques is the fraction of the nutrient in the plant derived from the (labeled) fertilizer and quantifying its use efficiency. Fertilizer use efficiency is a quantitative measure of the actual uptake of fertilizer nutrient by the plant in relation to the amount of nutrient added to the soil as fertilizer. A common form of expression of fertilizer use efficiency is plant recovery or "coefficient of utilization" of the added fertilizer.

$$\%\ \text{utilization of added fertilizer} = \frac{\text{Amount of nutrient in the plant derived from the fertilizer}}{\text{Amount of nutrient applied as fertilizer}} \times 100$$

The concept of fertilizer use efficiency, however, is much broader. It implies not only the maximum uptake of the applied nutrient by the crop but also the availability of the applied nutrient under variable climatic and edaphic conditions. Environmental issues, such as pollution resulting from the fertilizer application, should also be considered. It is important to study the efficient use of fertilizers because we are interested to obtain the highest possible yield with a minimum fertilizer application. The crop responds to the application of nutrients such as nitrogen and phosphorus when the soil is deficient in such nutrients. The objective should be to apply the fertilizer to the crop, not to the soil and to avoid the fertilizers becoming unavailable to the crop, i.e. sorbed in the case of phosphorus or lost as nitrate by leaching or as gaseous losses due to denitrification and/or volatilization in the case of nitrogen. It is, therefore, essential to ensure that the applied fertilizer is taken up by the crop to the highest possible extent. This is done in field trials by assessing the best fertilizer practices such as sources, timing, placement and their interactions in different farming systems (FAO, 1980; 1983a; 1985).The following methods can be utilized to assess the efficiency of fertilizer practices:

a) The classical or conventional agronomic approach based on yield

This measures the biological response or the effect of increasing fertilizer rates on crop yield. Yield is however dependent on a series of factors: some controllable, others not controllable. They all will influence, to a variable extent, the yield and quality of the product.

b) Methods based on nutrient uptake

Fertilizer nutrient uptake can be made by the non-isotopic difference method (difference method) as well as by the isotopic method.

i) Difference method

This indirect method is based on the differences in the nutrient uptake (nitrogen or phosphorus) between the control plot (without fertilizer application) and fertilized treatments. Recovery data estimated using the difference method are best referred to as 'apparent coefficient of utilization'.

Plots fertilized with N or P at different rates and control unfertilized plots are necessary to calculate the fertilizer derived from the source or for the nutrient use efficiency estimation. It is assumed that the nutrient uptake of the control plot measures the amount of nutrient available from the soil, whereas that of the fertilized treatments, the amount of nutrients available from soil and fertilizer. This method, furthermore, assumes that all nutrient transformations i.e. mineralization, immobilization and other processes in the case of nitrogen are the same for both fertilized and unfertilized soils. Obviously, this is an erroneous assumption, and can account for gross differences between recoveries calculated by non-isotope and isotopic methods (Broeshart, 1974, Westerman and Kurtz, 1974, Harmsen and Moraghan, 1988). It is important to note that the difference method allows to calculate nutrient use efficiency (NUE) for a single season, as nutrients in the crop roots that are incorporated in the soil organic matter pool is not taken in consideration which subsequently contributes to the succeeding crops.

ii) Isotopic method

The isotopic method is the only direct means of measuring nutrient uptake from the applied fertilizer is through the use of isotopes. The recovery data are known as the 'real coefficient of utilization'. Extensive work has been conducted using N-fertilizers labelled with the stable isotope ^{15}N and P-fertilizers labelled with the radioactive isotopes ^{32}P or ^{33}P. This does not mean that K and the other plant nutrients are not important. Initial work has been done with N and P utilizing isotopic methods, and many studies have also been conducted with the others, as researchers in developing countries became experienced and confident with the methodology (Broeshart, 1974, Fried, 1978, IAEA, 1980, 1983, FAO, 1980, Zapata and Hera, 1995). Isotopic techniques by means of radioactive tracers (^{32}P or ^{33}P) have been developed for a direct quantitative evaluation of phosphatic fertilizers. Use of radioisotope technique involves two methods depending upon the feasibility of proper labeling of P source to be tested.

(a) Direct method

When a phosphatic fertilizer is applied to soil, two sources of P (soil and fertilizer) are present in the growth medium for a crop. If the phosphatic fertilizer is labelled with ^{32}P, it is possible to differentiate between soil and fertilizer derived P in the plant. The amount of P taken up from the ^{32}P-labelled phosphatic fertilizer can be measured directly. This direct method is applied for the evaluation of phosphatic fertilizers, which can be labeled with a radio isotope (^{32}P or ^{33}P) during manufacturing. This includes superphosphate, diammonium phosphate and nitrophosphate.

(b) Indirect method

Natural source of P such as farmyard manure cannot be labeled with isotopes in similar way since the treatment involved in the labeling would drastically change their chemical and physical characteristics. Nevertheless an indirect method (three sources method) has been proposed to evaluate sources that cannot be labeled. In practice, the procedure is an extension of the direct method. It consists of determination of the uptake of P from a ^{32}P labeled source in presence and absence of the unlabeled nutrient source. In this technique ^{32}P carrier free solution or a ^{32}P labeled carrier solution with high enough specific activity is added to each plot to obtain sufficient activity (SA) in the plant material. By definition it is known that:

$$\%Pdff = \frac{\text{S.A. in plant material}}{\text{S.A. labeled fertilizer}} \times 100$$

Where, % Pdff is the P in plant derived from the labeled fertilizer, and S.A. is the specific activity, which is determined experimentally in the plant material and the labeled fertilizer and expressed in disintegration per minute (dpm). The following is the procedural steps in calculating the phosphorus use efficiency (PUE) from the applied fertilizer material.

1. Uptake of phosphorus from applied P source = $\frac{\%\ Pdff}{100} \times$ Total uptake of P

$$2.\ \%\ \text{P utilization of applied fertilizer material} = \frac{\text{P uptake from added fertilizer material}}{\text{P added through fertilizer material}} \times 100$$

C. Agricultural chemical residues studies using tracer techniques

Radioisotopes have proved to be a valuable tool in obtaining information on residues of agricultural chemicals including insecticide and weedicide. Many radioactive-labeled pesticides have been synthesized and used to obtain residue data needed for the establishment of safe limits or tolerances for pesticides. Radioisotopes have also proved useful in studying the mode of action of radioactive-labeled pesticides in plants and animals.

D. Use of Radioisotopes in SOM studies

The study of soil organic matter (SOM) is becoming increasingly important as world agriculture attempts to increase sustainability of soils while at the same time increase production to feed an ever-increasing population. Several attempts like addition of organic manures, green leaf manures, in-situ incorporation of green manures, crop residue return to soil and other organic amendments addition were often used to increase the SOM. Increase in soil organic matter content have multifold effect on soil physical and chemical properties favouring plant growth and ultimately increases crop productivity. Carbon is the energy source, which drives many of the nutrient cycles that occur in the soil. A continuous supply of accessible carbon is necessary for a steady supply of soil nutrients and to maintain soil structure. Due to the presence of large amount of background carbon in native soil, quantifying carbon contribution from the added organic source to different SOM pools and studying its process on soil fertility changes can be very a difficult procedure. However, with the help of carbon isotopes tracing the additions of different plant materials to SOM pools and its influence on soil properties such as soil nutrients and soil structure can easily be measured. Stevenson (1986) showed that by using ^{14}C labelled materials it was possible to identify plant C as it became incorporated into different fractions of the soil humus. Fu *et al.* (2000) used ^{14}C labeled corn residues to study the carbon dynamics and respiration of soil organisms in soils under different tillage regimes and were able to estimate the carbon budget and carbon partitioning during a short period after residue application. Trinsoutrot *et al.* (2000) used ^{13}C labelled oilseed rape to investigate the decomposition of the residues added to soil and to determine how the breakdown of the plant materials related to N content, soluble C compounds, and the cellulose and lignin content of the plant material.

Further, radioisotopic techniques have been used to study the SOM turnover rate using the atmospheric ratio of ^{13}C to ^{12}C and the relative carbon fixation by different plant species. Depending upon the physiographic parameter of a location such as altitude, latitude and temperature, the atmospheric ratio of ^{13}C to ^{12}C varies (Lefroy *et al.* 1995).When plants fix carbon during photosynthesis there is a degree of discrimination between the amount of ^{13}C and ^{12}C and such discrimination generally occurs during the carboxylation step in photosynthesis. In C_3 (Calvin cycle) plants greater discrimination against ^{13}C occurs than in C_4 (Hatch-Slack cycle) plants and as result lower ^{13}C:^{12}C ratio in C_3 plants than in C_4. CAM plants (crassulacian acid metabolism) plants show variable discrimination, but it is more often similar to C_4 plants. The ^{13}C:^{12}C ratio is generally measured as $\delta^{13}C$. A C_4 species such as maize will have a $\delta^{13}C$value of approximately –12‰ whereas in a C_3 species such as wheat or rice it will be approximately –26‰. The $\delta^{13}C$ of SOM is comparable to that of the source plant material (Schwartz *et al.,* 1986) and thus every change in vegetation between C_3 and C_4 plants results in a corresponding change in the $\delta^{13}C$ value of the SOM (Lefroy *et al.,* 1995). This means that when C_3 plants are grown in soils, which had previously been under C_4 vegetation (or vice versa) there is virtually an in situ labeling of the organic matter incorporated into the soil. Cerrie *et al.* (1985) first used this method in order to measure the turnover rate of organic matter in a 50-year-old sugarcane field, after forest clearing. Schwartz *et al.* (1986) used this principle to investigate changes in vegetation in the Congo while Skjemstad *et al.* (1990) studied the turnover of organic matter in pastures using this method. This principle has also been used by Balesdent *et al.* (1987) and Lefroy *et al.* (1993) to investigate changes in SOM as a result of cropping, while Bonde *et al.* (1992) used it to quantify maize root derived soil C.

E. Root Activity studies using isotopes techniques

Radioisotopes have been used to study the rooting pattern of different food crop and the response of roots to changes in the soil environment. Root studies are becoming increasingly important component of crop improvement and selection programs. The roots are responsible for plant water and nutrient uptake. Plant and microbes can also form associations at the root level, which may significantly affect crop productivity in different ways. There is a continuous development of new methods for studying root systems. Most classical methods (visual observations and/or physical separation of roots) are aiming at determining rooting pattern of crops but do not provide information on root activity, growth and physiological responses to environmental factors. Isotope techniques are unique tools to provide information about the soil-root interface interaction processes including nutrient mobility from soil to plant system. The study basically involves two approaches; one is through stem injection and another through soil injection. The first approach, involves the injection of a radioisotopes (^{32}P or ^{86}Rb) in the plant system, and the pattern of root distribution is determined by taking soil-root cores and measuring radioactivity in them. This approach doesn't reflect the true activity of the root system, because when the moisture is under limiting condition the radioisotopes is still translocated to the live, but dormant roots. The second approach involves the injection of the radioisotopes in the soil at various locations (lateral distances and vertical depths) to be tested for root activity, and the amount of radioisotopes is subsequently measured in the aerial parts of the plant. Radioisotopes (*e.g.*, ^{32}P), stable isotopes (*e.g.*, ^{15}N) and rare elements (*e.g.*, Sr) can be used to determine relative root activity distribution through this method by applying the tracer to different soil depths or distances from trees. The later approach is appropriate for delineating patterns of root activity in relation to the uptake of fertilizer nutrient under the soil and other environmental conditions prevalent during a specific period of time. Both these techniques have been applied extensively to field crops. Further the injection techniques for root activity studies may be used to study the effect of land preparation methods on root activity (tillage/plowing techniques for seedbed preparation and soil erosion control) and genotypic differences in response to environmental stress i.e. crop cultivars response to moisture stress.

The ^{32}P injection techniques for studying the root distribution of cereal grain roots was first described by Racz *et al.* (1964), and subsequently modified and improved by Rennie and Halstead (1965). Halstead and Rennie (1965) made studies on the movement and equilibrium distribution of the injected ^{32}P throughout the plant, and the effect of stage of growth and moisture stress on the net translocation of the carrier-free ^{32}P into the root system. Accurate quantification of P accumulated in the plant root systems is difficult since fine roots are often not recovered from soil. Foyjunnessa *et al.* (2014) demonstrated that wick-feeding ^{32}P via stem can effectively label P in roots in situ and facilitate quantitative estimation of total P accumulation by plant root systems.

F. Quantifying biological nitrogen fixation in legumes using tracer techniques

Nitrogen (N) is the most widely applied fertilizer because it is usually considered the main nutrient limiting factor in most agricultural systems (Dudal, 2002). The key role that fertilizer N has played in increasing crop yields, in particular of cereals, is widely recognized. The dynamics and interaction between various pools of nitrogen in the agricultural system including biological nitrogen fixation by legume crop and utilization of soil and fertilizers nitrogen is precisely studied with the help of ^{15}N (Jenkinson *et al.,* 1985, Hart *et al.,* 1986). Legumes are integrated components of any sustainable cropping systems. The most important characteristics of legumes is that their ability to form nodules and fix atmospheric nitrogen in symbiosis with rhizobium bacteria. Because of effective biological nitrogen fixation (BNF) capacity of the legume crop, it can be grown even in the absence of nitrogenous fertilizer application and thereby the fertilizer cost. For proper management and a full realization of the benefits of this plant-microbial association, it is necessary to estimate how much nitrogen is fixed under different field conditions. After quantification of nitrogen fixed by legume crop from the atmospheric source, other factors including fertilizer nitrogen management can be manipulated to increase the crop production. Further, with the help of this isotopic tracer method, the BNF efficiency can also be enhanced by targeting the organisms responsible for nitrogen fixation through manipulation in the soil management practices. Therefore for maximizing the biological nitrogen fixation, it is necessary to develop a suitable method to measure precisely the amount of nitrogen derived from the atmosphere by the crops. In N isotopic tracer studies, the system under investigation is supplied with materials containing $^{15}N/^{14}N$ ratios measurably different from the ^{15}N natural abundance. It is also essential that the nitrogen isotope ratio should again be measurably different from ^{15}N natural abundance at the time the system under investigation is sampled.

Further, the transfer of biologically fixed nitrogen from legume to non-legume which is grown in association with legume may be determined directly or indirectly using this isotopes ^{15}N methodology. In the direct method, the legume is labelled with ^{15}N without contaminating the soil during the labelling period. The nitrogen transfer from the legume to non-legume crop is evident from the tracer detected in the associated plant (non-legume). For this studies, several methods have been developed using ^{15}N, including N_2 labelling (Ta *et al.,* 1989), foliar labelling (Giller *et al.,* 1991), stem labelling by a cotton-wick method (Mayer *et al.,* 2003) and split-root labelling (Jensen, 1996, Vankessel *et al.,* 1985). However, the basic assumption in determination of the transferred nitrogen from legume to associated plant is that the labeled ^{15}N is distributed uniformly within the legume crop. Oghoghorie and Pate (1972) showed that some of the nitrate and symbiotically fixed N that is transported to the shoot is translocated back to the root and nodules. The longer the period of the growth cycle during which the legume is labelled, the better is the quantitative estimate. The following are the methods that are employed for quantitative estimation of BNF.

(i) Use of ^{15}N in nitrogen fixation studies

Burris and Miller (1941) are the pioneer's to use ^{15}N to study biological nitrogen fixation by legume crops. The advantage of this method is that it provides direct evidence for nitrogen fixation because the concentration of ^{15}N in the plants exposed to ^{15}N is extremely greater than the natural abundance (0.3663%) if fixation occurs. Therefore, the extent to which ^{15}N is detected in the plant provides an estimate of the proportion of the plant's N that was actually derived from fixation. This study is normally conducted in closed chamber filled with enriched gas of ^{15}N (Witty and Day, 1978). Since the environment within the chamber is likely to be different from that in a field situation, it is difficult to maintain plants in these chambers for long periods without affecting the growth conditions as compared to the field environment. Further, another drawback of this method of BNF quantification is that the enriched isotope gas (^{15}N) may get leaked over the period of plant growth cycle. As a result, such studies are normally conducted for short period and are subjected to errors while extrapolating the data from the short period studies to the entire growing season (Knowles, 1980).

ii) Quantification of BNF with the help of enriched fertilizer or substrates

Another method to quantify BNF by legume crop is the use of enriched fertilizer material. This method is basically on the principle of ^{15}N isotope dilution method. In this method, nitrogen fixing legume and associated plants (reference plants)are grown in a soil fertilized/tagged with ^{15}N enriched inorganic or organic fertilizers. It relies on differential dilution in the plant of ^{15}N-labelled fertilizer by soil and fixed nitrogen (McAuliffe *et al.,* 1958, Fried and Broeshart, 1975, Fried and Middelboe, 1977). The main advantage of this method over the earlier method is that it provides an integrated quantitative measurement of amount of fixed N accumulated by a crop over the crop growing season. The uptake of ^{15}N enriched fertilizer added to soil will result in a $^{15}N/^{14}N$ ratio greater than 0.3663% (natural abundance) within the plant, the extent of which is a reflection of uptake of the labeled ^{15}N fertilizer. A decrease in the atom %^{15}N excess of the fertilizer nitrogen within the plant is an indication of the extent to which the plant took up N from other unlabelled N sources, e.g. air. Like the previous method, this method is also based on the assumption that both non-fixing and fixing crops take up N from soil and fertilizer in the same ratio.

iii) The ^{15}N isotope dilution method (ID)

The third method to quantify BNF is ^{15}N isotopic dilution method. This is the case when both fixing and reference plants are grown in soil to which the same amount of fertilizer having the same ^{15}N enrichment, has been applied. Thus, in the absence of any supply of N other than soil and ^{15}N labeled fertilizer, a fixing plant and a non fixing reference plant will contain the same ratio of $^{15}N/^{14}N$, since they are taking up N of the same $^{15}N/^{14}N$ composition, but not necessarily the same total quantity of N. In both plants, the $^{15}N/^{14}N$ratio within the plant is lowered by the N absorbed from the unlabelled soil. However, in the presence of N_2, and because of atmospheric N_2 fixation by the fixing plant the $^{15}N/^{14}N$ ratio is lowered further. The extent to which the $^{15}N/^{14}N$ ratio in the fixing crop is decreased, relative to the non-fixing plant can be used to estimate N_2 fixed in the field. By using ^{15}N labeled fertilizer, 50% of the N in the reference crop was derived from the applied fertilizer. The assumption made in this method of ^{15}N isotope dilution is that either the fertilizer distribution is even with depth or the legume and reference crops have spatially similar nutrient uptake profiles, i.e. the root systems should be similar (Witty, 1984). It is implicit in the calculation that the enrichment of plant available soil N remains constant with time or that the legume and control have similar N uptake patterns. In practice when fertilizer N is added as a single application the enrichment of plant available soil N declines with time; and this decline can vary between the legume and the control plant. Depending on whether the control takes up soil nitrogen faster or slower than the legume, the calculated nitrogen fixation rate will be greater or less than the true value (Witty, 1983). Errors due to making this assumption may be reduced by the use of slow-release N fertilizer and by choice of a control plant which closely parallels the legume in its nitrogen uptake (Witty, 1984).

G. Applications of isotopes in soil physics research

Radioisotopes and radiation application in soil physics research has made a significant contribution in the areas of soil moisture measurement, tillage studies, soil and water conservation measures, soil density measurement, soil-water movement studies etc. For establishing accurate facts and findings in most of these studies is not possible without the application of radiation and isotopic techniques. A major factor in successful crop production is the presence of an adequate water supply. Soil moisture content can now be measured accurately and rapidly in the field using either a neutron scattering method or a gamma radiation source such as cobalt 60 or cesium 137. Also, irrigation efficiency studies are precisely studied by labeling water with radioisotopes of hydrogen (tritium). Precise soil density measurements also can be made using these gamma ray sources. It is clear that radioisotopes have greatly facilitated such investigations and are now being widely used in the field of soil physics research.

H. Application of radioisotopes techniques in soil erosion studies

Soil erosion by water and wind and its subsequent particle redistributions are natural process; however, it can be accelerated by several anthropogenic activities like poor soil and water management, and use changes, deforestation and overgrazing. As a result of soil losses, it not only affects the on-site crop productivity (Dercon *et al.,* 2003; 2006; Schmitter *et al.,* 2010), but also off-site problems of sediment deposition and associated contaminants that can find their way into dam/reservoirs and water bodies. Globally, soil erosion by water and wind is the predominant land degradation and the recent estimates shows that nearly three quarters of the total area of global agricultural land area affected by erosion is situated in the developing countries of Africa, Asia and Latin America. Therefore, to devise effective strategies for soil erosion control, there is an urgent need to obtain reliable quantitative data on the extent and rates of soil erosion/ redistribution of a specific location (Stroosnijder, 2005, Boardman, 2006). Traditional methods used to measure soil erosion are time consuming and the results obtained for an experimental plot are usually incomparable with one another. The use of radionuclides to measure soil erosion overcomes some of the limitations of the traditional methods and it is a valuable alternative compared to the traditional methods.

The task for alternative rapid and cost-effective techniques for assessing soil erosion to complement classical conventional methods has directed attention to the use of fallout radionuclides (FRN), as tracers to obtain quantitative estimates of soil erosion and deposition on agricultural landscapes (Ritchie and McHenry,

1990).The use of fallout radioisotopes to study soil erosion was started at the end of 1950s. The first fallout radioisotope which was used as a soil erosion and soil movement tracer was strontium-90 (Menzel, 1960, Frere and Roberts, 1963). They concluded that the ^{90}Sr loss from a field was due to soil erosion. Moreover, Frere and Roberts (1963) found no loss of ^{90}Sr from grassland. Later, studies are mainly focused on the use of FRN ^{137}Cs on soil erosion. The global fallout input of ^{137}Cs and $^{239-249}$PU from the atmosphere to the land surface occurred through wet (rainfall) in the 1950s and 1960s, due to the testing of atomic weapons (bomb fallout). Once ^{137}Cs is deposited, it is rapidly and strongly adsorbed by fine soil particles (clay and humus) and is lost when soils are eroded which subsequently re-accumulates at or near the soil surface. Tracing the FNR and documenting the radionuclide redistribution across the landscape is the effective tool for quantifying erosion extent and severity on the decadal time scale (Mabit *et al.,* 2008, Zapata and Nguyen, 2010). Because these transient tracers often result in characteristic depth profiles with a transition at subsurface depths, they offer some advantages over naturally produced tracers. The latter alone, however, potentially capable of long-term are averaged soil erosion rates. The ^{137}Cs isotope ($t^1/_2$ =30.5 years) is strongly fixed to the soil. By comparing the amount of isotope in an agricultural soil with nearby areas that have not been disturbed since the 1950s, it is possible to estimate how much soil had been lost in the past 50 years.

Several well established research organization have confirmed that the ^{137}Cs technique is recognized worldwide to be the primary and most widely used isotopic tracer for soil erosion/sedimentation investigations (Walling and He, 1999).The first attempts to use ^{137}Cs measurements for estimation of soil erosion were carried out in the 1960's (Yamagata *et al.,* 1963, Rogowski and Tamura, 1965) and since then ceasium-137 has been widely used to study soil erosion and deposition. Caesium-137 was also used to estimate tillage redistribution of soil (Lobb *et al.,* 1995, Quine *et al.,* 1999, Schuller *et al.,* 2003). However, the caesium method has also some limitations. Firstly, to obtain quantitative value of soil erosion it is important to know the reference value of ^{137}Cs inventory for the study area and sometimes it is difficult to find a reference place for the study area. Secondly, the calculation of value of soil erosion and deposition strongly depends on the model used; the models are sensitive to the parameter choice. In other words, it is essential to use models and determine the parameters for those models with great care. Probably to overcome the limitations of the method, the solution is to take into consideration other isotopes like ^{210}Pb or ^{7}Be together with ^{137}Cs. Later, other environmental radionuclides, such as lead-210 (^{210}Pb) and beryllium-7 (^{7}Be), find its applications in exploiting the essentially unique advantages provided by the FRN technique.

I. Application of isotopic tracer techniques in pollution studies

Sewage wastewater and sludge contains beneficial constituents such as macro and micronutrients essential for plant growth and rich in organic matter (Badawy and El-Motaium, 1999). Sewage wastewater irrigation and sludge application as organic manure for agriculture purpose improves soil fertility and sustains crop production (Kumazawa, 1997). On the other hand, sewage wastewater and sludge disposal to agricultural land without proper treatment causes a major environmental pollution globally. The presence of pathogens, toxic organic compounds and inorganic elements (heavy metals) in the sewage wastewater and sludge pose a serious threat to ecosystem through food chain contamination. In addition, most of these contaminants are not biodegradable and therefore, becoming dangerous to plant, animal and human health. Therefore, there is environmental and health concern when using sewage water and/or sludge in agriculture. Hence, necessary precautions and treatments like pathogen disinfection and toxic pollutants removal has to be considered prior to land application.

The conventional method of sewage water treatment includes primary treatment (precipitation), secondary treatment (biological), tertiary treatment (chlorination) and digestion. Chlorination may eliminate bacteria and amoeba cysts; however it can't influence enteric viruses and/or parasite eggs (Lund, 1975). Further, chlorine can react with the organic residues in the wastewater to form hazardous substances which is carcinogenic in nature. In certain countries like in Europe and USA, chlorination has been replaced by UV- irradiation to kill bacteria. However, eliminating pathogens from wastewater by this conventional method is cumbersome, costly and requires trained personnel's (Chang, 1997).

Nuclear techniques like gamma irradiation (cobalt-60 or Cesium-137) and electron beam has recently been used as an alternative to the conventional methods of treatments. Sufficient data are available for gamma radiation treatment of sludge, permitting its application on commercial scale. The possibility of beneficial and safe recycling of gamma-irradiated sludge in agricultural uses is documented. Ionizing radiation has been recognized as a fast and reliable means for pathogen removal (Kapila *et al.,* 1981, Pandya *et al.,* 1987 and El-Motaium *et al.,* 2000) and organic pollutant degradation (Abo-Elseoud *et al.,* 2004). Gamma radiation induces production of free radicals which in turn cause denaturation of cell protoplasm and damage of membranes and cell walls. These processes lead to lysis of organisms (Chang, 1997). However, the effect of ionizing radiation on the bioavailability of heavy metals was not well investigated. The Massachusetts Institute of Technology (MIT, 1980), found that electron beam significantly reduce the water-soluble fraction of several potentially toxic metals and thereby renders metals less available for plant uptake. In USA, studies have claimed that decreased solubility of metals in irradiated sludge is due to binding of water extractable heavy metals to sludge components (Sheppard and Mayoh, 1986).

Conclusion

To conclude, isotopic tracers of nutrient elements have proved a blessing par excellence due to their potentials in solving many mysteries in soil science. Also, radiation and isotopes techniques has made important contribution in the field of plant varietal development by inducing genetic variability in crop plant species; pest and disease infestation monitoring and management; fate of organic pollutants and pesticides in soil; and to preserve agricultural produce for higher shelf-life. Though, the knowledge concerning the use of tracer tools and techniques have developed very much and are widely used in health and medicinal science, but their applicability and critical evaluation for studying effects and fate of various agro-chemicals in soil-plant-environmental studies is inadequate. Hence, application of the current knowledge on tracer tools and techniques in soil-plant-environmental studies is more desirable and demanding under the current as well as future scenario so as to provide better understanding and to strengthen the knowledge on soil fertility and plant nutrition studies.

References

Abo El-Seoud, M., El-Motaium, R A., Mufeed, I., Batareh and Kreuzig R. (2004). Impact of gamma radiation on the degradability of polynuclear aromatic hydrocarbons in Egyptian sewage sludge. *Fresenius Environmental Bulletin,* 13 (1): 52-55.

Badawy, S. H. and El-Motaium, R. A. (1999). Effect of irradiated and non-irradiated sewage sludge application on some nutrients heavy metals content of soils and tomato plants. pp. 728-744. Proceeding of the 1st Congress on: Recent Technology in Agriculture. Bull. Fac. Agric., Cairo University, Special Edition.

Balesdent, J., Mariotti, A. and Guillet, B. (1987). Natural ^{13}C abundance as tracers for studies of soil organic matter dynamics. *Soil Biology and Biochemistry,* 19 (1): 25-30.

Boardman, J. (2006). Soil erosion science: reflections on the limitations of current approaches. *Catena,* 68 (2-3): 73-86.

Bonde, T A., Christensen, B. T. and Cerri, C C. (1992). Dynamics of soil organic matter as reflected by natural ^{13}C abundance in particle size fractions of forest and cultivated Oxisols. *Soil Biology and Biochemistry,* 24 (3): 275-277.

Broeshart, H. (1974). N tracer techniques for the determination of active root distribution and nitrogen uptake by sugarbeet. pp. 121-124. Proceedings of Symposium on Nitrogen and Sugarbeet. International Institute for Sugarbeet Research, Brussels.

Burris, R. H. and Miller, C. E. (1941). Application of ^{15}N to the study of biological nitrogen fixation. *Science,* 93: 114–115.

Cerri, C., Feller, C., Balesdent, J., Victória, R. and Plenecassagne, A. (1985). Application du traçage isotopique naturel en ^{13}C a l'étude de la dynamique de la matiére organique dans les sols. *Comptes Rendus Des Séances De L'Académie Des Sciences. Série II.* 300 (1): 423-428.

Chang, A. C. (1997). Land application of sewage sludge: pathogen issues. IAEATECDOC-971: 183-190.

Dercon, G., Deckers, J., Govers, G., Poesen, J., Sánchez, H., Vanegas, R., Ramirez, M. and Loaiza, G. (2003). Spatial variability in soil properties on slow-forming terraces in the Andes region of Ecuador. *Soil and Tillage Research,* 72 (1): 31- 41.

Dercon, G., Deckers, J., Poesen, J., Govers, G., Sánchez, H., Ramirez, M., Vanegas, R., Tacuri, T. and Loaiza, G. (2006). Spatial variability in crop response under contour hedgerow systems in the Andes region of Ecuador. *Soil and Tillage Research,* 86 (1): 15-26.

Dudal, R. (2002). Forty years of soil fertility work in sub-Saharan Africa. *Integrated Plant Nutrient Management in sub-Saharan Africa: From Concept to Practice.* pp. 7-22. (Eds) Vanlauwe, B., Diels, J., Sanginga, N. and Merckx. CAB International, Wallingford.

El-Motaium, R. A., Ezzat, H E M. and El-Batanony, M. (2000). Radiation Technology: a means of sewage water and sludge disinfection treatment in Egypt. *The Egyptian Journal of Medical Microbiology,* 9 (2): 387-392.

FAO. (1980). Maximizing the efficiency of fertilizer use by grain crops. FAO Fertilizer and Plant Nutrition Bulletin No. 3. Rome, Italy.

FAO. (1983a). Fertilizer use under multiple cropping systems. FAO Fertilizer and Plant Nutrition Bulletin No. 6. Rome, Italy.

FAO. (1983b). Micronutrients. FAO Fertilizer and Plant Nutrition Bulletin No. 7. Rome, Italy.

FAO. (1984). Fertilizer and Plant Nutrition guide. FAO Fertilizer and Plant Nutrition. Bulletin No. 9. Rome, Italy.

FAO. (1985). Efficient fertilizer use in acid upland soils of the humid tropics. FAO Fertilizer and Plant Nutrition. Bulletin No. 10. Rome, Italy.

Foyjunnessa, McNeill, A M., Doolette. A L., Mason, S. and McLaughlin, M. (2014). In situ ^{33}P-labelling of canola and lupin to estimate total phosphorus accumulation in the root system. *Plant and Soil*, 382 (1): 291-299.

Frere, M H. and Roberts, H J. (1963). The loss of strontium-90 from small cultivated watersheds. *Soil Science Society of America Proceedings,* 27 (1): 82-83.

Fried, M. and Broeshart, H. (1975). An independent measurement of the amount of nitrogen fixed by legume crops. *Plant and Soil,* 43 (1): 707–711.

Fried, M. and Middleboe, V.(1977). Measurement of amount of nitrogen fixed by a legume crop. *Plant and Soil,* 47 (3): 713–715.

Fried, M. (1978). Critique of "Field trials with isotopically labelled N fertilizer". *Nitrogen in the environment.* pp.43-62 (Eds) Nielsen D R and MacDonald J G. Academic Press, Inc. USA.

Frossard, E. and Sinaj, S. (1997). The isotope exchange kinetic technique: a method to describe the availability of inorganic nutrients. Applications to K, PO_4, SO_4 and Zn. *Isotopes in Environmental and Health Studies,* 33 (1-2): 61–77.

Fu, S., Coleman, D C., Schartz, R., Potter, R., Hendrix, P F. and Crossley, Jr. D A. (2000). ^{14}C distribution in soil organisms and respiration after the decomposition of crop residue in conventional tillage and no-till agroecosystems at Georgia Piedimont. *Soil and Tillage Research,* 57 (1-2): 31-41.

Giller Kenneth, E., Ormesher Judy and Awah Fru Martin. (1991). Nitrogen transfer from *Phaseolus* bean to intercropped maize measured using ^{15}N-enrichment and ^{15}N-isotope dilution methods. *Soil Biology and Biochemistry,* 23 (4): 229–346.

Halstead, E H. and Rennie, D A. (1965). The movement of injected ^{32}P throughout the wheat plant. *Canadian Journal of Botany,* 43 (11) 1359-1366.

Harmsen, K. and Moraghan J T. (1988). A comparison of the isotope recovery and difference methods for determining nitrogen fertilizer efficiency. *Plant and Soil,* 105 (1): 55-67.

Hart, P B S., Rayner, J H. and Jenkinson, D S. (1986). Influence of pool substitution on the interpretation of fertilizer experiments with ^{15}N. *European Journal of Soil Sciences,* 37 (3): 389–403.

IAEA. (1980). Soil nitrogen as fertilizer or pollutant. Panel Proceedings Series.Vienna, Austria.

IAEA. (1983). A guide to the use of nitrogen-15 and radioisotopes in studies of plant nutrition: calculations and interpretation of data. IAEA TECDOC-288. Vienna, Austria.

Jenkinson, D S., Fox, R H. and Rayner, J H. (1985). Interactions between fertilizer nitrogen and soil nitrogen-The so called priming effect. *European Journal of Soil Sciences,* 36 (3): 425–444.

Jensen, E S. (1996). Rhizodeposition of N by pea and barley and its effect on soil N dynamics. *Soil Biology and Biochemistry,* 28 (1): 65–71.

Kapila, S., Puranam, R S., Roa, N N., Modi, V V., Ekbote, R. and Sheh, M.N. (1981). Some aspects of hygienisation of sewage sludges. pp. 45-54. Proceeding of the national symposium on application of nuclear and applied techniques in public health and pollution control. Bahaba Atomic Research Center, Department of Atomic Energy, India.

Knowles, R. (1980). The measurement of nitrogen fixation. *Current Perspectives in Nitrogen Fixation.* Pp. 327-333. (Eds) Gibbson A H and Newton W E. Australian Academy of Science, Canberra.

Kumazawa, K. (1997). Use of sewage sludge for agriculture in Japan. pp. 111-127. Proceedings of consultants meetings on sewage sludge and wastewater for use in agriculture. 199 Joint FAO/IAEA Div. of Nuclear Techniques in Food and Agriculture, Vienna- Austria, International Atomic Energy Agency.

Larsen, S. (1952). The use of P-32 in studies on the uptake of phosphorus by plants. *Plant and Soil,* 4 (1): 1–10.

Lefroy, R D B., Blair, G J. and Conteh, A. (1995). Chemical fractionation of soil organic matter and measurement of the breakdown rates of residues. *Soil Organic Matter Management for Sustainable Agriculture.* pp. 149-158. (Eds) Lefroy R D B, Blair G J and Craswell E T. ACIAR Proceedings. No. 56. ACIAR Canberra, ACT., Ubon, Thailand.

Lefroy, R D B., Blair, G J. and Strong, W M. (1993). Changes in soil organic matter with cropping as measured by organic carbon fractions and ^{13}C natural isotope abundance. *Plant and Soil,* 155 (1): 399-402.

Lobb, D A., Kachanoski, R G. and Miller M H. (1995). Tillage translocation and tillage erosion on shoulder slope landscape positionsmeasured using ^{137}Cs as a tracer. *Canadian Journal of Soil Science,* 75 (2): 211-218.

Lund, E. (1975). Public health aspects of waste water treatment. IAEA-SM-194/104

Mabit, L., Benmansour, M. and Walling, D E.(2008). Comparative advantages and limitations of Fallout radionuclides (^{137}Cs, ^{210}Pb and ^{7}Be) to assess soil erosion and sedimentation. *Journal of Environmental Radioactivity,* 99 (12): 1799-1807.

Mayer, J., Buegger, F., Jensen, E S., Schloter, M. and Hess, J. (2003). Estimating N rhizodeposition of grain legumes using a ^{15}N in situ stem labelling method. *Soil Biology and Biochemistry,* 35 (1): 21–28.

McAuliffe, C D., Hall, N S., Dean, L A. and Hendricks, S B. (1948). Exchange reactions between phosphates and soils: hydroxylic surfaces of soil minerals. *Soil Science Society of America Proceedings,* 12 (C): 119–123.

McAuliffe, C., Chamblee, D S., Uribe-Arango, H. and Woodhouse, W W. (1958). Influence of inorganic nitrogen on nitrogen fixation by legumes as revealed by ^{15}N. *Agronomy Journal,* 50 (6): 334–337.

Menzel, R. (1960). Transport of strontium-90 in runoff. *Science,* 131 (3399): 499-500.

MIT. (1980). High energy electron treatment of wastewater liquid residuals, including selected. Proceedings of a seminar on "Electron disinfection of municipal sludge for beneficial disposal". June 23-24, Massachusetts Institute of Technology, Cambridge.

Oghoghorie, C.G.O. and Pate, J.S. (1972). Exploration of the nitrogen transport system of a nodulated legume using ^{15}N. *Planta,* 104 (1): 35–49.

Pandya, G.A., Kapila, S., Kelkar, V.B., Negi, S. and Modi, V.V. (1987). Inactivation of bacteria in sewage sludge by gamma radiation. *Environmental Pollution*, 43 (4): 281-290.

Quine, T.A., Walling, D.E., Chakela, Q.K., Mandiringana, O.T and Zhang X. (1999). Rates and patterns of tillage and water erosion on terraces and contour strips: evidence from caesium-137.

Racz, G.J, Rennie, D.A. and Huteheon W. I. (1964). ^{32}Phophorus injection method for studying the root system of wheat. *Canadian Journal of Soil Science,* 44 (1): 100-108.

Rennie, D.A. and Halstead, E.H. (1965). "A " P injection method for quantitative estimation of the distribution and extent of cereal grain roots", Isotopes and Radiation in Soil-Plant Nutrition Studies (Proc. Symp. Ankara, 1965), IAEA, Vienna.

Ritchie, J.C. and McHenry, J.R. (1990). Application of radioactive fallout cesium-137 for measuring soil erosion and sediment accumulation rates and patterns: a review. *Journal of Environmental Quality,* 19 (2): 215-233.

Rogowski, A.S. and Tamura, T. (1965). Movement of ^{137}Cs by runoff, erosion and infiltration on the alluvial captina silt loam. *Health Physics,* 11 (12): 1333-1340.

Schmitter, P, Dercon, G., Hilger, T., Thi Le Ha, T., Huu Thanh, N., Lam, N., Duc Vien, T. and Cadisch, G. (2010). Sediment induced soil spatial variation in paddy fields of Northwest Vietnam. *Geoderma,* 155 (3-4): 298-307.

Schuller, P., Ellies, A., Castille, A. and Salazar, I. (2003). Use of ^{137}Cs to estimate tillage- and water-induced soil redistribution rates on agricultural land under different use management in central-south Chile. *Soil and Tillage Research,* 69 (1-2): 69-83.

Schwartz, D., Mariotti, A., Lanfranchi, R. and Guillet, B. (1986). ^{13}C/^{12}C ratios of soil organic matter as indicators of vegetation change in the Congo. *Geoderma,* 39 (2): 97-103.

Sheppard, S.C. and Mayoh, K.R. (1986). Effects of irradiation of sewage sludge on heavy metal bioavailability. Atomic Energy of Canada Ltd., Pinawa, MB (Canada) Witeshell Nuclear Research Establishment AECL Oct., 55 p.

Skjemstad, J.O., Le Feuvre, R. P. and Prebble, R. E. (1990). Turnover of soil organic matter under pasture as determined by ^{13}C natural abundance. *Australian Journal of Soil Research,* 28 (2): 267-276.

Stevenson, F. J. (1986). *Cycles of Soil: Carbon, Nitrogen, Phosphorus, Sulfur, Micronutrients.* New York: John Wiley and Sons.

Stroosnijder, L. (2005). Measurement of erosion: is it possible? *Catena,* 64 (2-3): 162-173.

Ta, T.C., Faris, M.A. and Macdowall, F.D.H. (1989). Evaluation of ^{15}N methods to measure nitrogen transfer from alfalfa to companion timothy. *Plant and Soil,* 114 (2): 243–247.

Trinsoutrot, I., Recous, S., Mary, B. and Nicolardot, B. (2000). C and N flux of decomposing ^{13}C and ^{15}N Brassica napus L.: effect of residue composition and N content. *Soil Biology and Biochemistry,* 32 (11-12):1717–1730 .

Van Kessel, C., Singleton, P W. and Hoben H J. (1985). Enhanced N-transfer from a soybean to maize by vesicular arbuscular mycorrhizal (VAM) fungi. *Plant Physiology,* 79 (2): 562–563.

Walling, D E. and He Q. (1999). Improved models for estimating soil erosion rates from ^{137}Cs measurements. *Journal of Environmental Quality,* 28 (2): 611-622.

Wen, C., Blair, G J., Scott, J. and Lefroy, R. (1999). Nitrogen and sulfur dynamics in contrasting grazed pastures. *Australian Journal Agricultural Research,* 50 (8): 1381-1392.

Westerman, R L. and Kurtz, L T. (1974). Isotopic and non-isotopic estimation of fertilizer nitrogen uptake by Sudan-grass in field experiments. *Soil Science Society of America Proceedings,* 38 (1): 107-109.

Witty, J F. and Day, J M. (1978). Use of $^{15}N_2$ in evaluating asymbiotic N_2 fixation. *Isotopes in Biological Dinitrogen Fixation.* IAEA, Vienna.

Witty, J F. (1983). Estimating N_2-fixation in the field using ^{15}N-labelled fertilizer: Some problems and solutions. *Soil Biology and Biochemistry,* 15 (6): 631–639.

Witty, J F. (1984). The validity of some assumptions inherent in the application of the acetylene assay and the isotope dilution method. *Biological N_2 Fixation Newsletter,* 12: 1–3.

Yamagata, N., Matsuda, S. and Kodaira K. (1963). Run-off of caesium- 137 and strontium-90 from rivers. *Nature,* 200 (4907): 668-669.

Zapata, F. and Hera, C. (1995). Enhancing nutrient management through use of isotope techniques. *Nuclear Techniques in Soil-Plant Studies for Sustainable Agriculture and Environmental Preservation.* pp 83–105. IAEA, Vienna, Austria.

Zapata, F. and Nguyen, M. L. (2009). Soil erosion and sedimentation studies using environmental radionuclides. *Environmental Radionuclides: Tracers and Timers of Terrestrial Processes.* pp.295-322. (Ed) Froehlich K. Oxford, UK.

Advances in Nutrient Dynamics in Soil - Plant System, pp 367-375

Editors: R. Elanchezhian, A.K. Biswas, K. Ramesh and A.K. Patra

31

Scanning and Transmission Electron Microscopy for Ultra-structural Studies

K. Rajukumar and Manoj Kumar

ICAR- National Institute of High Security Animal Diseases
Anand Nagar, Bhopal – 462022, India

Knowledge of structure is a basic element of biological research. Understanding structures of various degrees of complexity contributes to a better understanding of function, ontogenetic and physiologic processes, as well as phylogenetic relationships. In biological sciences, ultrastructure refers to the nanostructure of a biological specimen, such as a cell, tissue, or organ, at scales smaller than can be viewed with light microscopy. Electron microscopy (EM) has a great impact on our knowledge and understanding of ultrastructure.

Principles of Electron Microscopy

The two integral aspects of microscopy are resolution and magnifying power. Resolution is a measure of capability of an image forming system to separate images of adjacent objects. Magnification can be defined as the ratio of the resolving power of the eye to the resolving power of the microscope. Magnification has no meaning without the simultaneous increase in resolution. In a light microscope (LM), a beam of light is directed through a thin object and a combination of glass lenses provide an image, which can be viewed by our eyes through an eye-piece. The image formed is realistic, since it used light in the visible spectrum (λ -=400 to 800nm). Since resolution cannot be less than half of wavelength, the ultimate resolution obtained using LM is 200nm. In order to gain higher resolution, the use of a light source with a far smaller wavelength is needed. Although the X- or g rays have lower wavelength, non-existence of high-performance lenses required to focus the beam to form an image limit their use in microscopy (Yaseen, 2014). In 1923, De Broglie showed that all particles have an associated wavelength linked to their momentum: λ=h/mv, where *m* and *v* are the relativist mass and velocity respectively, and h the Plank's constant. In 1927, Hans Bush showed that electric and magnetic fields (electromagnetic lenses) can focus an electron beam in the same way that a glass lens does for light. Combining these characteristics, the first electron microscope was built by Ernst Ruska and his mentor Max Knoll in 1931 as the project for his Ph.D. thesis (Ruska and Knoll, 1931).

When accelerated electrons strike an object, several events occur. If the electron is not interrupted by an atom in the sample, it will continue travelling in a straight line. If the electron comes in contact with the sample, it can either bounce off elastically (without losing energy), or inelastically (transferring some energy to the atom). This complex interaction between the accelerated electrons and the specimen results in various physical produces such as elastically scattered electrons, secondary electrons, x-rays, back scattered electrons, Auger electrons, etc (Murphy and Davidson, 2001). Contrast is obtained when interference occurs between electrons coming in from various angles. Based on the use of these physical products, different kinds of electron microscopes have been developed. Two of the most commonly used electron microscopes are Scanning Electron Microscope (SEM) and Transmission Electron Microscope (TEM).

Scanning Electron Microscopy

The SEM utilizes a beam of electrons by passing it through the sample in a raster pattern in order to image the surface of the sample, providing information about the sample such as topography, composition, and directionality. Usually, the beam of electrons is created by the passing of current through a tungsten filament in order to boil off electrons from the tungsten. The electrons are then accelerated (2-40 keV) and passed through an objective to focus into a beam with a diameter within the nanometer scale. When the beam hits the surface of the sample, electrons are ejected from the sample surface and collected by an analyzer. A driver translates the number of electrons counted and produces the image of the sample surface on the computer screen. Magnification in a SEM can be controlled over a range of about 10 to 500,000 times or more. The resolution depending on the instrument ranges between 1 and 20 nm (Bozzola and Russell, 1999).

However, for the imaging, the sample must be placed inside a chamber with a vacuum inside. As a result, the sample must be free of any moisture or foreign particles that may depart from the sample and cause damage to the microscope. This is accomplished through the use of dehydration protocols which effectively sterilize and rid the moisture from the sample. The sample size is only limited by the size of the microscope specimen chamber. The sample must also be electrically conductive. However, new developments in SEM design are the Low Vacuum and Environmental SEM which maintain the specimen chamber at low vacuum, enabling hydrated, uncoated samples to be imaged.

Hard, dry materials such as bone, wood, dried insects, seeds or teeth can be examined with little further treatment or directly. Living cells and tissues, soft-bodied organisms usually require fixation to preserve and stabilize their structure (Pollack and Tadmor, 2011; Echlin, 2009). Fixation is usually performed by incubation in a solution of a buffered chemical fixative or by cryofixation. The fixed cell or tissue is then dehydrated or cryo-substituted. Because air-drying causes collapse and shrinkage, this is commonly achieved by critical point drying. Critical point drying involves replacement of organic solvents such as ethanol or acetone, and replacement of these solvents in turn with a transitional fluid such as liquid carbon dioxide at high pressure. The carbon dioxide is finally removed while in a supercritical state, so that no gas-liquid interface is present within the sample during drying. The dry specimen is usually mounted on a specimen stub using colloidal silver and sputter coated with gold, carbon or gold/palladium alloy before examination in the microscope (Bozzola and Russell, 1999; Pollack and Tadmor, 2011).

Processing of biological material for SEM analysis (Murphy and Roomans, 1984)

1. **Fixation** – Immerse sample in 3% glutaraldehyde buffered with 0.1 M phosphate buffer at room temperature or 0-4 0C (2-4h, max. 24-48h).
2. **Washing** – Rinse tissue with 0.1 M phosphate buffer pH=7.2 - (3 x 10min.).
3. **Post-fixation** – Immerse sample in 1-2% osmium tetroxide in 0.1 M phosphate buffer pH=7.2 (2-4h) at room temperature and in a light tight container.

   1-2% osmium tetroxide solution: (1%) - 0.25g OsO_4 in 25 ml 0.1 M phosphate buffer (12.5 ml 0.2M phosphate solution + 12.5 ml distilled water) or **2% osmium tetroxide solution:** 0.25g OsO_4 in 12.5 ml 0.1 M phosphate buffer (6.25 ml 0.2 M phosphate solution + 6.25 ml distilled water).
4. **Washing** in 0.1 M phosphate buffer pH=7.2 (3 x 10 min.).
5. **Dehydration** in a graded ethanol or acetone solutions in water – 30%, 50%, 70% (can store tissue in 70% ethanol), 80%, 90%, 96%, 100% for 5-15 min each); 2 x 100% ethanol or acetone (15-30 min each).
6. **Critical Point Drying CPD**: This complicated process involves simply the replacement of liquid in the cells with gas in a chamber that is cooled and put under pressure. This process creates a completely dry specimen with minimal or no cellular distortion. The most commonly used medium used for CPD are CO_2 and Freon 13.
7. **Sample mounting:** The stub is often a small, flat, round piece of metal that has a stem – it looks a bit like a flattened mushroom. The basic method of attachment is to glue the specimen or bits of the specimen to the stub which has been covered with double sided sticky tape and a thin layer of foil. The glue is a special silver or carbon conductive glue.

8. **Metal coating:** The specimens must be gold coated because most material (but not gold) is transparent to the electron beam used by the SEM. There are two detectors in the SEM chamber which create a signal from electrons bouncing of the gold-coated specimen. These are used to make up an image of the specimen. If the specimen is not finely covered with an electron-opaque substance like gold, the electron beam would travel right through the specimen, creating no image and probably destroying the specimen too.
9. **Viewing specimens in the SEM:** The SEM has a monitor from which we as the operators view the specimen. The image is derived from the detection of excited electrons that are being bounced of the gold specimen at varying speeds and signals.

SEM applications

- SEMs have a variety of applications in a number of scientific and industry-related fields, especially where characterization of solid materials is beneficial.
- SEM allows us to visualize external morphological characteristics and is a very useful tool for obtaining data on systematic and taxonomic studies of parasites, fungi, etc.
- Scanning electron microscopy is a powerful tool to visualize the morphological variations occurring on the various cell, tissue and organ surfaces, embryos, epidermal characters of plants such as distribution pattern of different cell types with idioblastic elements such as trichomes, glands and stomata (Cordatella, *et al.,* 2012).
- SEM can be used for elemental analysis.
- In addition to topographical, morphological and compositional information, a Scanning Electron Microscope can detect and analyze surface fractures, provide information in microstructures, examine surface contaminations, reveal spatial variations in chemical compositions, provide qualitative chemical analyses and identify crystalline structures.
- SEMs have practical industrial and technological applications such as semiconductor inspection, production line of miniscule products and assembly of microchips for computers.

Transmission Electron Microscopy

Transmission electron microscope (TEM) employs a beam of electrons which is transmitted through an ultra-thin specimen, interacting with the specimen as it passes through. An image is formed from the interaction of the electrons transmitted through the specimen; the image is magnified and focused onto an imaging device, such as a fluorescent screen, on a layer of photographic film, or to be detected by a sensor such as a CCD camera., In the TEM, unlike the SEM, the electrons are accelerated at much high voltage (100-1000 kV), which makes it possible to achieve resolutions below 0.5 Ångströms and magnifications above 50 million times.

Sample preparation for TEM analysis

Sample preparation in TEM can be a complex procedure. The sample for TEM analysis must be made to withstand the electron beam and also the high vacuum inside the chamber that it is put into. For TEM observations, thin samples are required due to absorption of electrons in the material. High acceleration voltage reduces the absorption effects but can cause radiation damage. At these acceleration tensions, a thickness of about 60 nm is required for TEM. The sample preparation is difficult as a thin sample on a support grid must be prepared. The process can also be time consuming and expensive. For biological samples, fixation is the first step in sample preparation for EM. This is done to preserve the structure of cell with minimum alteration from living state and to protect them against alterations during embedding and sectioning procedures.

The general methods and procedures for plant specimen preparation for ultrastructural studies as well as the composition of fixatives, buffers, dehydration solvent and embedding media are similar to those of animal tissues. However, certain special characteristic features of plant tissues, such as a thick cellulosic cell wall, waxy substance in the cuticle, large amount of gases ion the intercellular spaces, the presence of vacuoles, etc have created fixation and resin filtration difficulties. Therefore, special modifications are required. The addition of chemicals such as caffeine in fixative can stabilize the phenol in the vacuole; however, the rupture of vacuoles caused by the fixatives still cannot be controlled, especially for plants with highly vacuolated cells. The application of vacuum infiltration during the initial fixation stage to remove gases from the tissues can be useful. Additional vacuum infiltration during resin infiltration procedure will facilitate the penetration of resin (Kuo, 2007).

Processing of plant specimens for TEM analysis

Materials required

0.2 M phosphate buffer (pH 7.4)

Solution A: 0.2M NaH_2PO_4

Solution B: 0.2M Na_2HPO_4

Add 19ml of solution A to 81 ml of solution B to make 100ml of 0.2M phosphate buffer.

To prepare 0.1 M working solution, add equal volume of distilled water.

2.5% glutaraldehyde solultion

25% glutaraldehyde	10 ml
Distilled water	40 ml
0.2M phosphate buffer	50 ml

2% Osmium tetroxide stock

Dissolve 1 g of osmium textroxide salt (supplied in ampoules) in 50 ml of double distilled water. Allow overnight to completely dissolve. Store at 4°C in tightly stoppered, amber coloured bottle.

1% OsO_4 working solution is prepared by adding equal volume of 0.2M phosphate buffer in 2% OsO_4.

Different grades of dehydrating solution (acetone or ethanol)

	30%	50%	70%	90%
Acetone or ethanol (ml)	30	50	70	90
Distilled water (ml)	70	50	30	10

Uranyl acetate stain

- Add excess of uranyl acetate to 10ml of filtered 50% ethanol in a 15 ml centrifuge tube.
- Vortex for 2 min and spin down hard to allow the excess UA to settle down.
- The solution is ready for use. It can be stoppered and stored at 4°C.
- Before use it is advisable to filter the stain.

Lead citrate solution

- Add one half a pellet of NaOH to 12 ml double distilled water in a centrifuge tube.
- Shake well till it dissolves.
- Add 50 mg of lead citrate, mix well for 2 minutes and centrifuge.
- The solution is ready for use. Stopper and store at 4°C.

2% phosphotungstic acid (PTA) (For negative staining)

PTA	1g
Distilled water	50 ml

Dissolve and adjust pH between 6.8 and 7.4 with 1N KOH.

Procedure

1. Primary fixation

- Fix 1 to 2 mm sized tissue samples in 2.5% glutaraldehyde overnight at 4°C preferably under vacuum.

2. Washing

- After fixation, wash the specimens three times in 0.1M phosphate buffer for 15 min.

3. Post-fixation or secondary fixation

- Fix specimens in 2% osmium tetroxide solution for 6 hours at 4°C.

4. Washing

- Wash the post fixed specimens thoroughly with phosphate buffer three times for 30 min at 4°C to wash off the excess fixative.

5. Dehydration

- Change the specimens serially though the following
- 30% ethanol (in water) for 30 min. at 4°C.
- 50% ethanol (in water) for 30 min. at 4°C.
- 70% ethanol (in water) for 30 min. at 4°C.
- 90% ethanol (in water) for 30 min. at 4°C.
- 100% ethanol (in water) for 30 min. at 4°C.
- 100% ethanol (in water) for 30 min at RT.
- Then in dry ethanol for 1 hour at RT.

6. Clearing

- Propylene oxide for 60 min at RT.

7. Infiltration

- 75% propylene oxide + 25% araldite mixture for 36 hour at RT under vacuum.
- 50% propylene oxide + 50% araldite mixture for 24 hour at RT under vacuum.
- 25% propylene oxide + 75% araldite mixture for 24 hour at RT.
- Rinse with pure araldite mixture at RT.

8. Embedding

Prepare Araldyte 502 mixture

Araldyte 502	54 ml
DDSA	46 ml
DMP-30	1.5-2% (added just before use)

- Embedding is done at RT in cavity moulds in pure araldite mixture.

9. Polymerization

- Keep the moulds for 24 hours at 50°C.
- Then raise the temperature of the incubator or oven to 60°C and allow 48 to 72 hours for hardening.
- After polymerization, remove the blocks from the mould by bending or cutting depending upon the mould type.

10. Sectioning

The blocks are trimmed to produce a pyramidal shape at the tip where the tissue is located. Trimmed blocks are fitted on to an ultramicrotome and ultrathin sections of 60 to 90 nm can be obtained using glass or diamond knives. Sections are picked up on specially made metal grids.

11. Staining of sections on grids

Double staining method using uranyl acetate and lead citrate (Reynolds, 1963) is described.

Staining procedure

Uranyl acetate staining

- Spread and secure a piece of parafilm inside a petridish.
- Place a drop (about 50ml) of uranyl acetate solution over the parafilm.
- Float the grid upon the stain drop with its section side facing down.
- Place the petridish cover and incubate in dark for 10 to 15 min.
- Take the grid and wash twice in each of 50% ethanol and double distilled water with continuous agitation. Dry carefully over filter paper without touching the section side on the grid.

Lead citrate staining

- Place a few pellets of NaOH over a piece of parafilm in a corner.
- Place a drop of lead citrate on the parafilm and stain the grid as indicated above for 5 to 10 min.
- Wash grid with 0.02M NaOH and then twice in double distilled water.
- Dry the grids and store in grid boxes.
- The stained sections are ready for viewing under TEM.

Negative Staining

Biologic structures, because of low mass density, interact weakly with electrons used for imaging, and therefore, show little contrast or detail. One of the most versatile ways to generate sufficient image contrast and resolution is negative staining with heavy metal ions, e.g., lead, tungsten, and uranium ions (Hazelton and Gelderblom, 2003). Negative staining is a rapid qualitative method for visualizing surface structures of particulate samples. Over the years, it has proved to be one of the most valuable and important techniques for examination of fine surface details of viruses, bacterial pili or flagella, protozoa, other biological membrane structures and proteins or protein aggregates. Basically, the method involves staining particulate samples in a layer of electron dense material on a carbon/formvar coated EM grid so that the specimen can be viewed as a light object against a dark background. Commonly used negative stains for TEM are phosphotungstic acid, ammonium molybdate, uranyl acetate, uranyl formate, osmium tetroxide, osmium ferricyanide and auroglucothionate. The major advantage with the negative staining technique is that it can be performed in a very short time and it requires no specialized equipments. But negative staining does not allow high resolution examination of samples and artefacts such as flattening of spherical or cylindrical structures are common, because negative staining involves heavy atom stains (Ohi *et al.,* 2004). This technique is not useful for tissue sections.

References

Bozzola, J J. and Russell, L D. (1999). Electron Microscopy Principles and Techniques for Biologists. Jones and Bartlet Publishers. Sudbury. M.A.

Cortadellas, N., Fernández, E. and Garcia, A. (2012). Biomedical and Biological Applications of Scanning Electron Microscopy. Handbook of instrumental techniques from CCiTUB, Spain.

Echlin, P. (2009). Handbook of sample preparation for scanning electron microscopy and X-ray microanalysis. Ed. Springer.

Hazelton, P. R. and Gelderblom, H. R. (2003). Electron Microscopy for Rapid Diagnosis of Emerging Infectious Agents. *Emerging Infectious Disease,* 9: 294–303.

Kuo. J. (2007). Electron Microscopy: Methods and Protocols, In: Methods in molecular biology 369, 2nd Edition, Humana Press.

Murphy, J. A. and Roomans, G. M. (1984). Preparation of biological specimens for scanning electron microscopy. Scanning Electron Microscopy, Inc., AMF O'Hare, II, 344 pp.

Murphy, D. B. and Davidson, M. W. (2001). Fundamentals of light microscopy and electronic imaging, 2nd edition, Wiley-Blackwell Publications.

Ohi, M., Li, Y., Cheng, Y. and Walz, T. (2004). Negative staining and image classification - powerful tools in modern electron microscopy. *Biological Procedures Online,* 6: 23-34.

Pollack, A. and Tadmor, T. (2011). Surface topography of hairy cell leukemia cells compared to other leukemias as seen by scanning electron microscopy. *Leukemia and Lymphoma,* 52 (S2): 14-17.

Ruska, E. and Knoll, M. (1931). The magnetic concentrating coil for fast electron beams. *Z. techn. Physik,* 12: 389-400.

Yaseen, M. J. (2014). The Resolution in the Electron Microscopy. *International Journal of Application or Innovation in Engineering and Management,* 3: 312 – 319.

Colour Figures

Chapter 7: Precision Nutrient Management for Improving Crop Productivity

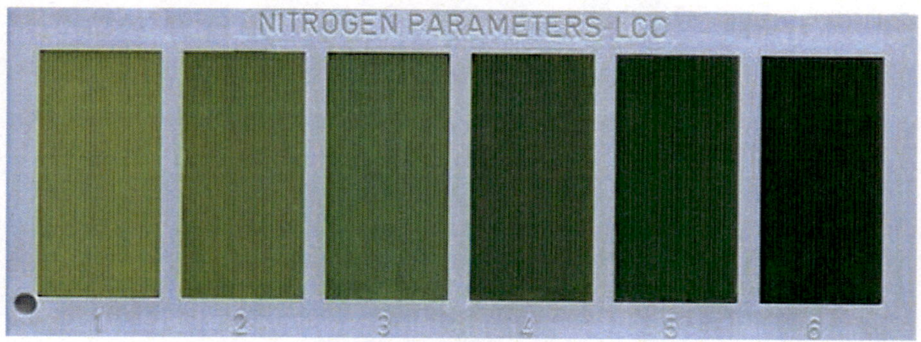

Six Panel Leaf Color Chart

Chapter 14: Nano Rock Phosphate Absorption and Utilization in Plants

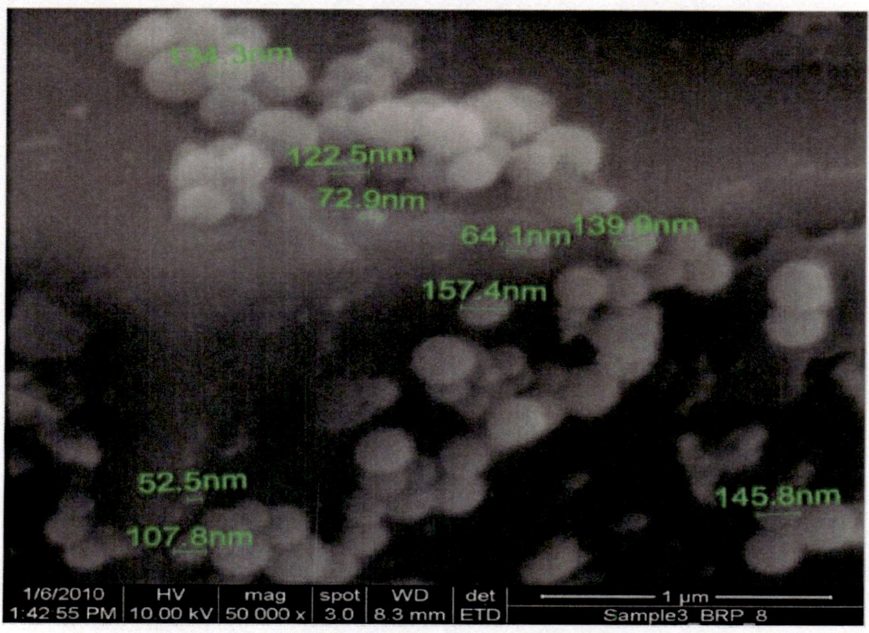

Fig 3. Scanning Electronic microscope photograph of BRP rock phosphate nano-particles

Fig. 4. Response of maize to HGRP-3, Stone-3 and Synthetic Hydroxy Apatite NPs (200nm)

Fig. 5. Growth of Maize roots under nano-particles of HGRP-3 and Stone-3 Rock phosphates

Chapter 17: Development of Nutrient Use Efficient Genotypes

Fig. 1: Screening of rice in low P plot

Chapter 19: Development of Nutrient Use Efficient Genotypes

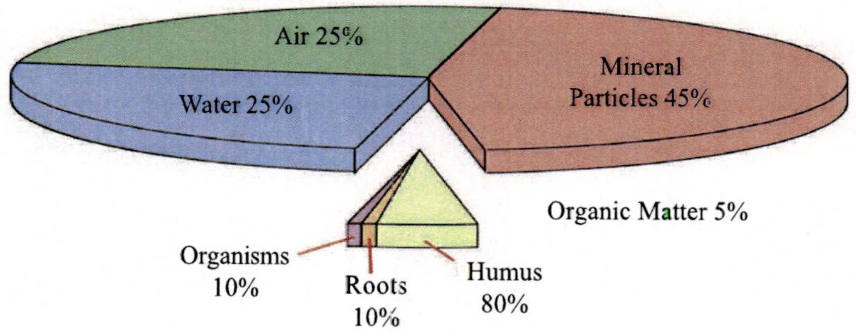

Fig. 1: Composition of soil *Source*: Pidwirny (2006)

Chapter 21: Simulation Modeling for Improving Nitrogen Use Efficiency in Crops and Cropping Systems

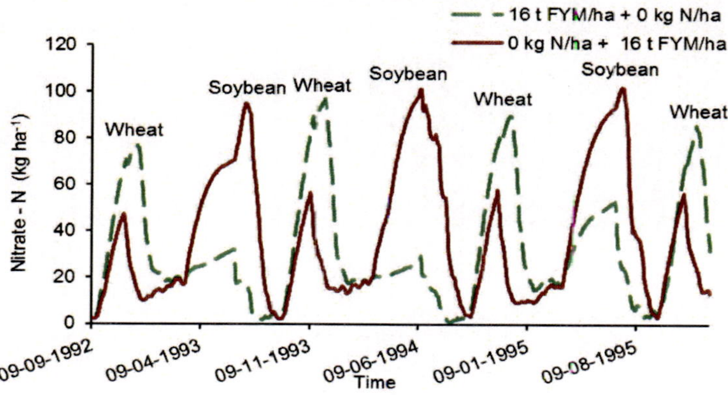

Fig. 2: Nitrate-N content in the profile (0-60 cm) as influenced by FYM application to soybean and wheat (16 t FYM ha^{-1} in each case).

Chapter 24: Impact of Conservation Agricultural Practices on Soil Organic Carbon and Nutrient Availability: An Overview

Fig 1: Residue retention under soybean-wheat system (left) and maize-gram system (right) in Vertisols of Central India

Fig. 2: Crop Residue Burnt by Farmers

Chapter 25: Impact of Conservation Agricultural Practices on Soil Organic Carbon and Nutrient Availability: An Overview

Fig. 1: ITK in biochar preparation (*Source*: http. biocharplus.blogspot.in)

Fig. 2: Raw material (a) and biochar (b) from subabool

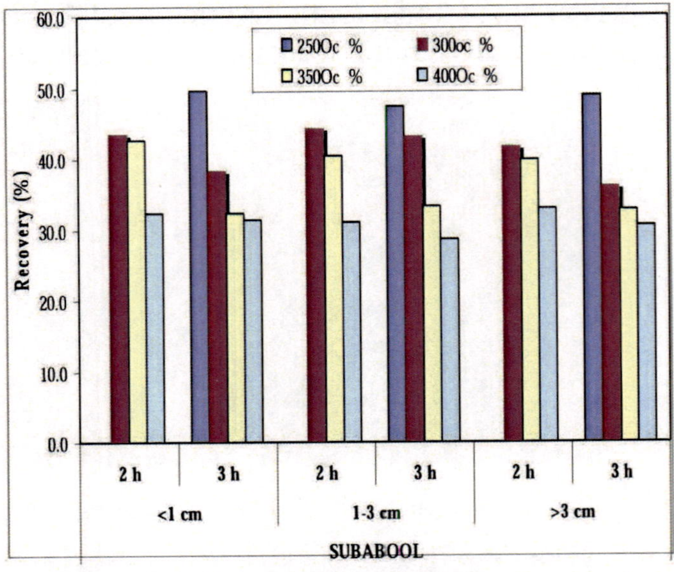

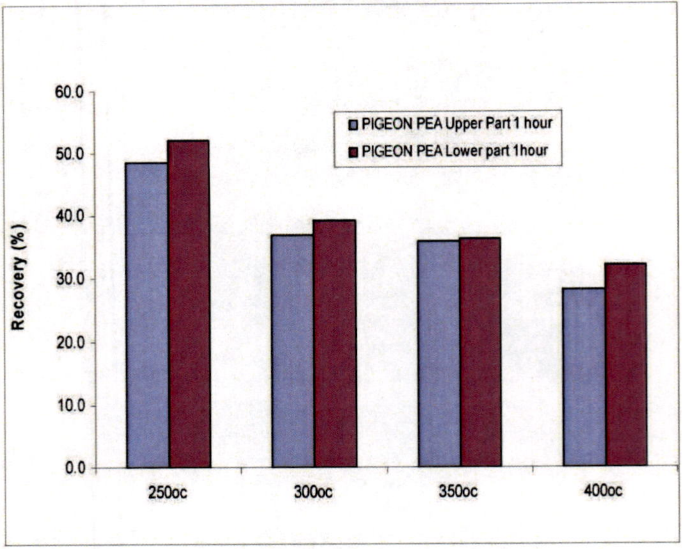

Fig. 3: Recovery of biochar from feedstock

About the Editors

Dr. R. Elanchezhian was born on 1971 in Vandal village in Sivaganga Distt. of Tamil Nadu. He studied B.Sc. (Agri.) during 1988 to 1992 from Tamil Nadu Agricultural University (Coimbatore), and M. Sc. (Plant Physiology) and Ph. D. (Plant Physiology) in 1995 and 1998, respectively, from IARI, New Delhi. He joined ARS (ICAR) in 1998 and started his career at ICAR-CIARI, Port Blair, Andaman & Nicobar Islands as a Scientist/ Scientist Sr. Scale (1998-2006). Then as a Sr. Scientist he worked at ICAR-RCER, Patna (2006-2012). Presently he is working as Principal Scientist at ICAR-IISS Bhopal, Madhya Pradesh, India, since February 2012. During his scientific career at ICAR-CIARI, Port Blair he worked on physiological and biotechnological management of abiotic/biotic stresses in rice, solanaceous vegetables with emphasise on salinity, drought and insect resistance. He has generated promising somaclones of rice and solanaceous vegetables with improved agro-morphological traits besides having moderate tolerance towards salinity and bacterial wilt and insect borer. He was involved in sustainable management of plant biodiversity of A&N Islands including collection, characterization, conservation and enhancement of ecologically and economically important plant species. He has got registered several indigenous germplasms of rice with NBPGR New Delhi. At ICAR Research Complex for Eastern Region, Patna he continued the work on biotechnological and physiological management of abiotic stress tolerance with special reference to submergence and drought stress conditions. Four drought tolerant rice lines with yield potential of 3.2 tons/ha have been identified for drought prone areas of eastern region of India. Besides three promising rice lines suitable for aerobic conditions with yield potential of 3.5 tons/ha were also identified. He has been associated with an ICAR network project on climate change and worked on modeling plant growth of major agricultural crops viz rice, wheat, maize and chickpea with respect to climate change in eastern region of India. He has undergone an international training on marker assisted selection at Clemson University, USA, wherein the genome of bitter melon has been mapped using AFLP and SSR markers. Besides the isolation and sequence characterization of anti-HIV (Map30) and anti-diabetic (Insulin) gene from bitter melon have also been done. Presently at ICAR-IISS, Bhopal, he is working on Nanoparticle delivery and internalization in plant systems for improving nutrient use efficiency of agricultural crops. Besides, he has been associated with projects on management for improving nutrient use efficiency of crops.

Dr Elanchezhian has contributed more than 100 publications, including 46 peer reviewed research papers, 8 research bulletins, 2 edited books, 9 book chapters, 7 popular articles, 24 conference proceedings and 4 training manuals which has been widely cited throughout the scientific community. He is a recipient of Dr. GS Sirohi Award (2002) and Dr. RD Asana Award for Young Scientist (2005) given by the Indian Society for Plant Physiology. He is bestowed with Fellowship of Indian Society for Plant Physiology for contribution in the field of Plant Physiology and Cognate sciences in the year 2008. He has guided several M.Sc. students for their thesis work. He has served as expert of several important committees.

Dr. A.K. Biswas was born in village Bagna in North 24 Parganas district of West Bengal on May 5, 1963. He obtained his B.Sc. (Ag.) Hons. degree in 1984 from B.C.K.V., Mohanpur, W.B. He received his M. Sc. (Gold Medalist) and Ph. D. Degrees from I.A.R.I., New Delhi and is presently working as Head of the Division of Soil Chemistry & Fertility at the Indian Institute of Soil Science, Bhopal; Madhya Pradesh.

The major areas of work of Dr. A.K. Biswas have been concerning the development of farmers' resource-based integrated plant nutrient supply system in soybean-wheat system, farmers' participatory technology development of integrated nutrient management coupled with soil moisture conservation for rainfed pulse-based cropping systems, environmental impact and risk assessment of recycling of distillery and sewage effluents in agriculture, determination of sink capacity of mineralogically and physico-chemically variant soils for some heavy metal pollutants and phyto-remedial efficiency of some flowering plants to decontaminate Cd, Pb and Cr contaminated soils, delineation of nitrate contamination of groundwater in heavily fertilized and intensively cultivated districts of the country, evaluation of the effect of climate change on soil carbon stock in different agro-ecoregions of the country, development of soil organic C and N turnover model in light of C saturation and stabilization theory, development of methodology to estimate relative soil quality index to assess soil health under different agro-ecoregions of the country, and development of nano-rock phosphate for commercial utilization of low-grade indigenous rockphosphates of the country and modified urea products like oleoresin coated urea, zeolite-impregnated and biochar coated urea. He is part of the team who developed Mridaparikshak - A Soil Testing Minilab for which patent has been applied.

Dr. Biswas has contributed more than 200 publications, including 90 research papers, 20 review papers, 20 popular articles, 3 edited books, 40 book chapters and several other publications which has been widely circulated throughout the scientific community.

Dr Biswas has been elected to the prestigious "Fellowship of National Academy of Agricultural Sciences (NAAS)" in 2016. He is also the recipient of the "Jawahar Lal Nehru Award" from ICAR for outstanding doctoral research in the field of Soil Science in 1992 and "The ISSS- Dr. J.S.P. Yadav Memorial Awards for Excellence in Soil Science- 2012" from the Indian Society of Soil Science, New Delhi. He has delivered lecture on "Maintaining soil health for evergreen revolution in the Agriculture and Forestry section of centenary Celebration of Indian Science Congress in Kolkata and got several best paper/ poster awards to his credit. He has guided several M.Sc. and Ph. D. students. He is a member of the Editorial Board of the Journal of the Indian Society of Soil Science since 2011, and a scientific reviewer of the Bioresource Technology, Journal of Hazardous Materials, Journal of the Indian Society of Soil Science, Agropedology, Legume Research, Annals of Soil and Plant Research and Journal of the Indian Society of Pulse Research and Development since 2003.

Dr. K. Ramesh was born on 1974 in Trichirappalli, Trichy Distt. of Tamil Nadu. He studied B. Sc. (Agri.) during 1992 to 1996 from Pandit Jawaharlal Nehru College of Agriculture, Karaikal (UT of Pondicherry) under Tamil Nadu Agricultural University (Coimbatore), and M. Sc. and Ph. D. in Agronomy in 1999 and 2002, respectively, from AC&RI, Madurai and AC&RI., Coimbatore under TNAU, Coimbatore. He served as Assistant Professor (Agronomy) in an Agricultural college in Tamil Nadu during 2001-2004 and has introduced innovative teaching methods in agronomy. Later joined as Scientist Gr.IV(2) (Agronomy) in 2004 at Council of Scientific and Industrial Research and started his career at CSIR - Institute of Himalayan Bioresource Technology, Palampur, Himachal Pradesh (2004-2008). Thereafter he moved to Indian Council of Agricultural Research as a Sr. Scientist at ICAR-Indian Institute of Soil Science, Bhopal. Presently, he is working as Principal Scientist at ICAR-IISS Bhopal, Madhya Pradesh, India, since September 2014.

During his scientific career at various institutions he has worked on the management of cropping systems for enhancing income of the farmers as well as soil health. He has developed promising *Sesbania roastrata* based cropping systems for sustaining rice yields under double cropped wet lands of western region of Tamil Nadu. As a cropping system agronomist at CSIR-IHBT,

Palampur, Himachal Pradesh, he has developed promising food crop+ natural sweetener (*Sesbania rebaudiana*) based cropping systems for Dhauladhar range of Kangra district of Palampur (HP). He was also involved in sustainable management of hilly areas with the introduction of Lavender (a medicinal aromatic perennial) in Himachal Pradesh and Arunachal Pradesh. He has developed molecule specific farming technology *Tagetes minuta* in western Himalayas. Presently at ICAR-IISS, Bhopal, he has been working on zeolites for enhancing nutrient use efficiency as well as Nanoparticle delivery in plant systems for improving nutrient use efficiency of agricultural crops and developing organic farming practices for agri-horticultural crops. Besides, he has been associated with projects on evaluation of new fertilizer products.

Dr K. Ramesh has contributed more than 80 publications, including 39 peer reviewed research papers, several book chapters, 20 popular articles, 25 conference proceedings and 4 training manuals which has been widely cited throughout the scientific community. He is a recipient of NAAS Young scientist award (1997-98), besides awards for bachelor and master's degree programs. He has served as Councilor of Indian Society of Agronomy for Madhya Pradesh state. He has served as expert of several important committees.

Dr. Ashok K. Patra is Director of the ICAR-Indian Institute of Soil Science located at Bhopal, Madhya Pradesh.

He did his early education at Araldihi High School and Dubrajpur Uttarayan Vidyayatan, Bankura, West Bengal. He studied B.Sc. (Agri.) during 1979-1983 from Banaras Hindu University (Varanasi), and M. Sc. and Ph. D. in the discipline of Soil Science & Agricultural Chemistry in 1985 and 1989, respectively, from Indian Agricultural Research Institute, New Delhi. He joined Agricultural Research Service (ICAR) in 1989 and started his career at the Indian Grassland & Fodder Research Institute (IGFRI), Jhansi as a Scientist/Scientist Sr. Scale (1990-1998). Then he moved to CIFE, Mumbai as a Senior Scientist (1998-1999), and to IARI, New Delhi as Senior Scientist (1999-2006) and Principal Scientist (2006-2014).

Under an ICAR-ICRISAT Collaborative programme, he was a postdoctoral scientist (1991-1993) at ICRISAT, Hyderabad, and under an Indo-UK Collaborative programme a Visiting Study Fellow (1996) at the Institute of Grassland and Environmental Research (IGER), Devon, UK. He was a recipient of the prestigious INRA Fellowship (2001-2003) of the French Research Ministry to work on molecular soil ecology in N cycling at the CNRS-Claude Bernard Université Lyon, France and made a significant contribution to unravel the complex processes of nitrogen cycling – its ecology, biodiversity and management in agro-ecosystems. For pursuing the frontier soil science research, DBT (GOI) awarded him the 'DBT Overseas Asssociateship'- 2008 for which he visited University of Notre Dame, USA during 2008-2009.

In addition to research, he was actively involved in teaching and guiding of postgraduate students of soil science at IARI, New Delhi (1999-2014). His research interest includes nitrogen cycling, ecology and microbial biodiversity, carbon sequestration and nanoparticles. His research work has been published in different leading professional journals and highly quoted. He contributed more than 220 publications, which includes refereed journals, reviews, books/book chapters, proceedings of seminars/conferences, etc. He is a recipient of several national and international awards/honours, namely British Council TCT Award 1996; DBT Overseas Associateship Award 2008; FAI Dhiru Morarji Memorial Award, 2011; Bharat Jyoti Award, 2012; Rajiv Gandhi Excellence Award, 2012; ISSS Dr. G.S. Sekhon Memorial Lecture Award of ISSS, 2012; IARI Hooker Award, 2013; Bioved Agri-innovation Award – 2015. He served as an Editor, Range Management and Agroforestry and Journal of the Indian Society of Soil Science. Currently he is Associate Editor, European Journal of Soil Science. He was Councillor, Indian Society of Soil Science, New Delhi (2005-2006); Secretary, Delhi Chapter of Indian Society of Soil Science (ISSS) 2005-2007; Member, Nature's Reader Panel 2009; President (Delhi Chapter), ISSS 2012-2014; Joint Secretary, ISSS 2013-2014, Vice President, Indian Society of Soil Science (2016-2017) and President, Agriculture & Forestry Sciences Section, Indian Science Congress Association (2016-2017). He is a life member of several professional societies and acted as Member/Chairman of several committees.

Dr Patra is a Fellow of the National Academy of Agricultural Sciences, Range Management Society of India and Indian Society of Soil Science.

Other books by IISS, Bhopal, India

Enhancing Nutrient Use Efficiency:
Concepts, Methods and Management Interventions
edited by K. Ramesh, A.K. Biswas, B.L. Lakaria S. Srivastava, A.K Patra

ISBN: 9789385516733
Price: INR 3600.00
Weight: 880gm
Pages: 482
Binding: Hardcover
Available: 2017

The book is intended to provide scientists, farmers, students and policymakers with insights into holistic approaches to nutrient use efficiency in major crops and cropping systems. Improving nutrient use efficiency is a prerequisite to reducing production costs in the wake of escalating cost of agricultural inputs in farming besides minimizing environmental contamination. Soil physical management is the foremost strategy to enhance the nutrient use efficiency. Besides describing the concepts and methods of nutrient use efficiency, management practices for dry land crops and cropping systems, cereal based, rainfed pulses, soybean, sugarcane, cotton, tobacco and oil seed, rapeseed-mustard and spices based cropping systems are also dealt in the book and would serve as a resource guide for enhancing nutrient use efficiency in various crops and cropping systems.

Climate Change and Natural Resources Management
edited by Lenka, S., N.K. Lenka & A. Subba Rao

ISBN: 9789381450673
Price: INR 2750.00
Weight: 820gm
Pages: 380
Binding: Hardcover
Available: 2013

This book addresses the important issues of food security and sustainability of natural resources of India in the context of the projected climate change. Agroecosystems being the sites of intense interaction between human beings and natural world, global climate change is likely to affect the resource base, the crop productivity, input use efficiency and overall the profitability of agricultural production systems to a great extent. However, the adverse effects of climate change can be alleviated through mitigation and adaptation strategies which carry importance due to the increasing population and food demand in India. Thus, this compilation covers possible sources and sinks of greenhouse gases in Indian context including the potentials of soil carbon sequestration, crop pest and soil management and scientific livestock management as mitigation and adaptation options. The likelihood of carbon credits and trading through best management practices can help Indian farmers earning carbon credits in future. The book is useful for researchers, farm managers, policy makers and also students engaged in climate change related studies.

Conservation Agriculture for Carbon Sequestration and Sustaining Soil Health
edited by J.Somasundaram et al.

ISBN: 9789383305322
Price: INR 4550.00
Weight: 1kg 235gm
Pages: 528
Binding: Hardcover
Available: 2014

This book comprises 41 dealing various issues, prospects and importance of conservation agriculture practices followed across different regions with special emphasis on rainfed regions. We hope this book on conservation agriculture will be highly useful to researchers, scientists, students, farmers and land managers for efficient and sustainable management of natural resources.

Nanotechnology in Soil Science and Plant Nutrition
edited by Adhikari, Tapan et al.

ISBN: 9789381450789
Price: INR 2750.00
Weight: 810gm
Pages: 320
Binding: Hardcover
Available: 2013

The book has 21 chapters addressing fundamentals and applied aspects of nanotechnology in soil science and plant nutrition research and written by explorers of a new frontier. The interpretation of subject matter in each is comprehensive, simple and lucid with relevant supporting data.

This book would offer a platform for basic, fundamental and advanced learning for students. It would also be useful and informative to researchers from SAUs and ICAR institutes.